RECENT ADVANCES IN GRAPHENE RESEARCH

Edited by **Pramoda Kumar Nayak**

Recent Advances in Graphene Research
http://dx.doi.org/10.5772/61909
Edited by Pramoda Kumar Nayak

Contributors

Yuancheng Fan, Fuli Zhang, Quanhong Fu, Kimal Chandula Wasalathilake, Cheng Yan, Godwin Ayoko, Biplab Sanyal, Soumyajyoti Haldar, Bhaghavathi Parambath Vinayan, Jan Smotlacha, Richard Pincak, Maria Del Prado Lavin Lopez, Jose Luis Valverde, Luz Sanchez-Silva, Amaya Romero, Fanyao Qu, Carlos Velasco-Santos, Ana Laura Martínez-Hernandez, Edgar Jimenez-Cervantes A., Juventino Lopez-Barroso, Sebastian Mackowski, Patrycja Łydżba, Janusz Jacak, Adam Rycerz, abdelkarim ouerghi, Claire Mathieu, Tevfik Onur Menteş, Emiliano Pallecchi, Andrea Locatelli, Gilles Patriarche, Rachid Belkhou

Published by InTech

Janeza Trdine 9, 51000 Rijeka, Croatia

Publishing Process Manager Edi Lipovic

Technical Editor SPi Global

Cover InTech Design team

First published October, 2016

Printed in Croatia

Additional hard copies can be obtained from orders@intechopen.com

Recent Advances in Graphene Research, Edited by Pramoda Kumar Nayak

p. cm.

Print ISBN 978-953-51-2638-6
Online ISBN 978-953-51-2639-3

For EU product safety concerns:
IN TECH d.o.o., Prolaz Marije Krucifikse Kozulić 3, 51000 Rijeka, Croatia,
info@intechopen.com or visit our website at intechopen.com.

Meet the editor

Pramoda Kumar Nayak is currently a senior researcher at the Centre for Low Dimensional Carbon Materials, Ulsan National Institute of Science and Technology, Republic of Korea. He received his doctoral degree in physics from the Indian Institute of Technology, Guwahati, India. After completion of his Ph.D, he worked as a postdoctoral researcher and visiting scientist in several institutions around the globe such as the Institute for Plasma Research, India, the National Cheng Kung University, Taiwan, and the National Tsing Hua University, Hsinchu, Taiwan, before coming to South Korea. His research interest is focused on two-dimensional materials, including transition metal dichalcogenides (TMD), graphene and hexagonal boron nitride (*h*-BN). He has published more than 40 international journals, 5 book chapters and 1 book.

Contents

Preface

Graphene is a perfectly two-dimensional (2D) material that exhibits fascinating physical, chemical and biosensing properties. Although, the scientists knew the existence of 2D crystal graphene, no one had worked out how to extract it from graphite. Since its isolation in 2004, it has captured huge attention among the scientists, researcher and industry worldwide. In recent years, the fascination with graphene has been growing very rapidly and it is considered as one of the most researched materials of the twenty-first century. Basically, graphene has redefined the limits of what a material can do: it boasts record thermal conductivity and the highest current density at room temperature ever measured (a million times that of copper); it is the strongest material known (a hundred times stronger than steel), but highly mechanically flexible; it is the least permeable material known (not even helium atoms can pass through it), the best transparent conductive film and the thinnest material known; and the list goes on. The vast amount of products, processes and industries that graphene could create a significant impact all stems from its amazing properties. When graphene is used as an improvement to existing materials or in a transformational capacity, its true potential could be realised.

The book "Recent Advances in Graphene Research" covers the latest advances in graphene research, including its synthesis, characterization, fundamental physical properties as well as potential applications, aiming to provide a compressive reference to the scientists in this cutting-edge field. This book brings together a team of experts to provide an overview of the most advanced topics in theory, experiments, spectroscopy and applications of graphene and its nanostructures. It contains 12 chapters divided into three sections.

Section 1 "Fundamentals of Graphene" contains four chapters. Chapter 1 introduces fractional quantum Hall effect in graphene, bilayer graphene and other Hall systems. Chapter 2 presents the electronic properties of less common forms of carbon nanostructures such as graphitic nanocone and graphitic wormhole. In Chapter 3, the authors discuss the electronic structure and topological quantum-phase transitions in strained graphene nanoribbons. Chapter 4 provides spectral statistics of weakly disordered triangular graphene flakes with zigzag edges.

In Section 2 "Graphene Synthesis", the optimized synthesis procedures of graphene and its derivatives are presented. It contains two chapters (Chapters 5 and 6). Chapter 5 contains optimization of the synthesis procedures of graphene and graphite oxide and Chapter 6 talks about Au intercalation in epitaxial graphene on SiC (0001).

The application of graphene and its nanostructured-based materials are described in Section 3 "Application of Graphene and its Nanostructures", which contains six chapters (Chapters 7–12). Energy transfer in graphene-based hybrid photosynthetic nanostructures is reported in Chapter 7. Chapter 8 emphasizes on heteroatom-doped graphene-based hybrid materials for hydrogen energy conversion. Chapter 9 talks about porous graphene materials for energy storage and conversion applications. Defects in graphene and its derivatives have been described in Chapter 10. Chapter 11 deals with plasmonic excitations in graphene for tuneable terahertz/infrared metamaterials and finally Chapter 12 reports graphene-based materials' functionalization with natural polymeric biomolecules.

I believe that this book will be very useful to a large number of researchers in various disciplines, both in academia and industry, seeking to gain up-to-date knowledge in the field of graphene.

I am very pleased to serve as the Editor of this book, which contains a wide variety of studies from authors all around the world. I would like to thank all the authors for their efforts in sending their best research papers to the attention of the audiences, including students, scientists and engineers, throughout the world.

I would also like to acknowledge the help given by InTech Open Access Publisher, in particular publishing process manager, Mr. Edi Lipović, for his assistance, patience and support throughout the whole process of this book project.

Dr. Pramoda Kumar Nayak
Low Dimensional Carbon Materials Centre,
Ulsan National Institute of Science and Technology,
Ulsan, Republic of Korea

Section 1

Fundamentals of Graphene

Chapter 1

Ultra-Quantum 2D Materials: Graphene, Bilayer Graphene, and Other Hall Systems—New Non-Local Quantum Theory of Hall Physics

Patrycja Łydżba and Janusz Jacak

Additional information is available at the end of the chapter

http://dx.doi.org/10.5772/64018

Abstract

We present a brief introduction of the fractional quantum Hall effect—a description of the phenomenon is provided and basic requirements for its formation are discussed. We recall assumptions of the standard composite fermion theory. Additionally, we present a list of the fractional quantum Hall effect puzzles. The chapter also introduces the non-local approach to quantum Hall physics, which is entirely based on a mathematical concept of braid groups and their reduction stimulated by an external magnetic field (in two-dimensional spaces). We emphasize the connection between a one-dimensional unitary representation of the system braid group and the particle statistics (unavoidable for any correlated Hall-like states). We implement our topological approach to construct hierarchies of FQHE fillings for various two-dimensional structures, including multi-layers. We show the remarkable agreement of our results with experimental findings.

Keywords: the fractional quantum Hall effect, graphene, bilayer graphene, braid groups, topology

1. Introduction to the fractional quantum Hall effect

In high magnetic fields and low temperatures, dips in a longitudinal resistivity ($\rho_{xx} \rightarrow 0$) and plateaus in a transverse one ($\rho_{xy} = h/\nu e^2$) appear for fractional fillings (ν) of Landau levels (LLs) —this transport feature is called the fractional quantum Hall effect (FQHE). Mentioned minima show an activated behavior, vanishing exponentially as temperature goes to zero and indicating the presence of a gap in the spectrum [1]. The latter phenomenon, despite its long history

[2], constantly receives a great amount of interest from the scientific society. Though many preliminary requirements for Hall-like states are commonly known, the comprehensive theory of this effect—capable of explaining all experimental findings in an elegant way—is still missing. One of the obvious facts is that the FQHE is impossible to obtain within a single-particle picture without interactions—where a partial filling of an elongated-states band immediately results in a nonzero value of the longitudinal resistivity. Among the explained necessity of strong inter-particle correlations in the system, a two-dimensional (2D) topology and a quantized kinetic energy (flat bands—as in the case of LLs) are also compulsory for evidencing the FQHE. Additionally, due to the fragility of these incompressible states, a high purity of the sample needs to be ensured too.

Let us emphasize that when the lowest Landau level (LLL) is partially occupied (or when a collectivization is restricted to one, arbitrary LL), kinetic energy remains constant ($E_k = \hbar\omega_c$, where $\omega_c = eB/mc$ is a cyclotron angular frequency). This applies also to the background potential energy. The ground state is, thus, expected to minimize the Coulomb repulsion. Since for a wide collection of magnetic fields the FQHE is actually observed, the latter requirement refers to collective Hall-like states (and not Wignier-crystal states with localized electrons [3], as it may seem at first).

The initial step towards an explanation of the FQHE was taken by Laughlin [4], who proposed the exact solution for a basic set of fillings from the LLL, $\nu = 1/q$ (q – odd),

$$\Psi_L = \prod_{i<j}^{N} \left(z_i - z_j\right)^q e^{-\frac{1}{4l_0^2}\sum_{i=1}^{N}|z_i|^2} \tag{1}$$

where N is the number of electrons, $l_0 = \sqrt{\hbar c/eB}$ stands for magnetic length and $z = x + iy$ is a complex position. Note that when two arguments of this wave function are swapped, Ψ_L gains an additional phase equal to $q\pi$. This detail of the solution (or more precisely—the quantum statistics of particles) is usually called the hallmark of Laughlin correlations and included in all theories of the FQHE—as a result of the Aharonov-Bohm effect [5] or the Berry phase acquired by a vortex [6–8]. However, the one-dimensional unitary representations (1DURs) of the full braid group (π_1)—which define the particle statistics—are periodic with a periodicity of 2π($e^{iq\pi} = e^{i\pi}$ for odd q). In order to deal with this issue, we suggest to associate FQHE particles (so-called composite fermions) with the appropriately constructed braid subgroups of π_1. This allows to differentiate them from ordinary fermions characterized with the full braid group. The latter idea is explained in detail in following sections and in authors' papers [9–11].

Unfortunately, despite its unbelievable accuracy confirmed by exact diagonalization methods (at least for a small amount of particles) [12], the Laughlin wave function was introduced as an ansatz or, if one prefers, as an inspired (educated) guess. Nonetheless, it might be educatory to recall his arguments [4, 13]. First, note that one can choose single-particle eigenstates within LLL (in a central gauge) to be eigenstates of a coordinate of the angular momentum ($\hat{l}_z$),

$$\varphi_m = z^m e^{-\frac{|z|^2}{4l_0^2}} \tag{2}$$

where m (integer) is an eigenvalue of $\hat{l}_z$. It is easy to establish the average area covered by an arbitrary electron, $\langle\varphi_m|\pi r^2|\varphi_m\rangle = 2\pi(m+1)l_0^2$. Thus, the degeneracy of a whole Landau level equals,

$$N_0 = m_{max} + 1 = \frac{S}{2\pi l_0^2} = \frac{SB}{hc/e} \tag{3}$$

where S is a sample surface and $hc/e = \phi_0$ is a magnetic field flux quantum. Additionally, m_{max} stands for a maximal value of m, which produces a state satisfying the $\langle\varphi_m|\pi r^2|\varphi_m\rangle \le S$ relationship. As a result, the general state of a multi-particle system from the LLL takes the form of,

$$\Psi = f(z_1, z_2, \ldots, z_N) e^{-\frac{1}{4l_0^2}\sum_{i=1}^N |z_i|^2} \tag{4}$$

where $f(z_1, z_2, \ldots, z_N)$ is an ordinary polynomial. A degree of this polynomial, D, cannot exceed the maximal $\hat{l}_Z$ eigenvalue realizable in the system. In other words, D is restricted by an inverse of a filling factor times a number of particles (or a degeneracy), $D \le m_{\max} \approx N_0 = \nu^{-1} N$.

Consider now a two-body problem. The eigenfunction for two particles with a relative angular momentum m and a centre of mass angular momentum M is [13],

$$\varphi_{m,M} = (z_1 - z_2)^m (z_1 + z_2)^M e^{-\frac{1}{4l_0^2}(|z_1|^2 + |z_2|^2)} \tag{5}$$

As long as m and M are assumed non-negative integers, the above state lies entirely in the LLL (it is constructed from linear combinations of one-body functions, φ_m) and, remarkably, appears to be the exact solution of a stationary Schrödinger equation with any central potential acting between two particles. Additionally, eigenvalues of the corresponding Hamiltonian are, simultaneously, eigenvalues of a Coulomb potential, $\hat{V} = \sum_{i<j}^N e^2/|Z_i - Z_j|$ (the fixed kinetic energy can be disregarded—it only adds even shifts to the energy levels),

$$v_m = \frac{\langle\varphi_{m,M}|\hat{V}|\varphi_{m,M}\rangle}{\langle\varphi_{m,M}|\varphi_{m,M}\rangle} \tag{6}$$

where v_m expectation values are so-called Haldane pseudopotentials, and they are independent from a central mass angular momentum eigenvalue. Hence, we obtained a discrete spectrum consisted of bound, repulsive potential states. This result is owed to the quenched kinetic energy, and it is possible to obtain only in two-dimensional spaces in the presence of a magnetic field. This finding should not be surprising—the high potential energy of interacting particles cannot be converted into the high kinetic energy (E_k is kept constant within one Landau level) and, hence, particles cannot "fly apart". Even classically, when a third (parallel to the magnetic field) dimension is lacking, the Lorentz force protects electrons from being pushed away from each other.

Let us now return to Laughlin's considerations, concerning the many-body ground-state wave function for filling factors, $v = 1/q$ (q – odd). Though we are currently unable to derive it analytically for $v < 1$, the lowest energy solution for a complete filling of the LLL ($v = 1$) is well known. It can be written as a Slater determinant of all accessible (indexed with an angular momentum eigenvalue) one-particle states [14],

$$\Psi_S = \begin{vmatrix} 1 & z_1 & z_1^2 \ldots & z_1^{N-1} \\ \ldots & \ldots & \ldots\ldots & \ldots \\ 1 & z_N & z_N^2 \ldots & z_N^{N-1} \end{vmatrix} e^{-\frac{1}{4l_0^2}\sum_{i=1}^{N}|z_i|^2} = \prod_{i<j}^{N}\left(z_i - z_j\right)e^{-\frac{1}{4l_0^2}\sum_{i=1}^{N}|z_i|^2} \tag{7}$$

where Ψ_S is the Slater function with the Vandermonde polynomial. Taking into account this conclusion, as well as bearing in mind the two-body eigenstate, it seems plausible that $f(z_1, z_2, \ldots, z_N)$ have a Jastrow-like form, $\prod_{i<j}^{N}(z_i - z_j)^q$. Another advantage of this choice lies in its tendency to keep electrons apart—there exists a considerable chance that it may reduce the Coulomb repulsion energy. Furthermore, the system is built from electrons, thus, only odd powers in the Jastrow factor are allowed. Finally, when we connect the q – solution with a $\nu = 1/q$ filling factor (this can supported by the extremal condition $D = m_{max}$), we arrive at the Laughlin wave function. Generally, it should be emphasized that all presented arguments are insufficient to prove that Ψ_L is a correct, lowest energy eigenfunction of the multi-particle Hamiltonian. However, its excellent compatibility with numerical results [12] entirely justifies a widespread use of the Laughlin ansatz.

Before skipping to another topic it is worth describing two other features of the Laughlin solution. First, the form of this wave function guarantees that every pair of particles has a relative angular momentum greater or equal to q. Thus, it needs to be an exact eigenfunction of the so-called hard-core potential with neglected long range part of the Coulomb repulsion ($v_m^{HC} = v_m$ for $m < q$ and $v_m^{HC} = 0$ for $m \geq q$) [15],

$$\sum_{m=0}^{\infty}\sum_{i<j}^{N} v_m^{HC} P_m(ij)\Psi_L = 0 \tag{8}$$

where the left-hand side equals to the hard-core potential operator (written in terms of projection operators, $P_m(ij)$, that selects states in which particles i and j have relative angular momentum equal to m) times Ψ_L . This implies that a finite amount of energy is needed to excite the system from its (Laughlin) ground state—there exists a gap in the spectrum (Δ). This gap is believed to be stable against perturbations, thus, all corrections arising from differences between the Haldane and the hard-core potential are expected to be small compared to Δ [13].

Finally, one can prove—with the use of a beautiful analogy developed by Laughlin [4]—that the Ψ_L state describes a uniform particle-density. Let us express the probability distribution in terms of the Boltzmann weight, $|\Psi_L|^2 = e^{-H_q} = e^{-(-2q\sum_{i<j}^{N}\ln(z_i - z_j) + \sum_{i=1}^{N}|z_i|^2/2l_0^2)}$ [14]. H_q turns out to be the potential energy of a one-component, two-dimensional classical plasma. Hence, we expect that electrons are distributed uniformly with a density $\rho = 1/(2\pi l_0^2 m)$ in a state described by $1/q$ filling of the LLL.

At this stage, readers should be warned that many incompressible Hall states—connected with filling factors falling out of the basic set ($1/q$ with odd q)—have already been discovered. Inspired by these results, scientists have been searching for the thorough and microscopic theory of the FQHE ever since. We are not going to present all models or ideas introduced by researchers over the years—this is not the aim of this paper. However, before proceeding to the topological explanation, it is worth becoming acquainted (at least briefly) with today's most widely accepted theory—he theory of composite fermions (CFs). In this very short description, we are going to focus on substantial advantages, as well as built-in problems of the approach.

In the CF model [16, 17], it is assumed that the FQHE and the IQHE (integer effect) are deeply connected and can be unified. The latter phenomenon, however, is usually identified with non-interacting particles, while strong correlations are necessary for the appearance of a fractional one. Nonetheless, the mapping can still be obtained if we assume that Coulomb interactions can be utilized—weakly interacting quasi-fermions can appear in the system—for fractional ν (this is, actually, a first tricky part—no one proved that Landau formulation [18] can be used in 2D spaces, where the spectrum of a Coulomb potential energy is discrete). Now, we only need these novel particles to play the same role for the FQHE as electrons for the IQHE.

Jain, who proposed this approach, postulated that quasi-particles are appearing as complexes of electrons with an even number of magnetic field flux quanta (currently identified with vortices [17], though this disambiguation is not entirely clear and might be misleading) and called them "composite fermions". Note that the many-body wave function embraces an additional Aharonov-Bohm phase, $\frac{e}{\hbar c}\oint A dl = \frac{e}{\hbar c}\int B ds = \frac{e}{\hbar c}\phi_0 = 2\pi$, when an electron (as an argument of Ψ_N) is encircling a single quantum [5]. The latter affects the quantum statistics of CFs—when two quasi-particles are exchanging, the wave function acquires a phase factor,

$$e^{i\alpha} = e^{i\pi}e^{i(p-1)\pi} \tag{9}$$

where the first term corresponds to the statistics of ordinary fermions, while the second term —to the half of an Aharonov-Bohm phase for $p - 1$ flux quanta pinned to each electron (p is an odd number). We have already mentioned that this is usually presented as an explanation of the Laughlin correlations.

To understand how the assertion of ϕ_0 results in the unification of IQHE and FQHE, it is useful to first consider non-interacting electrons in a completely filled Landau level, $\nu^*=n$ (with integer n). In this case an incompressible state is produced, thus, particles experience the integer (Hall) phenomenon. Furthermore, the evidenced magnetic field, $|B^*| = \phi_0 N/(S\nu^*)$, can be either positive or negative.

Now attach to every electron (e.g. by asserting an infinitely thin, massless solenoid) an even number of flux quanta pointing in +z direction. This converts electrons into composite fermions. Since these quasi-particles witness an unchanged magnetic field (B^*), we may expect them to also experience the IQHE (despite a different general structure of the energy spectrum compared to ordinary fermions). Finally, we have transformed an incompressible state of electrons into an incompressible state of CFs.

In the last step we adiabatically spread flux quanta pinned to electrons, until they become a part of the external magnetic field with an enhanced (or diminished) strength,

$$B = (p-1)\frac{\phi_0 N}{S} \pm \frac{1}{\nu^*}\frac{\phi_0 N}{S} \tag{10}$$

connected with a fractional filling factor of the form (the obtained hierarchy contains all fillings from the basic set – $n = 1$ and $p = q$ – but it includes many other ratios too),

$$\nu = \frac{1}{(p-1)\pm\frac{1}{\nu^*}} = \frac{n}{n(p-1)\pm \mathbf{1}} \tag{11}$$

The implemented assumption that a uniform magnetic field can be obtained from detaching localized ϕ_0 was justified—in the original paper [16]—by the uniform density of an initial (IQHE) state. Note that the minus sign in the above equation arises from the possibility of a negative initial field, B^*. Since $(p-1) > \frac{1}{\nu^*} = \frac{\mathbf{1}}{\boldsymbol{n}}$, the resulting B-field is always positive and points in a + z direction. Eventually, we have successfully connected the FQHE of electrons (ν) with the IQHE of composite fermions ($\nu^*=n$)—if and only if we assume that, during this smearing process, only quantitative changes occur in the energy spectrum. Thus, one can say that electrons effectively absorbed $p - 1$ flux quanta from the external magnetic field to transform into CFs, which evidence lower magnetic field and form integer quantum Hall state.

Popularity of the CF theory is owed to the fact that it is able to explain most of (experimentally observable) fractions from the LLL. It also gives a prescription for constructing ground-state (as well as excited-state) wave functions within the lowest Landau band. However, scientists have already stumbled upon a difficulty—the pyramids of fillings from higher levels are not counterparts of the Jain's hierarchy. It is, for example, impossible to justify the appearance of Hall plateaus in the vicinity of $\nu=5/2$ and 7/2. To deal with this problem, residual interactions between composite particles were introduced to the considerations and they were supposed to turn some of them into higher order composite fermions with different Landau levels. This approach, unfortunately, was rather unsuccessful but, due to the lack of a better explanation, it is still (quite) widely used [19–21].

Moreover, recent experiments on graphene bilayers confirmed that this quasi-particle approach, regardless of its undeniable advantages, cannot embrace the entire physics behind the FQHE [22–24]. This is due to the fact that the most robust plateau in a transverse resistance—in samples consisted of two weakly coupled sheets of atoms—develops near a half filling (and not a one-third filling, as expected from the CF). Additionally, the improvement of sample quality led to the discovery of novel incompressible states that fall out of the Jain's pyramid of fillings, even in the lowest Landau level of standard GaAs/AlGaAs structures. Among them, there are filling factors with odd denominators (like 4/11) and with even denominators (like 3/8).

As a result of the hunt for a model which can find a solution for all mentioned problems, the topological approach to quantum Hall effects was formulated [9]. In this chapter, we briefly recall its basic assumptions and we demonstrate that it provides a topological explanation of Jain's approach (Section 2). Additionally, we demonstrate that it can be used to explain all incompressible Hall-like states observed in conventional 2DEG, monolayer, and bilayer graphene samples (Section 3). We compare our predictions with experimental findings (Section 4).

2. Non-local theory of Hall systems: the origin and generalization of composite fermions

The non-local theory of Hall systems is entirely based on the well-known mathematical concept of braid groups [25]. The full braid group, $\pi_1(\Omega)$, is a homotopy group of an N-particle system configuration space [26],

$$\Omega=\left(M^N\setminus\Delta\right)/S_N \tag{12}$$

where M is a manifold on which particles are placed, S_N stands for the permutation group of N elements (its appearance results from the assumed indistinguishability of carriers) and Δ removes points (creates topological defects) for which positions of at least two classical particles, as M^N coordinates, are the same. Hence, it is consisted of closed trajectories—though the quotient group structure allows initial and final orderings of particles to differ by a permutation—performed in Ω and organized in homotopy classes. In the case of simply connected manifold, $\pi_1(\Omega)$ is generated by σ_i elements which describe simple exchanges of i'th and $(i+1)$'th carriers (classical—as M^N coordinates). It is worth emphasizing that *quantum particles do not travel braid trajectories*. However, a relationship between topological properties (reflected in a braid group form) with quantum properties of an arbitrary system cannot be ignored. Note that in the standard quantization method, multi-particle state vectors, Ψ_N, are selected as functions from Ω into complex numbers [26]. Thus, when arguments of Ψ_N are encircling a closed path in their configuration space, the multi-particle wave function gains a phase equal to a one-dimensional unitary representation (1DUR) of the corresponding braid from $\pi_1(\Omega)$. Additionally, in the Feynman path integral formulation of the propagator [27], an additional summation over classes of homotopical trajectories needs to be implemented for multiply connected Ω (one measure, $d\lambda$, in a whole path space cannot be defined). The weight factors emerging in this summation are defined by a one-dimensional unitary representation of $\pi_1(\Omega)$. Finally, the braid group shape and its 1DUR determine a quantum statistics of particles in the system (the allowed types of particles are settled by the topology of a manifold and by certain external factors—e.g. anyons may exist only in 2D spaces).

Let us now move back to Hall systems. It seems reasonable to assume that in the presence of a strong magnetic field, trajectories representing elements of the system braid group are of cyclotron orbit type—although, in general, they are not ordinary circles for highly interacting particles. In the topological approach, we define the surface of an LLL cyclotron orbit (as a representative of a braid from $\pi_1(\Omega)$ by the archetype of a correlated incompressible state—the IQHE state. Hence, a plaque which encircles an area equal to $\frac{S}{N_0}=\frac{\phi_0}{B}=\frac{hc}{eB}$ and embraces exactly one flux quantum can be identified with a cyclotron path in the lowest Landau band. In great magnetic fields (as for partial fillings of the LLL), this enclosed surface may be too small to allow an arbitrary particle (as an argument of Ψ_N or M^N coordinate) to reach its nearest neighbor—the Coulomb repulsion protects a uniform distribution. In this situation *simple exchanges, σ_i, are unenforceable and should be excluded from the braid group describing the system.* Note that it may be impossible to organize a $\pi_1(\Omega)$-subgroup (which satisfy group axioms) from remaining classes of trajectories. However, we have already established that the FQHE can be evidenced only for correlated many-body states and, thus, it requires a determined statistics of particles (and a determined system braid group). Seeing that this phenomenon is actually observed in transport measurements, the $\pi_1(\Omega)$ reduction needs to result in the emergence of a new group (*a cyclotron subgroup*), at least for selected fillings. Generators b_i of a latter subgroup stand for novel, multi-loop exchanges of carriers (loopless, σ_i, cannot be defined). Finally, each pair—a subgroup of the full braid group and a condition, a magnetic field strength, upon which this subgroup characterizes the multi-particle system—represents (and can be used to identify) a FQHE state in the LLL [28].

As we have already mentioned, cyclotron subgroups are always generated with multi-loop exchanges. Thus, trajectories (circled in M-space) representing b_i elements need to be open—they need to contain an integer number of closed loops, n, and exactly one half-loop. Simultaneously, cyclotron orbits (representatives of braids) are closed by the definition. As a result, the simplest non-trivial cyclotron paths are generated by b_i^2 elements and, hence, they consist an odd number of loops, $2(n+1/2)=2n+1$. We are going to demonstrate why multi-loop exchanges of particles (as arguments of Ψ_N or M^N coordinates) can be permissible in the system, even when loopless ones are not. For simplicity we restrict (at first) our considerations to $1/q$ (q – odd) fillings of the lowest Landau level. In this case, a single-loop path of a carrier experience q flux quanta per particle in the system and—since an LLL orbit embraces exactly one ϕ_0 – its surface is not large enough to reach a neighboring carrier (σ_i exchanges are unfeasible). In 3D manifolds this also applies to multi-loop trajectories (meaning that effective dimensions of loops are not raised; σ_i exchanges are, however, always accessible because particles can move freely in the direction parallel to the magnetic field)—each loop increases the total surface encircled by the path, as a circumvolution adds a new surface to the coil. Therefore, the latter also increases the magnetic field flux evidenced by the path—an individual loop experience q flux quanta per particle too. In 2D manifolds, however, an additional loop cannot enhance path's area and the evidenced flux remains unchanged (the whole trajectory, which represents an element of the system braid group, experiences q flux quanta per particle). As a consequence, all loops must share the total BS/N per particle quantity, which passes through a single-loop path—each loop receives a diminished portion of BS/N and experiences a lower effective magnetic field, B^*. Additionally, if the number of loops in a trajectory coincides with the inverse of an LLL filling factor, $\nu^{-1}=q$, then every loop evidences exactly one ϕ_0 per particle in the system ($BS/(qN)=BS/N_0=\phi_0$) and an effective magnetic field defined by the relation,

$$\frac{B^*S}{N}=\frac{1}{q}\frac{BS}{N}\rightarrow B^*=\frac{B}{q} \tag{13}$$

Since on the LLL cyclotron orbit falls a single magnetic field flux quantum, the surface encircled by an arbitrary loop (from the q-loop trajectory) is large enough to reach a neighboring particle (as an argument of Ψ_N or M^N coordinate). *Although the loopless exchanges are not permitted in the system, we have just demonstrated that* $\left(\frac{q-1}{2}\right)$*-loop ones are accessible and they generate the cyclotron subgroup of* $\pi_1(\Omega)$. The explicit form of b_i elements for a $1/q$ filling of the lowest Landau level is presented below,

$$b_i^{(q)}=\sigma_i^{\,q},\quad 1<i<N-1 \tag{14}$$

with 1DURs of the form ($\theta=\pi$ for composite fermions—we stick to this generally confusing name for history reasons),

$$b_i^{(q)}\rightarrow e^{iq\theta},\quad 1<i<N-1 \tag{15}$$

Note that the even-denominator rule (q– odd) follows immediately from the requirement of an open trajectory (as an M-space representative) for b_i. Additionally, Laughlin correlations seem to follow from this cyclotron subgroup formalism, rather than from quasi-particles formulation with auxiliary objects pinned to electrons [10].

Consider now a filling factor, which does not belong to the basic set of fillings ($\nu \neq 1/q$). In this situation, it is impossible for every loop from the q-loop trajectory to embrace exactly one flux quantum attributed to a single particle. We can, however, assume that all $q-1$ loops experience a single ϕ_0 per particle and only last loop evidences a residual (per particle) portion, $\pm\frac{1}{x}\phi_0$. Thus, the total magnetic field flux per particle, BS/N, can be written as follows,

$$\frac{BS}{N} = \frac{1}{\nu}\phi_0 = (q-1)\phi_0 \pm \frac{1}{x}\phi_0 \tag{16}$$

A fortunate situation occurs when x is an integer number. In this case the last loop can embrace an entire flux quantum, as the LLL cyclotron orbit does, when it fits perfectly to the separation of an electron (as an argument of Ψ_N or M^N coordinate) and its arbitrary x'th order neighbor. As a result, *the closing loop (as a representative of a winding of a cyclotron subgroup braid) defines the type of accessible exchanges, which are not consisted of q loopless σ_i (these are not accessible for the system), but are integral exchanges of multi-loop type*. Thus, for filling factors included in the Jain-like hierarchy [16, 17],

$$\nu = \left((q-1) \pm \frac{1}{x}\right)^{-1} \tag{17}$$

a cyclotron subgroup of the full braid group can be defined and it is generated by elements,

$$b_i^{(q)} = \sigma_i^{q-1}\sigma_i\sigma_{i+1}\ldots\sigma_{i+x-1}^{\pm 1}\ldots\sigma_i^{-1}\sigma_{i+1}^{-1} \tag{18}$$

The quantum statistics follows immediately from the 1DUR ($\theta = \pi$ for composite fermions),

$$b_i^{(q)} \rightarrow e^{i(q-1\pm 1)\theta}, \quad 1 < i < N-1 \tag{19}$$

In the above considerations, we assumed that the last loop of a q-looped path (as a representative of a trajectory circled in the Ω space) can be overwrapped in a direction opposite to one enforced by the external magnetic field. The latter results in an appearance of the minus sign in Eqs. (16)–(19).

It is reasonable to consider even more complex commensurability conditions (of an area encircled by each loop and a surface attributed to a single particle or a group of x particles)—

where $q-1$ loops, which constitute a simplest non-trivial cyclotron path, are divided into sets characterized by different integer numbers, $a \le b \le c \ldots$ (different evidenced fractions of flux quantum per particle, $\frac{1}{a}\phi_0 \ge \frac{1}{b}\phi_0 \ge \frac{1}{c}\phi_0 \ldots$). These sets need to be of even multitudes, $\alpha \ge \beta \ge \gamma \ldots$, to keep a rational character of the exchange braid,

$$b_i^{(q)} = \left(\sigma_i \sigma_{i+1} \ldots \sigma_{i+a-1} \ldots \sigma_i^{-1} \sigma_{i+1}^{-1}\right)^{\alpha} \cdot \left(\sigma_i \sigma_{i+1} \ldots \sigma_{i+b-1}^{\pm 1} \ldots \sigma_i^{-1} \sigma_{i+1}^{-1}\right)^{\beta} \cdot \ldots \tag{20}$$

q	ν	FQHE	
3	$\frac{1}{2 \pm 1/x}$	(−) $1, \frac{2}{3}, \frac{3}{5}, \frac{4}{7}, \frac{5}{9}, \frac{6}{11}, \frac{7}{13}, \frac{8}{15}, \frac{9}{17}, \frac{10}{19}, \ldots$	(+) $\frac{1}{3}, \frac{2}{5}, \frac{3}{7}, \frac{5}{9}, \frac{5}{11}, \frac{6}{13}, \frac{7}{15}, \frac{8}{17}, \frac{9}{19}, \frac{10}{21}, \ldots$
5	$\frac{1}{4 \pm 1/x}$	(−) $\frac{1}{3}, \frac{2}{7}, \frac{3}{11}, \frac{4}{15}, \frac{5}{19}, \frac{6}{23}, \ldots$	(+) $\frac{1}{5}, \frac{2}{9}, \frac{3}{13}, \frac{4}{17}, \frac{5}{21}, \frac{6}{25}, \ldots$
5	$\frac{1}{2 + 2 \cdot 1/2 \pm 1/x}$	(−) $\frac{2}{5}, \frac{3}{8}\left(\frac{5}{8}\right), \frac{4}{11}\left(\frac{7}{11}\right), \frac{5}{14}, \frac{6}{17}, \frac{7}{20}, \ldots$	(+) $\frac{2}{7}, \frac{3}{10}, \frac{4}{13}\left(\frac{9}{13}\right), \frac{5}{16}, \frac{6}{19}, \frac{7}{22}, \ldots$
5	$\frac{1}{2 - 2 \cdot 1/3 \pm 1/x}$	(−) $1, \frac{12}{13}, \frac{15}{17}, \frac{6}{7}, \frac{21}{25}, \ldots$	(+) $\frac{3}{5}, \frac{12}{19}, \frac{15}{23}, \ldots$
5	$\frac{1}{2 - 2 \cdot 1/4 \pm 1/x}$	(−) $\frac{4}{5}, \frac{10}{13}, \frac{3}{4}, \frac{14}{19}\left(\frac{5}{19}\right), \frac{8}{11}, \frac{18}{25}, \ldots$	(+) $\frac{4}{7}, \frac{10}{17}, \frac{3}{5}, \frac{14}{23}, \frac{5}{13}\left(\frac{8}{13}\right), \ldots$
5	$\frac{1}{4 \cdot 1/2 \pm 1/x}$	(−) $\frac{2}{3}, \frac{3}{5}, \frac{4}{7}, \frac{5}{9}, \frac{6}{11}, \frac{7}{13}, \frac{8}{15}, \frac{9}{17}, \frac{10}{19}, \ldots$	(+) $\frac{2}{5}, \frac{3}{7}, \frac{5}{9}, \frac{5}{11}, \frac{6}{13}, \frac{7}{15}, \frac{8}{17}, \frac{9}{19}, \frac{10}{21}, \ldots$
5	$\frac{1}{2 \cdot 1/2 + 2 \cdot 1/3 \pm 1/x}$	(−) $\frac{3}{4}, \frac{12}{17}\left(\frac{5}{17}\right), \frac{15}{22}, \ldots$	(+) $\frac{1}{2}, \frac{12}{23}, \ldots$
7	$\frac{1}{6 \pm 1/x}$	(−) $\frac{1}{5}, \frac{2}{11}, \frac{3}{17}, \frac{4}{23}, \ldots$	(+) $\frac{1}{7}, \frac{2}{13}, \frac{3}{19}, \frac{4}{25}, \ldots$
7	$\frac{1}{4 + 2 \cdot 1/2 \pm 1/x}$	(−) $\frac{2}{9}, \frac{3}{14}, \frac{4}{19}, \frac{5}{24}, \ldots$	(+) $\frac{2}{11}, \frac{3}{16}, \frac{4}{21}, \ldots$

Where q = the number of half loops in the exchange trajectory, ν = the hierarchy of fillings. Plus or minus signs denote the direction of a last loop in the multi-loop cyclotron path. Selected hole states are indicated in brackets. Experimentally observable ratios are highlighted—blue color for ν possible to explain within the CF model and red color for ν out of the Jain hierarchy. Additionally, filling factors of monolayer graphene Hall states (which develop in transport measurements) are underlined.

Table 1. Filling factors obtained from the non-local theory (typical semiconductor structures).

The elements above generate cyclotron subgroups for filling factors from the generalized hierarchy of the form,

$$\nu = (\alpha / a \pm \beta / b \pm \gamma / c \pm \cdots)^{-1} \tag{21}$$

1DURs of the resulting subgroups can also be easily estimated (the scalar representation of a subgroup can be identified with reduced representation of the full $\pi_1(\Omega)$),

$$b_i^{(q)} \rightarrow e^{i(\alpha\pm\beta\pm\gamma\pm...)\theta}, \quad 1<i<N-1 \tag{22}$$

Note that while constructing cyclotron subgroups of this type, one needs to keep an eye on the complexity of generators—highly structured braids are probably unfavorable and near corresponding fillings the Wigner crystal (and not the FQHE) phases may be in favor [29–31].

Filling factors obtained within the non-local theory of Hall systems are gathered in **Table 1**. Note that among these fillings one can find all famous and mysterious v, like $\frac{3}{8}$, $\frac{4}{11}$, $\frac{4}{13}$ and others (marked with red color), which cannot be predicted within the quasi-particle formulation (at least when unclear residual interactions are not implemented) [19, 32–35]. For this reason, the topological approach presented in this chapter is the first one to justify all incompressible (collective) states from the LLL, which appear in transport measurements [28].

Finally, it is also worth mentioning that the pyramid of fillings, established within the topological approach, is a generalized version of the well-known Jain's hierarchy (obtained when $\alpha=q-1$ and $a=1$ are implemented into Eq. 21). Thus, *two artificial flux quanta pinned to electron only model (in a very convenient way) an additional loop, which evidences exactly one ϕ_0 per particle and appears in the exchange trajectory (representing a generator of the system braid group).* The great dependability of the CF theory in the LLL regime can be explained by highlighting the simplicity of cyclotron subgroups generators in the case of Jain's hierarchy—as it was already emphasized, highly structured braids (usually encountered for different configurations of loops) are probably unfavorable and may not result in pronounced plateaus in Hall resistivity.

3. IQHE and FQHE in conventional 2DEG, monolayer and bilayer graphene

Before we proceed to the main aim of this section, it is worth discussing differences in the Landau level ladder between conventional 2DEG, monolayer and bilayer graphene samples (as well as their possible consequences). We first note that in both graphene structures, LLs are not distributed equidistantly on the energy axis, which was the case for typical quantum wells [36–39]. However, the latter is not reflected in the topological approach—an area encircled by a single-loop cyclotron orbit (A) is proportional to the bare kinetic energy of electrons (E_k) and not the general energy determined by LLs. Since a crystal field cannot affect the value of E_k, the embraced surface of an orbit (representing a $\pi_1(\Omega)$ element) varies in the same manner for monolayer and bilayer graphene as for conventional 2DEG. Explicitly, A is proportional to $2n+1$ (where n stands for an LL index). At the same time, this dependence carries another consequence—the commensurability conditions from higher LLs are different from those encountered in the lowest band (the single-loop cyclotron orbit for $n\neq0$ embraces exactly $(2n+1)\phi_0$ and not a single quantum as in the LLL case). Thus, we expect that pyramids of fillings for levels indexed by $n \neq 0$ are not counterparts of the well-known Jain's hierarchy, which has already been confirmed in experiments [40–42].

An important property of the Landau level ladder is its degeneracy (as a number of sublevels, rather than a number of one-particle states, in a single level). In typical 2DEG systems, the half-spin (usually marked with ↑ and ↓) stands for an exclusive degree of freedom and, thus, all Landau bands are doubly degenerated. In graphene materials, however, the existence of a so-called valley pseudospin (isospin)—due to the presence of identical Dirac cones in two non-equivalent corners of the Brillouin zone (usually marked with K' and K)—results in an additional degeneracy of the energy spectrum. As a result, for an arbitrarily picked range of factors $[\nu_1, \nu_2]$, the corresponding partially filled Landau level in monolayer and bilayer graphene samples is different than in conventional quantum-wells. We have already mentioned that commensurability conditions (and so the hierarchies of fillings) for distinct n are not identical. We may, thus, expect that some incompressible states, which are marked with v and occur in one of these structures, may not be permissible for the other one.

It should be emphasized that a nonzero Berry phase causes the monolayer graphene LLL to be placed exactly in the Dirac point—where the valence band meets the conduction band in a gapless energy spectrum [36, 37]. The lowest Landau level is, thus, equally shared between free electrons and free holes (only two, not four, spin-valley branches accessible for one type of particles). Since it is natural to define the filling factor in terms of an electronic density measured from the charge neutrality point, ν is counted with respect to the bottom of a conduction band (a third sublevel) and not the LLL as in typical semiconductors [36, 37]. Generally, this also applies to bilayer graphene; however, not only the LLL is placed exactly in the Dirac point, but also the first Landau band. Obviously, the latter affects and determines the convenient impact factor definition, as well as the number of spin-valley branches of the 1LL available for free electrons.

Finally, it is easy to notice that the non-local theory of Hall systems predicts identical hierarchies of LLL filling factors for typical semiconductor quantum-wells and monolayer graphene samples [31, 43–50]. The latter applies for both sublevels of the lowest band (also in graphene—we take into account only fillings accessible for free electrons). Additionally, collective Hall-like states from the second (spin or spin-valley) branch are described with similar filling ratios ($\nu_{0,2}$) as those from the first one ($\nu_{0,1}$). The only difference is obvious and lies in a constant shift of all fractions, $\nu_{0,2}=1+\nu_{0,1}$. As we have already mentioned, the cyclotron subgroup model can be used to describe the FQHE in higher LLs of these structures too—the recipe for its application has been included in subsection 3.1.

The case of bilayer graphene is absolutely unique. The appearance of an additional sheet of carbon atoms (additional surface) leads to completely different commensurability conditions. This, for example, results in a surprising form of the basic set, where $\nu=1/2$ corresponds to the most prominent incompressible state. We are going to describe this problem in detail in subsection 3.2.

Single-loop FQHE	Paired IQHE	q	ν	FQHE
$\frac{7}{3}, \frac{8}{3}$	$\frac{5}{2}$	3	$2+\frac{1}{3\cdot(2-1/x)}$	$\frac{7}{3}\left(\frac{8}{3}\right), \frac{20}{9}\left(\frac{25}{9}\right), \frac{11}{5}\left(\frac{14}{5}\right), \ldots$
			$2+\frac{1}{3\cdot(2+1/x)}$	$\frac{19}{9}, \frac{32}{15}, \frac{15}{7}, \ldots$
			$2+\frac{1}{3\cdot(2\cdot 1/2-1/x)}$	$\frac{8}{3}\left(\frac{7}{3}\right), \frac{5}{2}, \frac{22}{9}\left(\frac{23}{9}\right), \frac{29}{12}, \frac{12}{5}\left(\frac{13}{5}\right), \ldots$
			$2+\frac{1}{3\cdot(2\cdot 1/2+1/x)}$	$\frac{20}{9}\left(\frac{25}{9}\right), \frac{9}{4}\left(\frac{11}{4}\right), \frac{34}{15}, \ldots$
			$2+\frac{1}{3\cdot(2\cdot 1/3-1/x)}$	$3, \frac{14}{5}, \frac{19}{7}\left(\frac{16}{7}\right), \frac{8}{3}, \frac{29}{11}, \frac{34}{13}, \frac{13}{5}, \frac{44}{17}, \frac{49}{19}, \frac{18}{7}\left(\frac{17}{7}\right), \frac{59}{23}, \frac{64}{25}, \frac{23}{9}, \frac{74}{29}, \frac{79}{31}, \frac{28}{11}, \frac{89}{35}, \frac{94}{37}, \frac{33}{13}\left(\frac{32}{13}\right), \ldots$
$\frac{10}{3}, \frac{11}{3}$	$\frac{7}{2}$	3	$3+\frac{1}{3\cdot(2-1/x)}$	$\frac{10}{3}\left(\frac{11}{3}\right), \frac{29}{9}, \frac{16}{5}\left(\frac{19}{5}\right), \ldots$
			$3+\frac{1}{3\cdot(2+1/x)}$	$\frac{28}{9}, \frac{47}{15}, \frac{22}{7}, \ldots$
			$3+\frac{1}{3\cdot(2\cdot 1/2-1/x)}$	$\frac{11}{3}\left(\frac{10}{3}\right), \frac{7}{2}, \frac{31}{9}, \frac{41}{12}, \frac{17}{5}\left(\frac{18}{5}\right), \ldots$
			$3+\frac{1}{3\cdot(2\cdot 1/2+1/x)}$	$\frac{29}{9}, \frac{13}{4}\left(\frac{15}{4}\right), \frac{49}{15}, \ldots$
			$3+\frac{1}{3\cdot(2\cdot 1/3-1/x)}$	$4, \frac{19}{5}, \frac{26}{7}, \frac{11}{3}, \frac{40}{11}, \frac{47}{13}, \frac{18}{5}\left(\frac{17}{5}\right), \frac{61}{17}, \frac{68}{19}, \frac{25}{7}, \ldots$

Where q = the number of half loops in the exchange trajectory, ν = the hierarchy of fillings. Plus or minus signs denote the direction of a last loop in the multi-loop cyclotron path. Selected hole states are indicated in brackets. Experimentally observable ratios are highlighted—blue color for ν possible to explain within the CF model and red color for ν out of the Jain hierarchy. Additionally, results for different sublevels are separated by three lines.

Table 2. 1LL filling factors obtained from the non-local theory (typical semiconductor structures).

3.1. Higher Landau levels: graphene and conventional 2DEG

We consider a system subjected to external magnetic field, which leads to the partial filling of an arbitrary Landau band ($n>0$). To simplify further discussions, we introduce an additional parameter, m, that enumerates spin (or spin-valley) branches in each LL. The latter is two-valued ($m \in \{0, 1\}$) in the typical 2DEG case and four-valued ($m \in \{0, 1, 2, 3\}$) in the monolayer graphene case. We remind that the area embraced by a single-loop cyclotron orbit in the whole n'th Landau level equals to $(2n+1)\frac{hc}{eB} = \frac{(2n+1)\phi_0}{B}$ (it takes exactly $(2n+1)$ flux quanta and its surface is considerably enhanced, when compared to the LLL orbit). This change of an encircled area modifies—and allows for the existence of novel—commensurability conditions, all of which are listed below.

Single-loop FQHE	Paired IQHE	q	ν	FQHE
			$2+\frac{1}{3\cdot(2-1/x)}$	$\frac{7}{3}\left(\frac{8}{3}\right),\frac{20}{9},\frac{11}{5},\ldots$
			$2+\frac{1}{3\cdot(2+1/x)}$	$\frac{19}{9},\frac{32}{15},\frac{15}{7},\ldots$
$\frac{7}{3},\frac{8}{3}$	$\frac{5}{2}$	3	$2+\frac{1}{3\cdot(2\cdot1/2-1/x)}$	$\frac{8}{3}\left(\frac{7}{3}\right),\frac{5}{2},\frac{22}{9}\left(\frac{23}{9}\right),\frac{29}{12},\frac{12}{5}\left(\frac{13}{5}\right),\ldots$
			$2+\frac{1}{3\cdot(2\cdot1/2+1/x)}$	$\frac{20}{9},\frac{9}{4},\frac{34}{15},\ldots$
			$2+\frac{1}{3\cdot(2\cdot1/3-1/x)}$	$3,\frac{14}{5},\frac{19}{7},\frac{8}{3},\frac{29}{11},\frac{34}{13},\frac{13}{5},\frac{44}{17},\frac{49}{19},\frac{18}{7}\left(\frac{17}{7}\right),\ldots$
			$3+\frac{1}{3\cdot(2-1/x)}$	$\frac{10}{3}\left(\frac{11}{3}\right),\frac{29}{9},\frac{16}{5},\ldots$
			$3+\frac{1}{3\cdot(2+1/x)}$	$\frac{28}{9},\frac{47}{15},\frac{22}{7},\ldots$
$\frac{10}{3},\frac{11}{3}$	$\frac{7}{2}$	3	$3+\frac{1}{3\cdot(2\cdot1/2-1/x)}$	$\frac{11}{3}\left(\frac{10}{3}\right),\frac{7}{2},\frac{31}{9},\frac{41}{12},\frac{17}{5}\left(\frac{18}{5}\right),\ldots$
			$3+\frac{1}{3\cdot(2\cdot1/2+1/x)}$	$\frac{29}{9},\frac{13}{4},\frac{49}{15},\ldots$
			$3+\frac{1}{3\cdot(2\cdot1/3-1/x)}$	$4,\frac{19}{5},\frac{26}{7},\frac{11}{3},\frac{40}{11},\frac{47}{13},\frac{18}{5},\frac{61}{17},\frac{68}{19},\frac{25}{7}\left(\frac{31}{7}\right),\ldots$
			$4+\frac{1}{3\cdot(2-1/x)}$	$\frac{13}{3}\left(\frac{14}{3}\right),\frac{38}{9},\frac{21}{5},\ldots$
			$4+\frac{1}{3\cdot(2+1/x)}$	$\frac{37}{9},\frac{62}{15},\frac{29}{7},\ldots$
$\frac{13}{3},\frac{14}{3}$	$\frac{9}{2}$	3	$4+\frac{1}{3\cdot(2\cdot1/2-1/x)}$	$\frac{14}{3}\left(\frac{13}{3}\right),\frac{9}{2},\frac{40}{9},\frac{53}{12},\frac{22}{5}\left(\frac{23}{5}\right),\ldots$
			$4+\frac{1}{3\cdot(2\cdot1/2+1/x)}$	$\frac{38}{9},\frac{17}{4},\frac{64}{15},\ldots$
			$4+\frac{1}{3\cdot(2\cdot1/3-1/x)}$	$5,\frac{24}{5},\frac{33}{7},\frac{14}{3},\frac{51}{11},\frac{60}{13},\frac{23}{5},\frac{78}{17},\frac{87}{19},\frac{32}{7}\left(\frac{31}{7}\right),\ldots$

Where q = the number of half loops in the exchange trajectory, ν = the hierarchy of fillings. Plus or minus signs denote the direction of a last loop in the multi-loop cyclotron path. Selected hole states are indicated in brackets. Experimentally observable ratios are highlighted. Additionally, results for different sublevels (only three, accessible in measurement, are presented) are separated by three lines.

Table 3. 1LL filling factors obtained from the non-local theory (monolayer graphene).

1. We first consider a single-loop path (as a representative of a braid from the appropriate braid group). The simplest situation—in which this trajectory fits perfectly to the separation of two neighboring particles—corresponds to $2n+1$ flux quanta attributed to a single electron from the partially filled level,

$$\begin{cases} \dfrac{N_0}{N-(2n+m)N_0} = 2n+1 & \textit{for conventional} \\ \dfrac{N_0}{N-(4n-2+m)N_0} = 2n+1 & \textit{for graphene} \end{cases} \tag{23}$$

where a denominator of the left-hand side expression counts the number of electrons lying in the n'th LL (in the bottom expression a "−2" factor needs to be omitted, while investigating n=0). Furthermore, the whole fraction determines the number of ϕ_0 per particle from the highest available level. Simultaneously, the right-hand side is also connected to the number of ϕ_0, however, grasped by the single-loop cyclotron orbit. One can easily determine filling factors, which fulfil the above equation,

$$\begin{cases} \nu = \dfrac{N}{N_0} = 2n+m+\dfrac{1}{2n+1} & \textit{for conventional} \\ \nu = \dfrac{N}{N_0} = 4n-2+m+\dfrac{1}{2n+1} & \textit{for graphene} \end{cases} \tag{24}$$

If the system is described with these ν, then exchanges of nearest neighbors, σ_i, are accessible and they generate $\pi_1(\Omega)$. Something surprising should already be noticed—the full braid group is coupled to the IQHE only for the zeroth Landau band. For other levels, the corresponding filling is fractional (e.g. $\nu=7/3$ is obtained, in both structures, when the $m=0$ branch of the 1LL is investigated). As a result, a collective Hall-like state is described with the fractional quantization of a transverse resistivity (just like for the ordinary FQHE state in the LLL), but the Laughlin correlations are described with a $q=1$ power in the Jastrow polynomial and loopless elements generate the system braid group (just like for the ordinary IQHE state in the LLL). We have, thus, stumbled across a novel phenomenon —the single-loop FQHE [51, 52]—which can be obtained only in higher Landau bands ($n>0$). We may also expect that this effect is very robust (simple, not structured, generators construction), which was confirmed in transport measurements of typical 2DEG structures and monolayer graphene samples [20, 40–42, 53].

In higher LLs it is possible that the cyclotron orbit surface is greater than the separation of neighboring particles (classical—as arguments of Ψ_N or M^N coordinates). Thus, for selected filling factors a single-loop path (as a representative of an element from the system braid group) can be generated as a double exchange of x-order neighbors (with integer x), $(\sigma_{i+1}\cdots\sigma_{i+x-1}\cdots\sigma_i^{-1}\sigma_{i+1}^{-1})^2$. This corresponds to the situation when exactly ϕ_0/x is attributed to a single electron from the partially filled level,

$$\begin{cases} \dfrac{N_0}{N-(2n+m)N_0} = \dfrac{2n+1}{x} \to \nu = 2n+m+\dfrac{x}{2n+1} & \textit{for conventional} \\ \dfrac{N_0}{N-(4n-2+m)N_0} = \dfrac{2n+1}{x} \to \nu = 4n-2+m+\dfrac{x}{2n+1} & \textit{for graphene} \end{cases} \tag{25}$$

Note that complete filling of the n'th LL is achieved when $x=2n+1$ and the braid group is generated by the loopless exchanges of particles and their $(2n+1)$-order neighbors.

2. In higher Landau bands also exists a possibility that the cyclotron orbit is too small to reach a neighboring (classical) carrier and, hence, the loopless exchanges are not allowed and need to be excluded from $\pi_1(\Omega)$. Fortunately, we can consider—similarly as in the LLL case—more structured, multi-loop exchanges to provide generators of the braid group describing the system. Commensurability conditions, which allow to establish the corresponding filling factors (connected with subgroups generated by b_i elements), resemble ones introduced for the zeroth Landau band. However, while possible ν for Hall-like states, we need to remember about a different number of flux quanta grasped by the cyclotron orbit ($(2n+1)\phi_0$ in the n'th LL). Additionally, we cannot take into account all particles in the system, but only those, which are placed in a partially filled level. Finally, we remind that if we consider a q-looped trajectory encircled in a 2D space, then all q loops must share the total magnetic field flux per particle, which is evidenced by a single-loop path.

To obtain a hierarchy—which resembles Jain's pyramid of fillings—we assume that $q-1$ loops experience exactly $(2n+1)\phi_0$ per particle, while the last one embraces a reduced portion, $(2n+1)\phi_0/x$, attributed to a single carrier. Thus, the appropriate commensurability conditions (with different, integer x) take the form of,

$$\begin{cases} \dfrac{N_0}{N-(2n+m)N_0} = (2n+1)\cdot(q-1) \pm (2n+1)\dfrac{1}{x} & \textit{for conventional} \\ \dfrac{N_0}{N-(4n-2+m)N_0} = (2n+1)\cdot(q-1) \pm (2n+1)\dfrac{1}{x} & \textit{for graphene} \end{cases} \tag{26}$$

with filling factors, which satisfy the requirements above, belonging to the set,

$$\begin{cases} \nu = 2n+m+\dfrac{1}{(2n+1)\cdot\left[(q-1)\pm 1/x\right]} & \textit{for conventional} \\ \nu = 4n-2+m+\dfrac{1}{(2n+1)\cdot\left[(q-1)\pm 1/x\right]} & \textit{for graphene} \end{cases} \tag{27}$$

Note that—despite the general similarity to the CF hierarchy—a supplementary factor, $2n+1$, is included in the denominators of all fractions from Eq. 27 (as a consequence of an enhanced kinetic energy). The last winding of a q-looped trajectory, for these fillings, fits perfectly to the area embraced by x classical particles. As a consequence, exchanges of x'th order neighbors are allowed and, thus, elements

$$b_i^{(q)} = \sigma_i^{q-1}\sigma_i\sigma_{i+1}\ldots\sigma_{i+x-1}^{\pm 1}\ldots\sigma_i^{-1}\sigma_{i+1}^{-1} \tag{28}$$

generate the cyclotron subgroup for the system.

3. Finally, the pairing of electrons—which occurs as a result of the Fermi sea instability—can also be investigated. In this case the number of particles is reduced by half and it is reasonable to consider the IQHE formation (for pairs),

$$\begin{cases} \dfrac{N_0}{N-(2n+m)N_0} = 2 \rightarrow \nu = 2n+m+\dfrac{1}{2} & \textit{for conventional} \\ \dfrac{N_0}{N-(4n-2+m)N_0} = 2 \rightarrow \nu = 4n-2+m+\dfrac{1}{2} & \textit{for graphene} \end{cases} \tag{29}$$

The equation above is constructed in such a manner that implementing a decreased number of particles—$\frac{N-(2n+m)N_0}{2}$ or $\frac{N-(4n-2+m)N_0}{2}$—leads to the exact IQHE commensurability condition.

We have presented hierarchies that gather all filling factors (from all LLs) in which particles can experience quantum Hall effects in typical semiconductor and monolayer graphene samples. We have also explicitly presented these results, for the first Landau band, in **Tables 2** and **3**.

3.2. Bilayer graphene

The FQHE in bilayer systems (not only bilayer graphene) is exceptional, as noticed earlier by Eisenstein [54]. The unparalleled basic set of fillings – with $\nu=1/2$ being the most robust incompressible state – has its origin in the appearance of an additional surface [28, 55]. The supplementary sheet of carbon atoms, coupled to the primary one by a nonzero hopping integral [36–39], leads to the electron density located in both graphene planes. As a result, bilayer graphene samples are not strictly two-dimensional. Classically (trajectories, which represent elements of the system braid group, are classical), this means that particles can move freely between opposite layers of the structure. Thus, while considering whether the cyclotron orbit area is sufficiently large to enable the existence of particle exchanges in the system, graphene planes should not be investigated separately. For example, the $\nu=1$ state—corresponding to the integer phenomenon—is realized when the single-loop cyclotron path encircles a surface equal to S/N, and not $2S/N$. However, while examining states described by fractional fillings of the LLL, even more interesting feature is revealed. Since multi-looped trajectories can be partly located in both graphene layers, the surface (and flux) provided by the additional plane needs to be taken into account. It is expected that the most energetically efficient trajectory is realized, when only one loop embraces (utilizes) the supplementary surface and magnetic field flux. As a consequence, its dimensions are not raised—they are equal to those of a single-loop path. Simultaneously, remaining loops ($q-1$ in the case of a q-looped trajectory) share the total per particle quantity, BS/N, associated with the primary layer. Latter results in a novel form of the commensurability conditions,

$$\frac{BS}{N}=\frac{1}{\nu}\phi_0=(q-2)\phi_0\pm\frac{1}{x}\phi_0 \tag{30}$$

with the hierarchy of fillings being a modified version of the composite fermion pyramid,

$$\nu=\left((q-2)\pm\frac{1}{x}\right)^{-1} \tag{31}$$

Here we have assumed that one of $q-2$ loops evidences a diminished (per particle) portion, $\pm\frac{1}{x}\phi_0$. Thus, it fits perfectly to the separation of an arbitrary particle and its x-order neighbors (as Ψ_N arguments) and allows them to exchange. Additionally, we speculate that the latter loop can be overwrapped in the direction opposite to one enforced by an external magnetic field (a minus sign in Eq. 31).

Even-denominator filling factors form a basic set—they are expected to be associated with most prominent Hall-like states [22–24] However, ratios with odd denominators are also included in the above hierarchy. The latter can be seen in **Table 4**, which gathers results from the whole LLL (and 1LL). In this paper we assumed that spin-valley branches of zeroth and first Landau bands, accessible for free electrons, are filled alternately. In other words, these sublevels are placed on the energy axis in the following order, $0K'\uparrow$, $1K'\uparrow$, $0K'\downarrow$, $1K'\downarrow$. Thus, filling factors responsible for the FQHE plateaus—n the $2<\nu<3$ range—are defined by the hierarchy from Eq. 33 (with a constant "+2" shift).

Although an even number of loops, $q-1$, is explicitly included in the commensurability conditions, it is an odd number, q, that constitutes a multi-looped trajectory (which represents a square of the cyclotron subgroup generator). This results in the Laughlin correlations with an odd power in the Jastrow-like polynomial—the multi-particle wave function is antisymmetric even for $\nu=1/2$ state.

In bilayer graphene, similarly as in typical 2DEG and monolayer graphene structures, kinetic energy of particles increases with Landau level index, n. Hence, the area encircled by a single-loop cyclotron orbit (representing a braid group element) is also enhanced, $(2n+1)hc/eB$, and it embraces precisely $(2n+1)$ quanta of the magnetic field flux. This leads to commensurability conditions, which resemble ones already presented in this section and other authors' papers [10, 52, 55]. We investigate them very briefly,

Sublevel	Single-loop FQHE	Paired IQHE	q	ν	FQHE
			3	$\frac{1}{1+1/x}$	$\frac{1}{2}, \frac{2}{3}, \left(\frac{1}{3}\right), \frac{3}{4}, \frac{4}{5}, \ldots$
$0K'\uparrow$			5	$\frac{1}{3-1/x}$	$\frac{1}{2}, \frac{2}{5}, \frac{3}{8}, \frac{4}{11}, \ldots$
			5	$\frac{1}{3+1/x}$	$\frac{1}{4}, \frac{2}{7}, \frac{3}{10}, \frac{4}{13}, \ldots$
			3	$1+\frac{1}{3\cdot(1-1/x)}$	$\frac{5}{3}, \left(\frac{4}{3}\right), \frac{3}{2}, \frac{13}{9}, \ldots$
$1K'\uparrow$	$\frac{4}{3}, \frac{5}{3}$	$\frac{3}{2}$	3	$1+\frac{1}{3\cdot(1+1/x)}$	$\frac{7}{6}, \frac{11}{9}, \frac{5}{4}, \frac{19}{15}, \ldots$
			3	$1+\frac{1}{3\cdot(1/2-1/x)}$	$\frac{4}{3}, \frac{7}{5}, \frac{13}{9}, \ldots$
			3	$2+\frac{1}{1+1/x}$	$\frac{5}{2}, \frac{8}{3}, \frac{11}{4}, \frac{14}{5}, \ldots$
$0K'\downarrow$			5	$2+\frac{1}{3-1/x}$	$\frac{5}{2}, \frac{12}{5}, \frac{19}{8}, \frac{26}{11}, \ldots$
			5	$2+\frac{1}{3+1/x}$	$\frac{9}{4}, \frac{16}{7}, \frac{23}{10}, \frac{30}{13}, \ldots$
			3	$3+\frac{1}{3\cdot(1-1/x)}$	$\frac{11}{3}, \left(\frac{10}{3}\right), \frac{7}{2}, \frac{31}{9}, \ldots$
$1K'\downarrow$	$\frac{10}{3}, \frac{11}{3}$	$\frac{7}{2}$	3	$3+\frac{1}{3\cdot(1+1/x)}$	$\frac{19}{6}, \frac{29}{9}, \frac{13}{4}, \frac{49}{15}, \ldots$
			3	$3+\frac{1}{3\cdot(1/2-1/x)}$	$\frac{10}{3}, \frac{17}{5}, \frac{31}{9}, \ldots$

Where q = the number of half loops in the exchange trajectory, ν = the hierarchy of fillings. Plus or minus signs denote the d irection of a last loop in the multi-loop cyclotron path. Selected hole states are indicated in brackets. Experimentally observable ratios are highlighted - a blue colour for ν possible to explain within the CF model and a red colour for ν out of the Jain hierarchy. Additionally, results for different sublevels are separated by three lines.

Table 4. LLL and 1LL filling factors obtained from the non-local theory (bilayer graphene).

1. First, it is possible that the system braid group is generated by loopless exchanges of x-order neighbors. This corresponds to the situation where exactly $(2n+1)/x$ flux quanta are attributed to a single particle,

$$\frac{N_0}{N-\varepsilon N_0}=\frac{2n+1}{x}\rightarrow \nu=\varepsilon+\frac{x}{2n+1} \tag{32}$$

   where ε counts completely filled Landau sublevels and $x=1, 2, \ldots, 2n+1$. Note that the IQHE is realized in the system when x equals $2n+1$ and the single-loop cyclotron orbit fits perfectly to the separation of a particle and its $(2n+1)$-order neighbor. For other values of x electrons experience the so-called single-loop FQHE – the phenomenon allowed only in higher Landau bands ($n>0$).

2. The $(q-1)/2$-loop exchanges of x-order neighbors (as arguments of Ψ_N or M^N coordinates) can become generators of the system braid group (a cyclotron subgroup), when loopless ones are not accessible. This is achievable when q-looped path is arranged as follows: all $q-2$ loops experience precisely $(2n+1)\phi_0$ per particle, while the last loop evidences only a residual portion attributed to a single electron, $\frac{(2n+1)}{x}\phi_0$,

$$\frac{N_0}{N-\varepsilon N_0}=(2n+1)(q-2)\pm\frac{(2n+1)}{x}\rightarrow \nu=\varepsilon+\frac{1}{(2n+1)[q-2\pm 1/x]} \tag{33}$$

   It should be emphasized that—in the above commensurability condition—we have not considered an additional per particle quantity, $\frac{BS}{N}$, supplied by the second layer of the structure. The latter can be performed because a remaining loop of the trajectory utilizes the supplementary magnetic field flux—it can be, thus, omitted in Eq. 33.

3. Finally, the pairing of electrons—which occurs as a result of the Fermi sea instability—can also be investigated. In this case the number of particles is reduced by half and it is reasonable to consider the IQHE formation (for pairs),

$$\frac{N_0}{N-\varepsilon N_0}=2\rightarrow \nu=\varepsilon+\frac{1}{2} \tag{34}$$

   The above equation is constructed in such a manner that implementing a decreased number of particles, $\frac{N-\varepsilon N_0}{2}$, leads to the exact IQHE commensurability condition.

We have presented hierarchies that gather all filling factors (from all LLs) in which particles can experience quantum Hall effects. We have also explicitly presented these results—for lowest and first Landau bands—in **Table 4**.

A careful reader probably has already noticed that an application of the non-local theory of Hall systems to structures with a greater (than two) number of layers is straightforward. The most important modification concerns the commensurability conditions for q-looped cyclotron trajectories (when only multi-loop exchanges are accessible and generate the braid group describing the system). Each additional layer of atoms supplies an additional surface (and magnetic field flux) that needs to be embraced by this path. The most energetically efficient trajectory seems to be achieved when every added plane is utilized by a single loop. Dimensions of these loops are not raised and they are not included, at least explicitly, in the commensurability conditions. As a result, only remaining $q-\gamma$ ones (where γ stands for the number of supplementary layers) must share the total flux per particle, $BS/(N-\varepsilon N_0)$, which is experienced by the single-loop path.

4. Comparison between theory and experiment

As mentioned previously, scientists are currently unable to understand the entire physics behind the fractional quantum Hall effect. Experimenters constantly conduct novel analyses to gain an insight into this non-trivial phenomenon. As a result, many measuring techniques have been developed—the local compressibility measurements or experiments in Hall-bar and two-terminal geometries. Selected results, for all structures considered in this chapter, are presented in **Figure 1**. It is also worth emphasizing that experiments carried out in monolayer and bilayer graphene samples are exceptional. For example, it is possible to modify the carrier density (with a lateral gate voltage) in a fixed magnetic field strength (**Figure 1b** and **c**).

Figure 1a presents a very famous transport measurement conducted on a typical GaAs/AlGaAs quantum well sample (and published in the Pan et al. paper [19]). Its uniqueness is owed to the well-developed plateaus in the longitudinal resistance for LLL fillings which are impossible to obtain within the quasi-particle approach (at least without implementing residual and unclear interactions between CFs) [16, 17]. The appearance of these incompressible states can be, however, explained by the non-local theory of quantum Hall effects. Consider a multi-loop cyclotron trajectory of a particle (as M^N coordinate) that represents a square of the system braid group generator. It is probable that additional loops, belonging to this path, fit to the separation of higher order (not nearest) neighbors. In this case the incompressible Hall-like state can also be formed—which was confirmed by Pan's experiment—but it cannot be captured by Jain's model. The latter is owed to the fact that every flux quanta pinned to an electron is a convenient model of a supplementary loop in the cyclotron trajectory only when it fits perfectly to the inter-particle distance.

A surprising form of the basic set in bilayer graphene samples—where $\nu=1/2$ is the most prominent incompressible state—can be observed in transport measurements depicted in **Figure 1b**. We have already mentioned that this feature follows immediately from the necessity to embrace (by the multi-loop cyclotron path) a flux supplied by the additional layer (surface). One loop is, thus, wasted on the utilization and falls out from the commensurability conditions—the obtained hierarchy contains even-denominator filling factors. However, the exchange trajectory (representing a braid group generator) consists of an odd number of half loops that protects the antisymmetric character of a particle statistics.

Finally, another interesting feature is worth to be noted—the remarkable stability of incompressible Hall-like states for $\nu = 7/3$, $8/3$ and 5/2. The unexpected robustness of related plateaus makes "7/3 and 8/3 states unlikely to be the analogues of the 1/3, 2/3 Laughlin correlated states" [40]. This conclusion seems to agree with the non-local theory predictions. The braid group describing the system for these fillings (connected with the paired IQHE and the single-loop FQHE) is generated with loopless exchanges. Thus, the Laughlin correlations are described by a $p = 1$ power in the Jastrow polynomial, despite the fractional quantization of a transverse resistivity. Hence, the robustness of these states is expected to exceed one for the ordinary FQHE states, which was confirmed experimentally (**Figure 1c** and **d**).

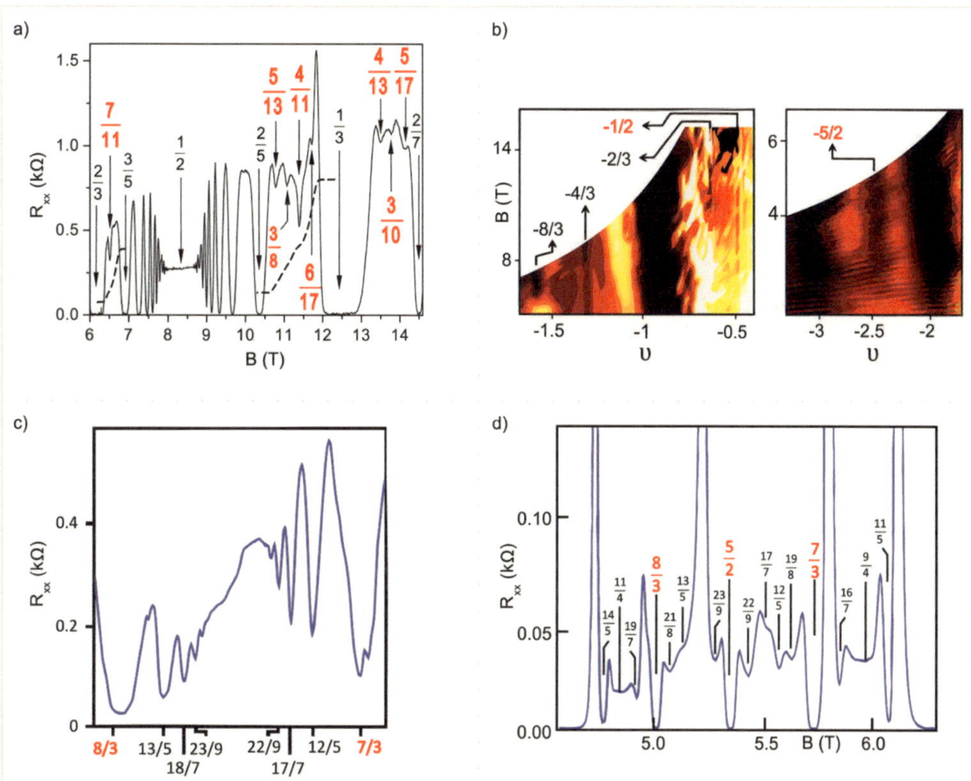

Figure 1. (a) Typical semiconductor quantum well. A longitudinal resistance as a function of external magnetic field. The figure is based on the Pan et al. paper [19]. (b) Bilayer graphene. A longitudinal resistance as a function of a magnetic field and a filling factor. $R_{xx} \approx 0.1[\Omega]$ is colored dark red and $R_{xx} \approx 4[k\Omega]$ is colored bright yellow. The figure is based on the Ki et al. paper [23]. (c) Monolayer graphene. A longitudinal resistance as a function of a filling factor. The figure is based on Amet et al.'s [53] paper. (d) Typical semiconductor quantum well. A longitudinal resistance as a function of external magnetic field. The figure is based on the Choi et al.'s [40] paper. Important filling factors (e.g. falling out of the CF hierarchy or corresponding to the single-loop FQHE) are highlighted.

5. Concluding remarks

In conclusion, the topological approach to quantum Hall effects is based on a concept of braid groups and their reduction stimulated by an external magnetic field (in 2D spaces). This model can be used to derive hierarchies of FQHE fillings for various two-dimensional structures (including multi-layers), which fit perfectly to the experimental findings. Additionally, it can be used to explain many issues that are believed to be mysterious—the puzzles of quantum Hall structures.

Acknowledgements

Support from the NCN Project UMO-2011/02/A/ST3/00116 is acknowledged.

Author details

Patrycja Łydżba* and Janusz Jacak

*Address all correspondence to: patrycja.lydzba@pwr.edu.pl

Faculty of Fundamental Problems of Technology, Wroclaw University of Science and Technology, Wrocław, Poland

References

[1] Boebinger GS, Chang AM, Stormer HL, Tsui DC. Magnetic field dependence of activation energies. Physical Review Letters. 1985; 55(15):1606–1609.

[2] Tsui DC, Stormer HL, Gosard AC. Two-dimensional magnetotransport in the extreme quantum limit. Physical Review Letters. 1982; 48(22):1559–1562.

[3] Bonsall L, Maradudin AA. Some static and dynamical properties of a two-dimensional Wigner crystal. Physical Review B. 1977; 15(4):1959–1973.

[4] Laughlin RB. Anomalous quantum hall effect: an incompressible quantum fluid with fractionally charged excitations. Physical Review Letters. 1983; 50(18):1395–1398.

[5] Aharonov Y, Bohm D. Significance of electromagnetic potentials in the quantum theory. The Physical Review. 1959; 115(3):485–491.

[6] Berry MV. Classical adiabatic angles and quantal adiabatic phase. Journal of Physics A: Mathematical and General. 1985; 18:15–27.

http://dx.doi.org/10.5772/64018

[7] Arovas D, Schrieffer JR, Wilczek F. Fractional statistics and the quantum Hall Effect. Physical Review Letters. 1984; 53(7):722–723.

[8] Read N. Theory of the half-filled Landau level. Semiconductor Science and Technology. 1994; 9:1859–1864.

[9] Jacak J, Jóźwiak I, Jacak L. New implementation of composite fermions in terms of subgroups of a braid group. Physics Letters A. 2009; 374:346–350.

[10] Jacak J, Łydżba P, Jacak L. Fractional quantum Hall effect revisited. Physica B. 2015; 475:122–139.

[11] Jacak J, Łydżba P, Jacak L. Homotopy approach to fractional quantum Hall Effect. Applied Mathematics. 2015; 6:345–358.

[12] Jain JK, Kamilla RK. Quantitative study of large composite-fermion systems. Physical Review B. 1997; 55(8):417–420.

[13] Girvin M. Introduction to the fractional quantum Hall effect. Seminaire Poincare. 2004; 2:53–74.

[14] Chakraborty T, Pietilainen P. The fractional quantum Hall effect, properties of an incompressible quantum fluid. Springer-Verlag, Berlin-Heidelberg-New York-London-Paris-Tokyo; 1988 .

[15] Haldane FDM. Fractional quantization of the Hall effect: a hierarchy of incompressible quantum fluid states. Physical Review Letters. 1983;51(7):605–608.

[16] Jain JK. Composite-fermion approach for the fractional quantum Hall effect. Physical Review Letters. 1989; 63:199–202.

[17] Jain JK. A note contrasting two microscopic theories of the fractional quantum Hall effect. Indian Journal of Physics. 2014; 88:915–929.

[18] Landau LD. The theory of a Fermi liquid. Soviet Physics JETP. 1957; 3(6):920–925.

[19] Pan W et al. Fractional quantum Hall effect of composite fermions. Physical Review Letters. 2003; 90(1):016801.

[20] Pan W et al. Experimental studies of the fractional quantum Hall effect in the first excited Landau level. Physical Review Letters. 2003; 90:075307.

[21] Mandal SS and Jain JK. Theoretical search for the nested quantum Hall effect of composite fermions. Physical Review B. 2002; 66:155302.

[22] Kou A et al. Electron-hole asymmetric integer and fractional quantum Hall effect in bilayer graphene. Science. 2014; 345(6192):55–57.

[23] Ki D, Fal'ko VL, Abanin DA, Morpurgo AF. Observation of even denominator fractional quantum Hall effect in suspended bilayer graphene. Nano Letters. 2014; 14:2135–2139.

[24] Bao W et al. Magnetoconductance oscillations and evidence for fractional quantum Hall states in suspended bilayer and trilayer graphene. Physical Review Letters. 2010; 105:246601.

[25] Birman JS. Braids, links, and mapping class groups. Princeton University Press. 1975; (No. 82).

[26] Sudarshan ECG, Imbo TD, Imbo CS. Topological and algebraic aspects of quantization: symmetries and statistics. Annales de l'Institut Henri Poincaré, Section A. 1988; 49(3): 387–396.

[27] MacKenzie R. Path integral methods and applications. arXiv:quant-ph/0004090. 2000 .

[28] Łydżba P, Jacak J. Topological origin of all incompressible states from the lowest Landau level: Preliminary results on wave functions. Forthcoming.

[29] Wexler C, Ciftja O. Novel liquid crystalline phases in quantum Hall systems. International Journal of Modern Physics B. 2006; 20(7):747–778.

[30] Williams FIB. Experiments on melting in classical and quantum two dimensional electron systems. In: International Conference on Physics in Two Dimensions, 1991. Helvetica Physica Acta, Switzerland; 1992.

[31] Du X et al. Fractional quantum Hall effect and insulating phase of Dirac electrons in graphene. Nature. 2009; 462:192–195.

[32] Pan W et al. Fractional quantum Hall effect at Landau level filling ν=4/11. Physical Review B. 2015; 91:041301.

[33] Samkharadze N et al. Observation of incompressibility at ν=4/11 and ν=5/13. Physical Review B. 2014; 91:081109.

[34] Stormer HL. Nobel lecture: the fractional quantum Hall effect. Reviews of Modern Physics. 1999; 71(4):875–889.

[35] Du RR et al. Experimental evidence for new particles in the fractional quantum Hall effect. Physical Review Letters. 1993; 70(19):2944–2947.

[36] Goerbig MO. Electronic properties of graphene in a strong magnetic field. Reviews of Modern Physics. 2011; 83:1193–1243.

[37] McCann E. Electronic properties of monolayer and bilayer graphene. NanoScience and Technology. 2012; 57:237–275. DOI: 10.1007/978-3-642-22984-8_8.

[38] Novoselov KS. Unconventional quantum Hall effect and Berry's phase of 2π in bilayer graphene. Nature Physics. 2006; 2:177–180.

[39] Fal'ko VL. Electronic properties and the quantum Hall effect in bilayer graphene. Philosophical Transactions of the Royal Society A. 2008; 366:205–219.

[40] Choi HC et al. Activation gaps of fractional quantum Hall effect in the second Landau level. Physical Review B. 2008; 77:081301.

[41] Dean CR et al. Contrasting behavior of the 5/2 and 7/2 fractional quantum Hall effect in a tilted field. Physical Review Letters. 2008; 101:186806.

[42] Eisenstein JP. New physics in high Landau levels. Physica E: Low-dimensional Systems and Nanostructures. 1999; 6: 29–35.

[43] Lee DS et al. Transconductance fluctuations as a probe for interaction-induced quantum Hall states in graphene. Physical Review Letters. 2012; 109:056602.

[44] Bolotin KI. Observation of the fractional quantum Hall effect in graphene. Nature Letters. 2009; 462:196–199.

[45] Skachko I. Fractional quantum Hall effect in suspended graphene probed with two-terminal measurements. Philosophical Transactions of the Royal Society A. 2010; 368:5403–5416.

[46] Feldman BE. Fractional quantum Hall phase transitions and four-flux states in graphene. Physical Review Letters. 2013; 111:076802.

[47] Feldman BE. Unconventional sequence of fractional quantum Hall states in suspended graphene. Science Reports. 2012; 337:1196–1199.

[48] Ghahari F. Measurement of the ν=1/3 fractional quantum Hall energy gap in suspended graphene. Physical Review Letters. 2011; 106:046801.

[49] Peterson MR, Nayak C. Effects of Landau level mixing on the fractional quantum Hall effect in monolayer graphene. Physical Review Letters. 2014; 113:086401.

[50] Dean CR. Multicomponent fractional quantum Hall effect in graphene. Nature Physics. 2011; 7:693–696.

[51] Łydżba P, Jacak L, Jacak J. Hierarchy of fillings for the FQHE in monolayer graphene. Scientific Reports. 2015; 5:1–16.

[52] Jacak J, Jacak L. The commensurability condition and fractional quantum Hall effect hierarchy in higher Landau levels. Pisma v ZhETF. 2015; 102:19–25.

[53] Amet F et al. Composite fermions and broken symmetries in graphene. Nature Communications. 2015; 6:1–7.

[54] Eisenstein JP. New fractional quantum Hall state in double-layer two-dimensional electron systems. Physical Review Letters. 1992; 68(9):1383–1386.

[55] Jacak J, Jacak L. Difference in hierarchy of FQHE between monolayer and bilayer graphene. Physical Letters A. 2015; 379:2130–2134.

Electronic Properties of Carbon Nanostructures

Jan Smotlacha and Richard Pincak

Additional information is available at the end of the chapter

http://dx.doi.org/10.5772/63633

Abstract

The carbon nanostructures are perspective materials for the future applications. This has two reasons: first, the hexagonal atomic structure, which enables a high molecular variability by placing different kinds of the defects, and second, good electronic properties that can be modified for the purpose of the concrete applications with the help of the defects and of the chemical ingredients. Many kinds of the nanostructures are investigated. Here, the properties of less common forms are examined—the graphitic nanocone and graphitic wormhole.

Keywords: graphene, graphitic wormhole, graphitic nanocone, spin-orbit coupling, zero modes

1. Introduction

The carbon nanostructures are the materials whose molecular structure is derived from graphene—the hexagonal carbon plain lattice (**Figure 1**). Because of their electronic structure, they are the promising materials for the construction of nanoscale devices (quantum wires, nonlinear electronic elements, transistors, molecular memory devices, or electron field emitters) and the inventions in the material science.

The planar geometry of the molecular surface is disrupted by the disclinations in the molecular structure that are most often presented by the pentagons and the heptagons in the hexagonal lattice. This change of the geometry is manifested by the positive or the negative curvature, respectively, that can be enlarged by the supply of higher number of the defects. In this way, by the supply of 1–5 pentagonal defects, we get conical structures with different values of the vortex angle (**Figure 2**).

One more defect can be added and a nanotube is created. This nanostructure can be considered closed as well as opened, i.e., without the cap that contains the pentagonal defects. The second case is more common (**Figure 3**, left part). The number of the defects can be increased up to 12, and in this way, a completely closed, spherical nanostructure arises (fullerene—**Figure 3**, middle part).

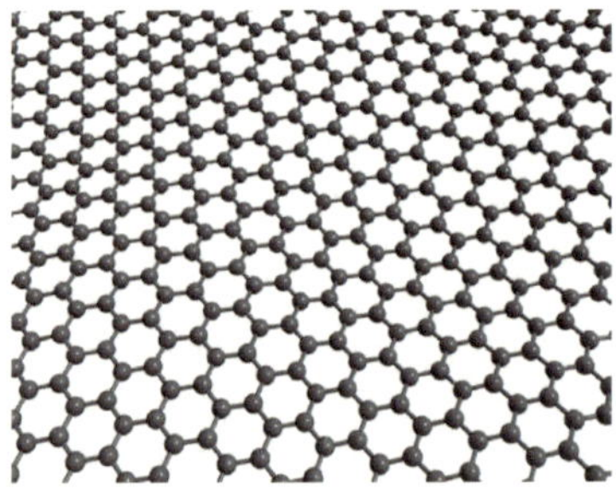

Figure 1. Hexagonal carbon plain lattice.

Figure 2. Conical nanostructures with different numbers of pentagonal defects in the tip.

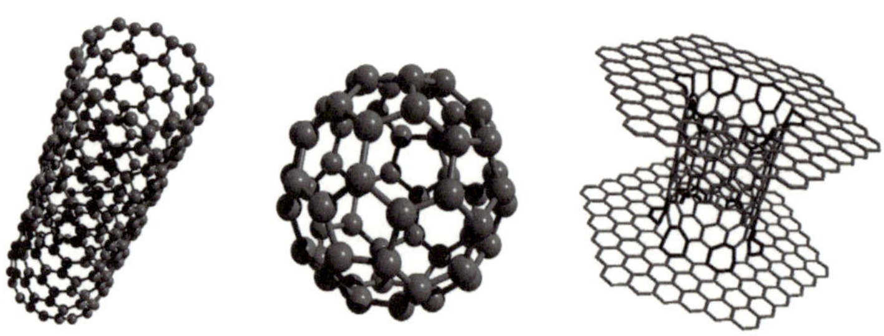

Figure 3. Different kinds of graphene nanostructures: nanotube (left), fullerene (middle), wormhole (right).

Analogical manipulations with the graphene lattice can be made by the supply of the heptagonal defects (**Figure 4**). For the case of 12 heptagonal defects, if they are placed appropriately, the wormhole structure is created (**Figure 3**, right part).

http://dx.doi.org/10.5772/63633

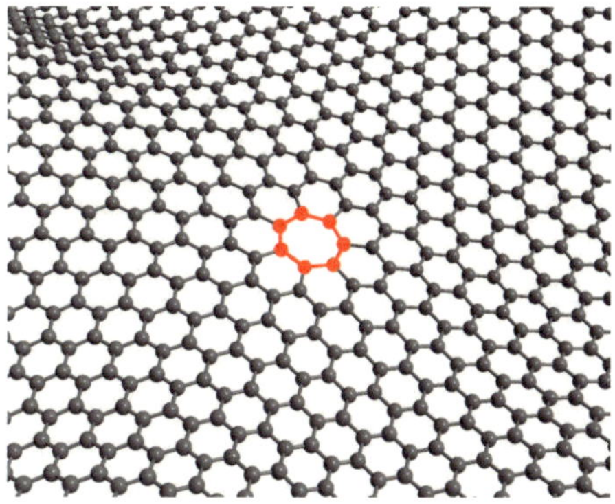

Figure 4. Hexagonal lattice disclinated by one heptagonal defect.

A lot of other variants of the graphitic nanostructures can be created using different combinations of the pentagonal and the heptagonal defects. Some of them are presented in **Figure 5**.

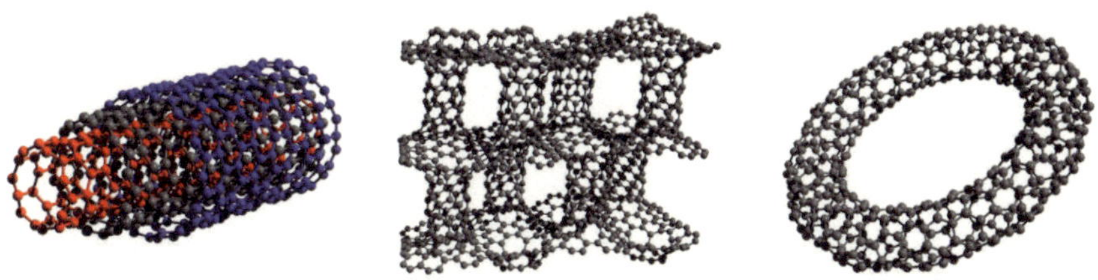

Figure 5. Less common forms of the graphene nanostructures: triple-walled nanotube (left), pillared graphene (middle), nanotoroid (right).

We investigate the electronic properties of several kinds of the carbon nanostructures. After the explanation of the computational methods, we demonstrate how to utilize these methods for the purpose of the investigation of graphene and some simple forms of the nanostructures—different kinds of nanoribbons and their modifications. Then, we will concentrate on the calculation of the properties of more complicated forms—the graphitic nanocone [1–3] and the graphitic wormhole [4–6]. In the first case, we consider the influence of the additional effects like the spin-orbit coupling (SOC) and the boundary effects coming from the finite size and from the extreme curvature of the surface geometry in the tip. In the second case, we investigate the effects that arise in the place of the wormhole bridge. Here, two additional effects appear: first, the SOC arising in the connecting nanotube and second, the increase of the electron mass due to relativistic effects coming from the extreme curvature of the surface geometry. As a result, the chiral massive electrons should be observed.

2. Computational formalism

The electronic structure can be characterized by the density of states (DOS)—the number of the electronic states per the unit interval of energies. This quantity can be used as the measure of the density of the electrons and generally we can say that the higher value of DOS, the higher conductivity. With the help of DOS, the electric field can be calculated as well. Besides DOS, one more quantity is defined—the local density of states (LDOS). It is DOS related to the unit area of the molecular surface or to the unit area of the surface in the space of the wave vector $\vec{k}$. Then, the quantities depend on the variables as follows:

$$DOS = DOS(E), \quad LDOS = LDOS(E,\vec{r}), \quad \text{resp.} \quad LDOS = LDOS(E,\vec{k}). \tag{1}$$

Two methods are used for the calculation of LDOS. The first one is used for the periodical structures with the planar geometry (plain graphene, nanotubes, nanoribbons, etc.), the second one is used for the structures that are aperiodical or that have the curved geometry (fullerene, nanocone, wormhole, nanotoroid, etc.). We outline here the base of these methods. Both the methods start on solving the Schrödinger equation for the electron bounded on the molecular surface

$$\hat{H}\psi = E\psi. \tag{2}$$

Here,

$$\hat{H} = \frac{\hat{k}^2}{2m} + \hat{V}(\vec{r}) + \hat{U}(\vec{r}), \tag{3}$$

$\hat{V}$ representing the potential of the periodic crystal, $\hat{U}$ representing the external potential that is responsible for the curvature.

2.1. Periodical structures with planar geometry

For the periodical structures, the external potential in Eq. (3) is zero, and the carbon lattice can be divided into several sublattices, each containing equivalent atomic sites. We can denote the sites corresponding to the different sublattices as $A, B, C, ...$ or $A_1, A_2, A_3, ...$. We can find a unit cell—the smallest possible cell in the structure that contains all possible inequivalent atomic sites (**Figure 6**).

In the case of graphene, the wave function, which solves Eq. (2), can be expressed as in references [7, 8]

$$\psi = C_A\psi_A + C_B\psi_B, \tag{4}$$

where the components ψ_A, ψ_B correspond to the particular sublattices. In the tight-binding approximation, we postulate the solution in the form

$$\psi_{A(B)} = \sum_{A(B)} \exp[i\vec{k} \cdot \vec{r}_{A(B)}] X(\vec{r} - \vec{r}_{A(B)}), \tag{5}$$

where $X(\vec{r})$ is the atomic orbital function. The overlap is zero, i.e.,

$$\int X^*(\vec{r} - \vec{r}_A)X(\vec{r} - \vec{r}_B)\mathrm{d}\vec{r} = 0. \tag{6}$$

By the substitution of the solution in Eq. (5) into the Schrödinger equation (Eq. (2)), multiplying it by $\psi^{\dagger}$ and making the integration over $\vec{r}$, we create the expressions

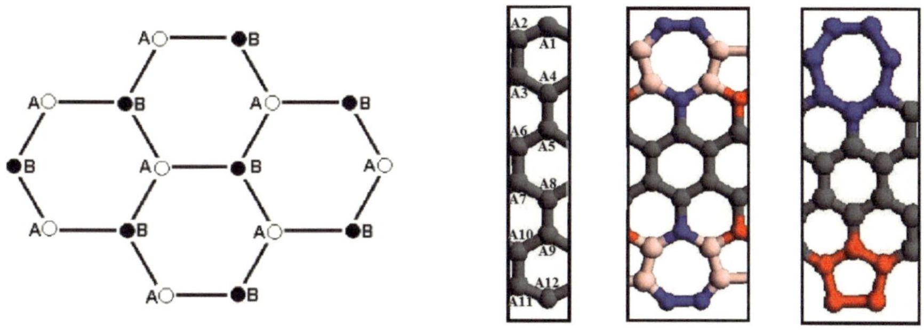

Figure 6. Unit cells for different periodical structures: graphene (left), different kinds of nanoribbons (right). The graphene structure is considered to be infinite in 2D, the nanoribbons are considered to have final width and the second size is infinite as well.

$$H_{ab} = \int \psi_a^* H \psi_b \mathrm{d}\vec{r}, \quad S = \int \psi_A^* \psi_A \mathrm{d}\vec{r} = \int \psi_B^* \psi_B \mathrm{d}\vec{r}, \quad a,b \equiv A,B. \tag{7}$$

If we suppose that the functions X are normalized so that $\int X^*(\vec{r}-\vec{r}_{A(B)})X(\vec{r}-\vec{r}_{A(B)})\mathrm{d}\vec{r}=1$, then S gives the number of the unit cells in the nanostructure. Now, the Schrödinger equation is transformed into the matrix form

$$\begin{pmatrix} H_{AA} & H_{AB} \\ H_{BA} & H_{BB} \end{pmatrix}\begin{pmatrix} C_A \\ C_B \end{pmatrix} = ES\begin{pmatrix} C_A \\ C_B \end{pmatrix}. \tag{8}$$

The eigenvalues of the matrix in this equation are the energy eigenvalues, and they create the electronic spectrum. For this purpose, first, we need to determine the values of the matrix elements H_{ab}. From their definition follows

$$\begin{aligned} H_{ab} &= \int \psi_a^* H \psi_b \mathrm{d}\vec{r} = \sum_{a,b} \exp\left[-\mathrm{i}\vec{k}\cdot(\vec{r}_A - \vec{r}_B)\right] \int X^*(\vec{r} - \vec{r}_a) H X(\vec{r} - \vec{r}_b)\mathrm{d}\vec{r} \\ &= \sum_{a,b} \exp\left[-\mathrm{i}\vec{k}\cdot(\vec{r}_A - \vec{r}_B)\right]\gamma_{ab}, \end{aligned} \tag{9}$$

where γ_{ab} ab denotes the corresponding hopping integral. The eigenvalues are labeled by $\vec{k}$ and in the nearest-neighbor approximation, they can be expressed as

$$E(\vec{k}) = \pm\gamma_0\sqrt{1 + 4\cos^2\frac{k_y a}{2} + 4\cos\frac{k_y a}{2}\cos\frac{k_x a\sqrt{3}}{2}}, \tag{10}$$

where $\gamma_0=\gamma_{AA}=\gamma_{BB}$ is the hopping integral for the nearest neighboring atoms and a=2.46A is the distance between the next-nearest atomic neighbors. DOS and LDOS are then defined as

$$DOS(E) = -\frac{1}{\pi}\lim_{\omega\to 0}\mathrm{Im}\int\frac{d\vec{k}}{E - E(\vec{k}) + i\omega}, \qquad LDOS(E,\vec{k}) = \delta(E - E(\vec{k})). \qquad (11)$$

The corresponding graphs of electronic spectrum and DOS are given in **Figure 7**. We see that for zero energy, the density of states has zero value. This property is typical for the semimetallic nanostructures. For the metallic nanostructures, a peak appears for zero energy. In the first case, a gap is present around zero in the electronic spectrum. Its width can be influenced by the additional defects in the hexagonal structure or by the chemical admixtures and in this way, a material with the predefined properties can be synthesized. In the second case, the gap around zero energy is absent.

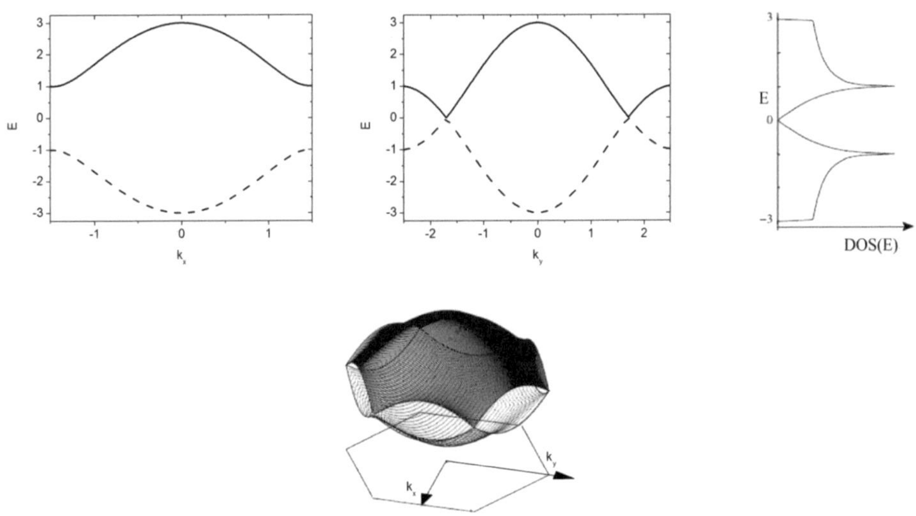

Figure 7. The electronic spectrum for k_y = 0 (left) and k_x = 0 (middle) and the density of states (right) of graphene. The solid and dashed lines correspond to the positive and the negative energy values, respectively. In the bottom, we see the form of the electronic spectrum for an arbitrary value of the wave vector.

In a similar way, but with a more complicated structure of the wave function in Eq. (4) and for a larger size of the matrix in Eq. (8), the electronic spectrum and DOS can be found for other nanostructures like the nanoribbons in the right side of **Figure 6** [9].

The results we see in **Figure 8**. In the left part, the shape of the segment of the concrete nanoribbon is present. The plot of the electronic spectrum and DOS are given in the middle and in the right part, respectively. The direction of the wave vector $\vec{k}$ is considered longitudinal.

The upper part corresponds to the nanoribbon with zigzag edges [10] and with 12 atomic sites in the unit cell (see **Figure 6**). That is why the size of the appropriate matrix in Eq. (8) would be 12 × 12 and its spectrum contains 12 eigenvalues. This corresponds to 12 lines in the graph of the electronic spectrum. The graph of DOS shows a zero energy peak that signalizes the metallicity of this kind of nanostructure. It is a typical property for the zigzag nanoribbons unlike the armchair nanoribbons [10].

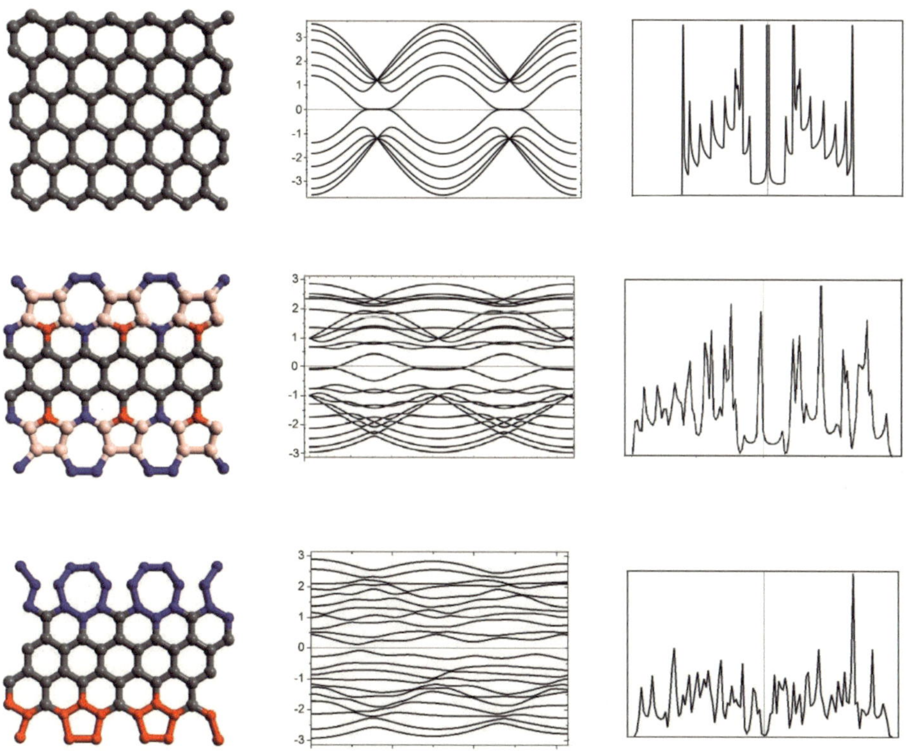

Figure 8. Electronic spectrum and DOS of different kinds of nanoribbons.

The middle and the bottom part correspond to some modifications of the previous form—the nanoribbon with the reconstructed edges. This causes the enlargement of the unit cell (**Figure 6**) and, consequently, a more complicated structure of the electronic spectrum. The metallic properties depend on the concrete kind of the modification: for the nanostructure in the middle part, the zero energy peak in DOS is preserved, whereas it disappears for the nanostructure in the bottom part. Furthermore, in the first case, the electronic spectrum gets a more complicated structure—the number of the Dirac points, where the lines are crossing, is doubled. This feature remains and strengthens for the larger width: in **Figure 9**, the form of the electronic spectrum is depicted for the same kind of the nanostructure and its width is three times larger.

2.2. Structures with curved geometry

For the curved structures, the nontrivial geometry is described by the external potential $U(\vec{r})$ in Eq. (3). Because of the aperiodicity, the eigenvalues cannot be labeled by the wave vector $\vec{k}$. Nevertheless, the solution of the Schrödinger equation (Eq. (2)) can be expressed with the help of the solutions for the previous case as so labeling by the wave vector will still play a key role in the following procedure.

$$\Psi(\vec{r}) = \int \mathrm{d}\vec{k}\psi_{\vec{k}}(\vec{r}), \tag{12}$$

For the purpose of the calculations, we express the wave function that solves Eq. (2) in the case of zero external potential in the form of the so-called "Luttinger–Kohn base" [11]:

$$\psi_{\vec{k}}(\vec{r}) = f_A(\vec{\kappa})e^{i\vec{\kappa}\cdot\vec{r}}\psi_A(\vec{K},\vec{r}) + f_B(\vec{\kappa})e^{i\vec{\kappa}\cdot\vec{r}}\psi_B(\vec{K},\vec{r}), \tag{13}$$

where $\vec{\kappa} = \vec{k} - \vec{K}$, $\vec{K}$ being the Dirac point and $E(\vec{k})$ is the appropriate eigenvalue for the zero external potential. After the substitution of this expression into Eq. (12), we get

$$\Psi(\vec{r}) = \int \mathrm{d}\vec{\kappa}\left(f_A(\vec{\kappa})e^{i\vec{\kappa}\cdot\vec{r}}\psi_A(\vec{K},\vec{r}) + f_B(\vec{\kappa})e^{i\vec{\kappa}\cdot\vec{r}}\psi_B(\vec{K},\vec{r})\right). \tag{14}$$

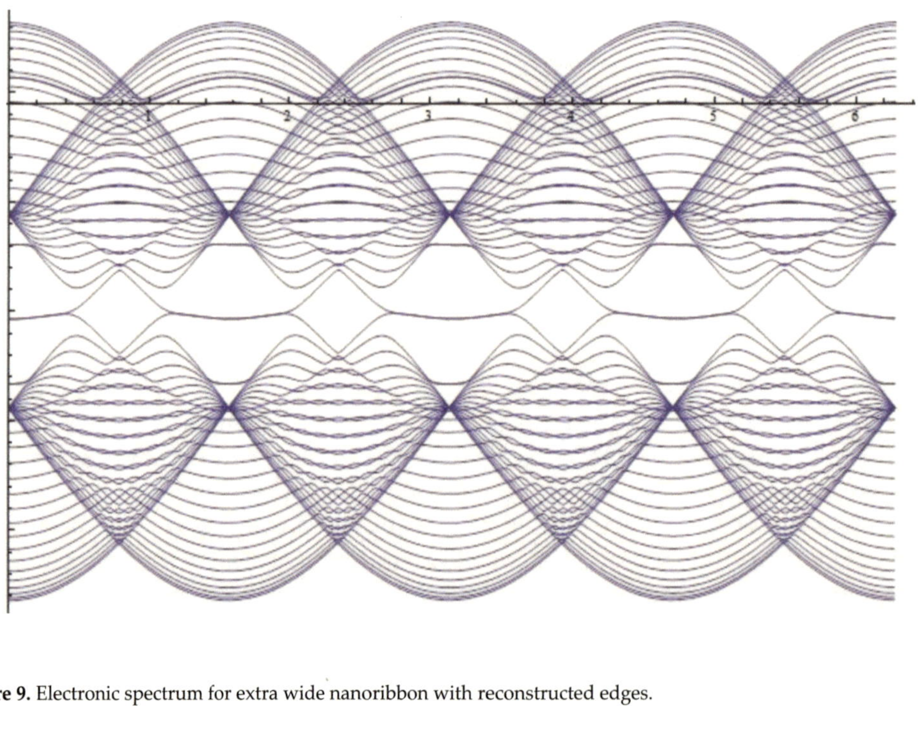

Figure 9. Electronic spectrum for extra wide nanoribbon with reconstructed edges.

This wave function will be substituted into Eq. (3). In reference [12], a sequence of steps is described whose result is the transformation of this equation to a two-dimensional (2D) Dirac-like equation for the massless fermion. In the practical calculations, a suitable choice of the coordinates is useful. In our case, we will suppose the rotational symmetry. Then, we perform the transformation of the coordinates: $(x, y, z) \to (\xi, \varphi)$, where φ is the angular coordinate. Then, the resulting Dirac-like equation will have the form [13, 14]

$$\mathrm{i}\sigma^{\alpha} e^{\mu}_{\alpha}[\partial_{\mu} + \Omega_{\mu} - \mathrm{i}a_{\mu} - \mathrm{i}a^{W}_{\mu} - \mathrm{i}A_{\mu}]\Psi = E\Psi. \tag{15}$$

The meaning of the particular terms is the following: e^{μ}_{α}, the zweibein, is connected with metric and using the corresponding tensor, it can be defined as

$$g_{\mu\nu}(x) = e^{\alpha}_{\mu}(x) e^{\beta}_{\nu}(x) \eta_{\alpha\beta}. \tag{16}$$

Here, $\eta_{\alpha\beta}$ is the metric of the plain space without curvature. Next term, Ω_{μ}, which is the spin connection in the spinor representation, is defined as $\Omega_{\mu} = \frac{1}{8}\omega_{\mu}^{\alpha\beta}[\sigma_{\alpha}, \sigma_{\beta}]$. Here, ω_{μ} is a more usual form of the spin connection. For its definition, we have to stress that the rotational symmetry is supposed. Then, it has the form

$$\omega^{12}_{\varphi} = -\omega^{21}_{\varphi} = 1 - \frac{\partial_{\xi}\sqrt{g_{\varphi\varphi}}}{\sqrt{g_{\xi\xi}}} = 2\omega, \qquad \omega^{12}_{\xi} = \omega^{21}_{\xi} = 0. \tag{17}$$

It remains to explain the sense of the gauge fields a_{μ}, a_{μ}^{W}, A_{μ}. First two of them ensure the circular periodicity. Their form is

$$a_{\varphi} = N/4, \qquad a^{W}_{\varphi} = -\frac{1}{3}(2m+n), \tag{18}$$

where N is the number of defects and (n,m) is the chiral vector. The last term, A_{μ}, represents one possible additional effect—the magnetic field.

The rotational symmetry enables to find the solution of Eq. (15) with the help of the substitution

$$\Psi^{T} = \begin{pmatrix} F_1 \\ F_2 \end{pmatrix} = \frac{1}{\sqrt[4]{g_{\varphi\varphi}}} \begin{pmatrix} u_j(\xi) e^{i\varphi j} \\ v_j(\xi) e^{i\varphi(j+1)} \end{pmatrix}, \qquad j = 0, \pm 1, \ldots, \tag{19}$$

from which we get the system

$$\frac{\partial_{\xi} u_j}{\sqrt{g_{\xi\xi}}} - \frac{\tilde{j}}{\sqrt{g_{\varphi\varphi}}} u_j = Ev, \qquad -\frac{\partial_{\xi} v_j}{\sqrt{g_{\xi\xi}}} - \frac{\tilde{j}}{\sqrt{g_{\varphi\varphi}}} v_j = Eu. \tag{20}$$

Here,

$$\tilde{j} = j + 1/2 - a_{\varphi} - a_{\varphi}^{W} - A_{\varphi}. \tag{21}$$

From the solution, LDOS is defined as the square of the absolute value of the wave function. In this case, we get it as the sum of squares of the absolute values of its ξ-components:

$$LDOS(E,\xi) = |u(E,\xi)|^2 + |v(E,\xi)|^2 . \tag{22}$$

3. Properties of the graphitic nanocone

The graphitic nanocone is a nanostructure that can be created from the plain graphene by the insertion of the pentagonal defects into the hexagonal structure. The number of these defects can vary from one to five. In this way, the conical tip arises and its smoothness and the vortex angle are given by the number of the defects and their placement. Then, the real geometry of the graphitic nanocone and the pure conical geometry are different (**Figure 10**, left part). The value of the vortex angle φ for the purely conical geometry can be calculated as

$$\sin\frac{\varphi}{2} = 1 - \frac{N}{6}, \tag{23}$$

where N is the number of the pentagonal defects in the conical tip.

The electronic structure of the graphitic nanocone with purely conical geometry and without any additional effects was investigated in reference [15]. There, the solution of Eq. (15) for this case is derived. Here, using the gauge field theory approach, we introduce the results of the calculations in different approximations: the nanocone with purely conical geometry influenced by the SOC [1] and the same case with the additional effect of the Coulomb interaction coming from the charge placed into the conical tip [3]. The reason is following: it is one of the possibilities how to simulate the real geometry in the conical tip.

3.1. Electronic structure influenced by the spin-orbit coupling

In the case of the purely conical structure, this form is assigned to the Hamiltonian in the Schrödinger equation (Eq. (2)) [15]:

$$\hat{H}_0 = \begin{pmatrix} H_1 & 0 \\ 0 & H_{-1} \end{pmatrix} \qquad \hat{H}_{0s} = \mathrm{i}\hbar v_F \left\{ \tau^y \partial_r - \tau^x r^{-1} \left[(1-\eta)^{-1} \left(s\partial_\varphi - \frac{3}{2}\mathrm{i}\eta \right) - \frac{1}{2}\tau^z \right] \right\}. \tag{24}$$

In this equation, v_F is the Fermi velocity, $s = \pm 1$ denotes the value of the K spin, $\eta = N/6$, τ^x, τ^y, τ^z are the Pauli matrices—these matrices have nothing to do with SOC. The points on the surface are described by the coordinates r and φ. The value of r is given by the distance from the tip (see **Figure 10**, right part).

SOC is incorporated using the substitutions [1]

$$\partial_r \to \partial_r - \frac{\delta\gamma'}{4\gamma R}\sigma_x(\vec{r}), \qquad \mathrm{i}\partial_\varphi \to \mathrm{i}\partial_\varphi + s(1-\eta)A_y\sigma_y, \tag{25}$$

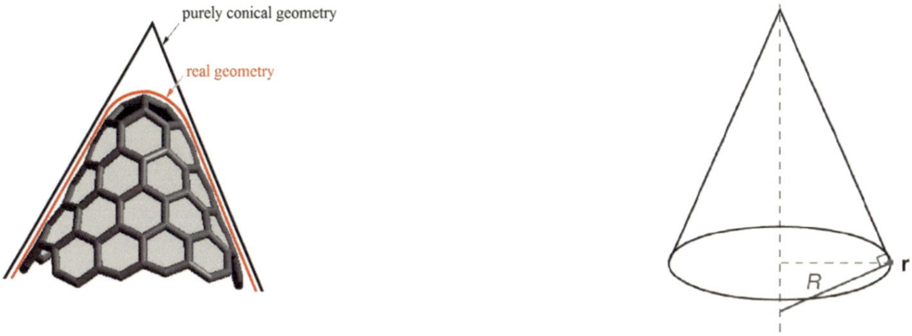

Figure 10. Left: the deviation of the geometry of the graphitic nanocone from the geometry of the real nanocone. Right: The notation of the variables in the nanocone.

where $R=\frac{(1-\eta)r}{\sqrt{\eta(2-\eta)}}$, $A_y=s\frac{2\delta p}{(1-\eta)}\sqrt{\eta(2-\eta)}$. Here, in the analogy to the case of the nanotube, R represents the curvature [16] (the geometrical meaning is also presented in **Figure 10**, right part), A_y is connected with the curvature of the carbon bonds. Next, γ and γ' represent the nearest neighbor and the next-nearest neighbor hopping integrals, respectively, and $\sigma_x(\vec{r})=\sigma^x\cos\varphi-\sigma^z\sin\varphi$. Here, σ^x, σ^y, σ^z are the Pauli matrices, but (unlikely the τ matrices) this time they are connected with SOC. The parameters p and δ are connected with the hopping integrals and the atomic potential, respectively. Their closer explanation can be also found in reference [16]. The following choice of their values was used: δ is of the order between 10^{-3} and 10^{-2}, $\frac{\gamma'}{\gamma}\sim\frac{8}{3}$, $p\sim 0.1$. After making the transformation $\hat{H}_s \to e^{\mathrm{i}\frac{\sigma_y}{2}\varphi}\hat{H}_s e^{-\mathrm{i}\frac{\sigma_y}{2}\varphi}$, which transforms the coordinate frame into the local coordinate frame, the final form of the Hamiltonian is

$$\hat{H}_s = \hbar v_F \begin{pmatrix} 0 & \partial_r - \mathrm{i}\frac{1}{r}\xi_x\sigma_x(\vec{r}) - \frac{\mathrm{i}s\partial_\varphi}{(1-\eta)r} - \frac{A_y}{r}\sigma_y - \frac{3\eta}{2(1-\eta)r} + \frac{1}{2r} \\ -\partial_r + \mathrm{i}\frac{1}{r}\xi_x\sigma_x(\vec{r}) - \frac{\mathrm{i}s\partial_\varphi}{(1-\eta)r} - \frac{A_y}{r}\sigma_y - \frac{3\eta}{2(1-\eta)r} - \frac{1}{2r} & 0 \end{pmatrix}. \tag{26}$$

It includes the strength of SOC through the parameters ξ_x, ξ_y:

$$\xi_x = \frac{\delta\gamma'\sqrt{\eta(2-\eta)}}{4(1-\eta)\gamma}, \qquad \xi_y = A_y + \frac{1}{2(1-\eta)}. \tag{27}$$

Now, the equation

$$\hat{H}_s\psi(r,\varphi) = E\psi(r,\varphi) \tag{28}$$

will be solved for the calculation of LDOS. Similarly as in Eq. (19), we can do the following factorization due to the rotational symmetry:

$$\psi(r,\varphi) = e^{ij\varphi}\begin{pmatrix} f_{j\uparrow}(r) \\ f_{j\downarrow}(r) \\ g_{j\uparrow}(r) \\ g_{j\downarrow}(r) \end{pmatrix} \tag{29}$$

It changes the equation into the form

$$\begin{pmatrix} 0 & 0 & \partial_r + \frac{F}{r} & -\frac{\mathrm{i}}{r}C \\ 0 & 0 & -\frac{\mathrm{i}}{r}D & \partial_r + \frac{F}{r} \\ -\partial_r + \frac{F-1}{r} & \frac{\mathrm{i}}{r}D & 0 & 0 \\ \frac{\mathrm{i}}{r}C & -\partial_r + \frac{F-1}{r} & 0 & 0 \end{pmatrix}\begin{pmatrix} f_{j\uparrow}(r) \\ f_{j\downarrow}(r) \\ g_{j\uparrow}(r) \\ g_{j\downarrow}(r) \end{pmatrix} = E\begin{pmatrix} f_{j\uparrow}(r) \\ f_{j\downarrow}(r) \\ g_{j\uparrow}(r) \\ g_{j\downarrow}(r) \end{pmatrix}. \tag{30}$$

Next parameters appearing in this equation are

$$F = \frac{sj}{1-\eta} - \frac{3}{2}\frac{\eta}{1-\eta} + \frac{1}{2}, \qquad C = \xi_x - \xi_y, \qquad D = \xi_x + \xi_y. \tag{31}$$

In reference [1], a numerical method is introduced in detail that helps to find the solution of this system. Using a modified version of Eq. (22) (we sum up the squares of absolute values of four components instead of two components), we calculate LDOS from this solution. For different numbers of the defects in the conical tip, we see the resulting LDOS in **Figure 11**. It involves the modes $j = -1, 0, 1, 2, 3$ with the same weight. While for the case of one and two defects in the tip, arbitrary energy and $r = 0$, LDOS grows to infinity, in the case of three defects in the tip, this effect appears close to zero energy only. Using a more thorough analysis, one could find out that the peak for the case of three defects corresponds to the case of the mode $j = -1$ and for other modes and the same number of defects, the behavior is the same as in the case of one and two defects.

From these results follows that there could be a strong localization of the electrons in the tip, especially near zero energy in the case of three defects. Now, we will be interested, if this behavior remains the same after the inclusion of some boundary effects that should simulate the real geometry of the nanostructure. Furthermore, we would like to ensure in this way the quadratical integrability of the solution.

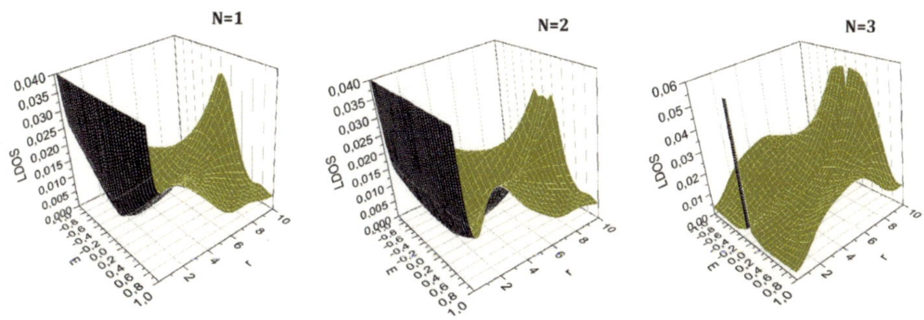

Figure 11. Three-dimensional graphs of LDOS of the graphitic nanocone influenced by SOC. Here, LDOS corresponds to the sum of the solutions corresponding to $j = 1,0,1,2,3$. The number of the defects in the tip in the particular cases: $N = 1$ (left), $N = 2$ (middle) and $N = 3$ (right).

3.2. Incorporation of the boundary effects by a charge simulation

The influence of the charge considered in the conical tip is expressed in the Hamiltonian by the presence of the diagonal term $-\frac{\kappa}{r}$, where $\kappa = 1/137$ is the fine structure constant. This term substitutes the diagonal zeros in Eq. (30):

$$\begin{pmatrix} -\frac{\kappa}{r} & 0 & \partial_r + \frac{F}{r} & -\frac{i}{r}C \\ 0 & -\frac{\kappa}{r} & -\frac{i}{r}D & \partial_r + \frac{F}{r} \\ -\partial_r + \frac{F-1}{r} & \frac{i}{r}D & -\frac{\kappa}{r} & 0 \\ \frac{i}{r}C & -\partial_r + \frac{F-1}{r} & 0 & -\frac{\kappa}{r} \end{pmatrix} \begin{pmatrix} f_{j\uparrow,C}(r) \\ f_{j\downarrow,C}(r) \\ g_{j\uparrow,C}(r) \\ g_{j\downarrow,C}(r) \end{pmatrix} = E \begin{pmatrix} f_{j\uparrow,C}(r) \\ f_{j\downarrow,C}(r) \\ g_{j\uparrow,C}(r) \\ g_{j\downarrow,C}(r) \end{pmatrix}. \quad (32)$$

In this way, the parallel influence of both the Coulomb interaction and SOC is considered [3]. To solve the resulting equation, we use the analogy of the numerical method used in [1]—this analogy is presented in reference [3]. From the calculated results, LDOS is calculated using Eq. (22) again.

The graphs of LDOS based on the found solution are sketched in **Figure 12** for the same modes and numbers of the defects as in **Figure 11**, i.e., $-1 \leq j \leq 3$. In spite of our expectations, this time the behavior of the found result is the same for arbitrary number of the defects, i.e., the appearance and the uniqueness of the peak in the case of three defects are distorted.

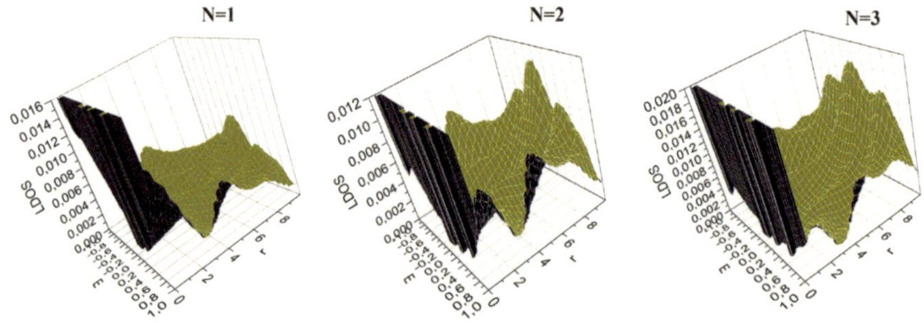

Figure 12. Graphs of LDOS of the graphitic nanocone influenced by the Coulomb interaction (including the influence of SOC) for different distances $\vec{r}$ from the tip, $-1 \leq j \leq 3$ and for different numbers of the defects.

3.3. Comparison of the results

Now, we would like to verify the possible quadratic integrability of the solution found for the case of the additional effect coming from the charge simulation. In **Figure 13**, we see the dependence of LDOS on $\vec{r}$ variable close to zero energy for the case of the influence of SOC only and of the simultaneous influence of SOC and the Coulomb interaction. We see here that in comparison with the first case, in the second case, the decrease of LDOS close to $r = 0$ is much faster and one could suppose that the quadratic integrability of the acquired solution is achieved here. To gain confidence with our conclusion, we have to do the integration of LDOS in the investigated interval close to $r = 0$ in all the outlined cases. This task is still in progress.

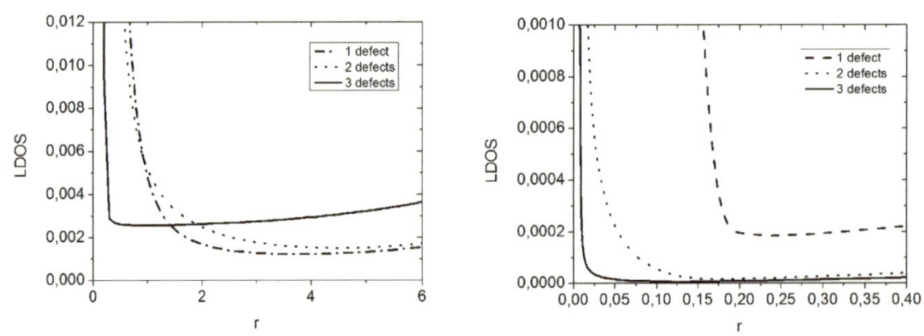

Figure 13. Behavior of LDOS for zero energy close to $r = 0$ for different numbers of defects in the conical tip: influence of SOC only (left) and the simultaneous influence of SOC and the Coulomb interaction (right).

4. Properties of the graphitic wormhole

The wormhole is understood as a form that arises when two graphene sheets are connected together with the help of the connecting nanotube. This can be achieved by the supply of the heptagonal defects on both sides of the given nanotube. The number of the defects can vary from 1 to 12. The composition of the graphitic wormhole is depicted in **Figure 14**: it consists of the connecting nanotube and two (perturbed or unperturbed) graphene sheets. The places of the connections are called the wormhole bridges. Because of the physical limitations, the radius of the nanotube must be much larger than its length (this fact is ignored in **Figure 14** for the better illustration of the composition). The limit case of 12 defects is described in references [4, 17], in the other cases, we speak about the so-called "perturbed wormhole" Here, using the formalism of the subsection II B, we derive the electronic structure for both cases, and we will find out the form of the zero modes on the wormhole bridge. Furthermore, we investigate the influence of the additional effects that could appear here due to extreme curvature in the place of the wormhole bridge—the relativistic mass acquisition of the present electrons. This effect together with the effect of SOC that appears in the carbon nanotubes [16] could lead to the appearance of the zero modes of the chiral massive electrons in the place of the wormhole bridge. This could serve as a useful instrument for the detection of the wormhole structures in the graphene bilayer during the process of the synthesis of the corresponding material.

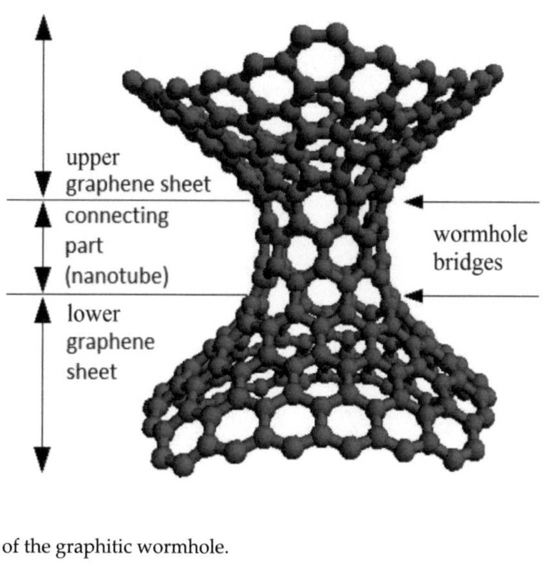

Figure 14. The composition of the graphitic wormhole.

4.1. Electronic structure

We will solve Eq. (15) in the subsection II B. In this case, the metric tensor has the form

$$g_{\mu\nu} = \Lambda^2(r_\pm)\begin{pmatrix} 1 & 0 \\ 0 & r_\pm^2 \end{pmatrix}, \qquad \Lambda(r_\pm) = (a / r_\pm)^2 \theta(a - r_\pm) + \theta(r_\pm - a). \tag{33}$$

Here, θ is the Heaviside step function, r_-, r_+ are the polar coordinates corresponding to the lower and the upper graphene sheet, respectively and $a=\sqrt{r_- r_+}$ is the radius of the wormhole.

The values of the components of a_μ depend on the chiral vector [18] of the connecting nanotube. For our purpose, this vector is (6*n*, 6*n*) and (6*n*, 0). In most cases, a_μ has then the components

$$a_\varphi = \frac{3}{2}, \qquad a_r = 0. \tag{34}$$

The only exception is when the chiral vector is (6*n*, 0), where *n* is not divisible by 3. Then,

$$a_\varphi = \frac{1}{2}, \qquad a_r = 0. \tag{35}$$

Knowledge of the spin connection is also needed—the values of the components have the form

$$\Omega_\varphi = -\frac{\mathrm{i}}{2}\sigma_3\left(r\frac{\Lambda'(r)}{\Lambda(r)} + 1\right), \qquad \Omega_r = 0. \tag{36}$$

All these expressions we substitute into Eq. (15). The resulting equation is

$$\mathrm{i}v_F\sigma^\mu(\partial_\mu + \Omega_\mu \mp \mathrm{i}a_\mu)\psi^\pm = \varepsilon\psi^\pm. \tag{37}$$

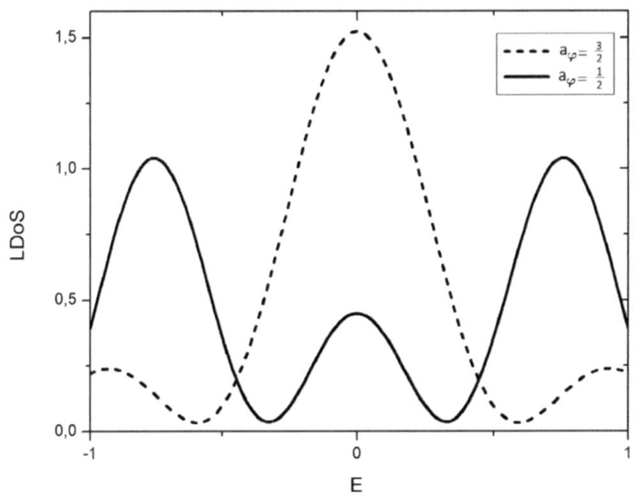

Figure 15. Local density of states on the bridge of the graphitic wormhole for different values of a_φ.

Here, each sign ± corresponds to a different Dirac point (the corner of the reciprocal unit lattice). We get these four possibilities: for $r \geq a$,

$$-\mathrm{i}v_F\left(\partial_r+\frac{1}{r}\mathrm{i}\partial_\theta\mp\frac{a_\varphi}{r}+\frac{1}{2r}\right)\psi_B^\pm=\varepsilon\psi_A^\pm,\qquad -\mathrm{i}v_F\left(\partial_r-\frac{1}{r}\mathrm{i}\partial_\theta\pm\frac{a_\varphi}{r}+\frac{1}{2r}\right)\psi_A^\pm=\varepsilon\psi_B^\pm \tag{38}$$

and for $0<r\leq a$,

$$\mathrm{i}v_F\left(\frac{r}{a}\right)^2\left(\partial_r-\frac{1}{r}\mathrm{i}\partial_\theta\pm\frac{a_\varphi}{r}-\frac{1}{2r}\right)\psi_B^\pm=\varepsilon\psi_A^\pm,\qquad \mathrm{i}v_F\left(\frac{r}{a}\right)^2\left(\partial_r+\frac{1}{r}\mathrm{i}\partial_\theta\mp\frac{a_\varphi}{r}-\frac{1}{2r}\right)\psi_A^\pm=\varepsilon\psi_B^\pm. \tag{39}$$

In the first case, the solution is

$$\begin{aligned}\psi^\pm=\begin{pmatrix}\psi_A^\pm(r,\varphi)\\ \psi_B^\pm(r,\varphi)\end{pmatrix}=c_1\begin{pmatrix}J_{j\mp a_\varphi-1/2}(kr)\\ -\mathrm{isgn}\varepsilon\, J_{j\mp a_\varphi+1/2}(kr)\end{pmatrix}\\ +c_2\begin{pmatrix}Y_{j\mp a_\varphi-1/2}(kr)\\ -\mathrm{isgn}\varepsilon\, Y_{j\mp a_\varphi+1/2}(kr)\end{pmatrix}.\end{aligned} \tag{40}$$

Here, $J_j(x)$ and $Y_j(x)$ are the Bessel functions of the integer order j and the energy $\varepsilon=\pm v_F k$. To calculate LDOS, similarly as in the previous section, Eq. (22) is used. In **Figure 15**, different behavior of LDOS, depending on the gauge field a_φ, is manifested.

4.2. Zero modes

For the presented possibilities, we investigate the zero modes—solutions of the Dirac equation for the zero energy. For this purpose, we consider zero values of the component $\psi_A^\pm$ of the solution. Then, from Eqs. (49) and (50) follows: for $r\geq a$,

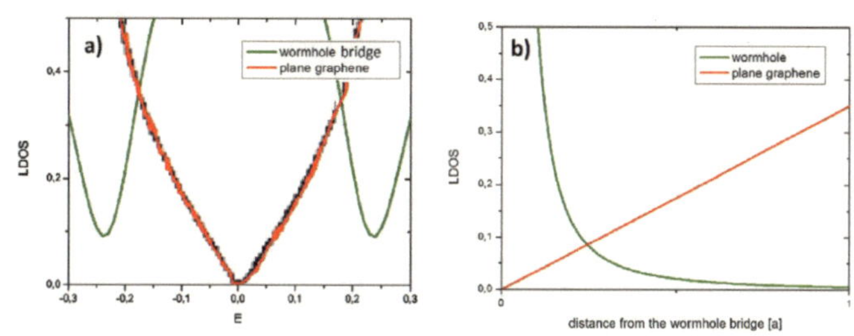

Figure 16. Comparison of the properties of the wormhole and the plain graphene: (a) local density of states, (b) zero modes.

$$\left(\partial_r-\frac{1}{r}\mathrm{i}\partial_\theta\mp\frac{a_\varphi}{r}+\frac{1}{2r}\right)\psi_B^\pm=0 \tag{41}$$

and for $0<r\leq a$,

$$\left(\partial_r-\frac{1}{r}\mathrm{i}\partial_\theta\pm\frac{a_\varphi}{r}-\frac{1}{2r}\right)\psi_B^\pm=0. \tag{42}$$

If $a_\varphi=\frac{3}{2}$ and $r\geq a$, the solution is

$$\psi_B^-(r,\varphi) \sim r^{-j-2} e^{\mathrm{i}\, j\varphi}. \tag{43}$$

The second possibility for this value of a_φ is $0<r\le a$, the corresponding solution is then

$$\psi_B^-(r,\varphi) \sim r^{-j+2} e^{\mathrm{i}\, j\varphi}. \tag{44}$$

Both solutions are strictly normalizable only for $j=0$. Analogous solution holds for ψ_B^+ and for $\psi_A^\pm$ if the components $\psi_B^\pm$ are chosen as zero.

There are no strictly normalizable solutions for the value $a_\varphi = \frac{1}{2}$. It means that in this case, the zero modes do not exist.

On the base of these results, one could expect a strong localization of LDOS near Fermi energy on the wormhole bridge. It is demonstrated in **Figure 16a**, where LDOS of the plain graphene is supplied for the comparison. It could be experimentally observed. In **Figure 16b**, we see the comparison of the zero modes of these two structures at different distances from the wormhole bridge.

4.3. Case of massive fermions

In the continuum gauge field theory, zero mass of the fermions in the Dirac equation is considered (in other words, it is very small in comparison with energy). On the other hand, the extreme curvature of the investigated structure leads to such values of the Fermi velocity that cause the appearance of the relativistic effects. The changes of the Fermi velocity due to curvature and other effects were demonstrated in references [13, 19]. As a result, the mass of the fermions becomes considerable, similarly as in the bilayer graphene [20, 21]. This effect is strengthened by the effective mass acquisition during the motion along the tube axis that happens due to the extreme size difference between the graphene sheets and the wormhole radius. This change of the space topology of graphene from 2D to 1D is similar to the string theory compactification. It means that we can image the wormhole connecting nanotube as the 1D object.

So we need to incorporate a mass term into the Dirac equation (Eq. (15)). To solve this problem, we go through the system of the corresponding equations (Eq. (20)) and transform it into the following differential equation of the second order:

$$\left(\partial_{\xi\xi} - \frac{1}{2g_{\xi\xi}} \partial_\xi g_{\xi\xi} + \frac{\tilde{j}}{2} \sqrt{\frac{g_{\xi\xi}}{g_{\varphi\varphi}^3}} \partial_\xi g_{\varphi\varphi} - \tilde{j}^2 \frac{g_{\xi\xi}}{g_{\varphi\varphi}} + E^2 g_{\xi\xi} \right) u_j = 0. \tag{45}$$

To simplify the calculations, the cylindrical geometry is supposed: the radius vector of the point at the surface changes as

$$\vec{R} = (R\cos\varphi, R\sin\varphi, \xi), \tag{46}$$

where R is the radius of the cylinder. In this case, Eq. (45) is considerably simplified:

$$\left(\partial_{\xi\xi} + E^2 - \frac{\tilde{j}^2}{R^2}\right)u_j = 0, \tag{47}$$

which is solved by [22]

$$u_j(\xi) = Ae^{k\xi} + Be^{-k\xi}. \tag{48}$$

Here,

$$k = \sqrt{\frac{\tilde{j}^2}{R^2} - E^2}. \tag{49}$$

In references [23, 24], in a very similar form the dispersion relation is given for the massive 1D Dirac equation:

$$k = \sqrt{M^2 - E^2}, \tag{50}$$

where M is the mass of the corresponding fermion. From reference [22], an analogy indeed follows between the 2D massless and 1D massive case. On this base, we rewrite Eq. (45) into the form

$$\left(\partial_{\xi\xi} - \frac{1}{2g_{\xi\xi}}\partial_\xi g_{\xi\xi} + \frac{\tilde{j}}{2}\sqrt{\frac{g_{\xi\xi}}{g_{\varphi\varphi}^3}}\partial_\xi g_{\varphi\varphi} - \tilde{j}^2\frac{g_{\xi\xi}}{g_{\varphi\varphi}} + (E^2 - M^2)g_{\xi\xi}\right)u_j = 0, \tag{51}$$

in this case, the mass M corresponds to the fermion in the altered conditions. Now we find the corrections of LDOS of the graphitic wormhole for different values of M. It is shown in **Figure 17**. Our prediction is that these massive particles arising in the wormhole nanotubes could create energy bulks on the wormhole bridge and in the close area that should be experimentally measured by the STM or by the Raman spectroscopy [25]. Moreover, this effect could be strengthened by the effect of SOC present in the connecting nanotube that was described in reference [16] for the nanotubes and in Section 3 for the nanocone. This effect causes next energy splitting and as the result, the aforementioned chiral massive electrons could appear.

Another possibility to identify the wormhole structure comes from the fact that the massive particles could create strain solitons and topological defects on the bridge of the bilayer graphene that should propagate throughout the graphene sheet. These are almost macroscopic effects and should be caught by the experimentalists [26].

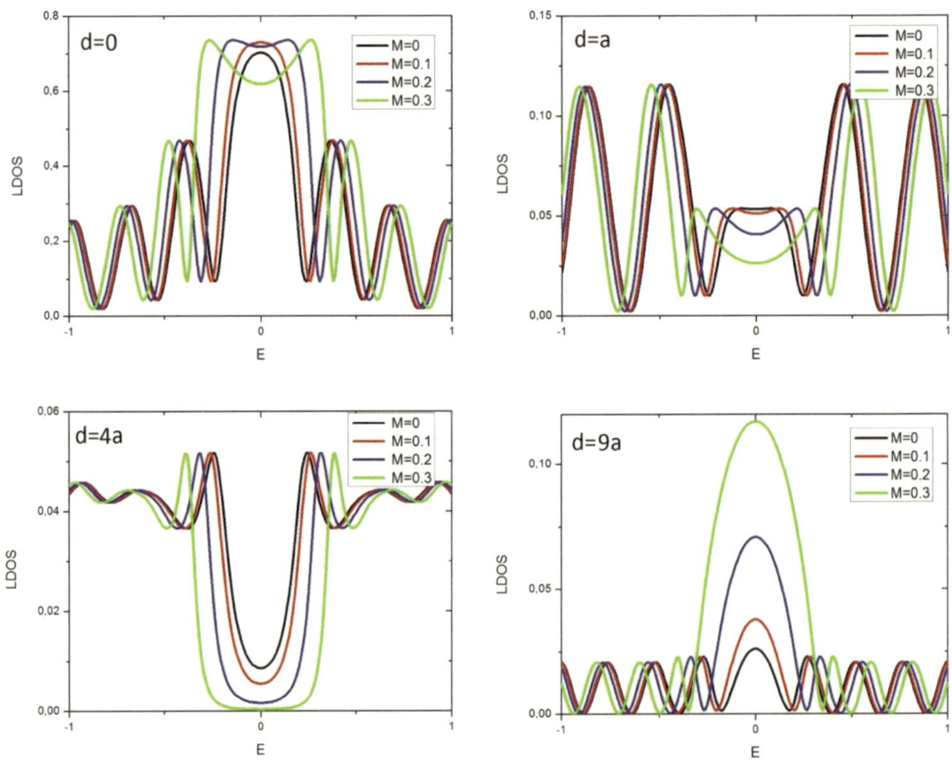

Figure 17. Comparison of LDOS for different masses of fermions at different distances from the wormhole bridge.

4.4. Case of perturbed wormhole

Now we will investigate how the electronic structure changes if the number of the heptagonal defects on the wormhole bridge is lowered—in this way, the perturbed wormhole is created.

In **Figure 18**, the possible forms of this structure are depicted. Due to symmetry preservation, only the even numbers of the defects, i.e., 2, 4, 6, 8, or 10, are considered.

The metric of the sheets can be draught by the radius vector

$$\vec{R}(z,\varphi) = \left(a\sqrt{1+\vartriangle z^2}\cos\varphi, a\sqrt{1+\vartriangle z^2}\sin\varphi, z\right), \tag{52}$$

where $\vartriangle$ is a positive real parameter; its value is derived from the number of the defects of the wormhole. In the case of $N = 2$ defects, we can say that the value of this parameter is negligible, so $\vartriangle << 1$. Then, the nonzero components of the metric are

$$g_{zz} = 1 + \frac{a^2 \vartriangle^2 z^2}{1 + \vartriangle z^2} \sim 1 + a^2 \vartriangle^2 z^2, \qquad g_{\varphi\varphi} = a^2(1 + \vartriangle z^2). \tag{53}$$

The nonzero components of the gauge fields are

$$a_\varphi = N/4, \qquad a_\varphi^W = -(2m+n)/3, \tag{54}$$

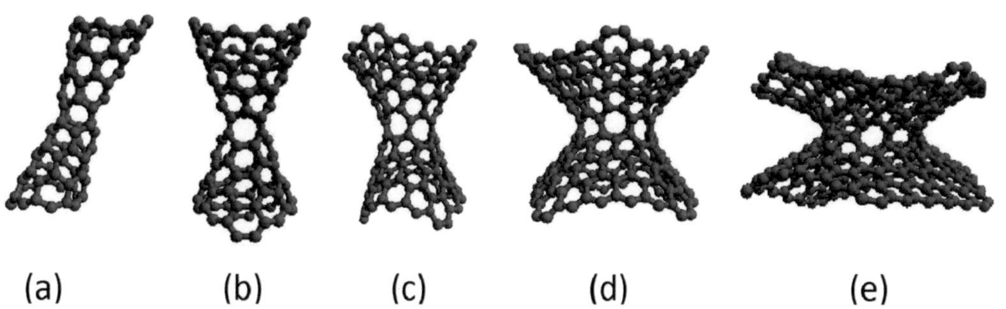

Figure 18. Different forms of the perturbed wormhole: (a) Two defects, (b) Four defects, (c) Six defects, (d) Eight defects, (e) Ten defects.

where (n, m) is the chiral vector of the connecting nanostructure. Then, regarding the form of the spin connection and by the substitution into Eq. (15), we get the solution

$$\psi_A(z) = C_{\vartriangle 1} D_{\nu_1}(\xi(z)) e^{ij\varphi} + C_{\vartriangle 2} D_{\nu_2}(i\xi(z))) e^{ij\varphi}, \tag{55}$$

$$\begin{aligned} \psi_B(z) = {} & \frac{C_{\vartriangle 1}}{E}\left(\partial_z D_{\nu_1}(\xi(z)) - \frac{\tilde{j} D_{\nu_1}(\xi(z))}{a}(1 - \frac{1}{2}\vartriangle^2 z^2) \right) e^{-ij\varphi} \\ & + \frac{C_{\vartriangle 2}}{E}\left(\partial_z D_{\nu_2}(i\xi(z)) - \frac{\tilde{j} D_{\nu_2}(i\xi(z))}{a}(1 - \frac{1}{2}\vartriangle^2 z^2) \right) e^{-ij\varphi}, \end{aligned} \tag{56}$$

where

$$\nu_1 = i\frac{a^2\Delta - 4a^2E^2 + 4ia\sqrt{\Delta}\tilde{j} + 4\tilde{j}^2}{8a\sqrt{\Delta}\tilde{j}}, \qquad \nu_2 = -i\frac{a^2\Delta - 4a^2E^2 - 4ia\sqrt{\Delta}\tilde{j} + 4\tilde{j}^2}{8a\sqrt{\Delta}\tilde{j}}, \tag{57}$$

$$\xi(z) = (-\Delta)^{1/4}\left(\sqrt{\frac{a}{2\tilde{j}}} + \sqrt{\frac{2\tilde{j}}{a}}z\right), \tag{58}$$

$D_\nu(\xi)$ being the parabolic cylinder function. The functions $C_{\Delta 1}=C_{\Delta 1}(E)$, $C_{\Delta 2}=C_{\Delta 2}(E)$ serve as the normalization constants. We see the graph of the local density of states in **Figure 19** (left part).

In the case of more than two defects, the value of Δ is not negligible, and we can get only the numerical approximation of LDOS. The derivation of the value of the parameter Δ follows from **Figure 19** (right part). From this figure it follows that in the middle part, the upper branch of the graphene sheet converges to the line $z=x\cdot\tan\alpha$, where we can suppose that the angle α depends on the number of the defects N linearly, i.e., $\alpha=\frac{\pi}{2}-N\cdot\frac{\pi}{24}$. (In this case, $\alpha=\frac{\pi}{2}$ corresponds to 0 defects and $\alpha=0$ corresponds to 12 defects.) Simultaneously, from Eq. (52) follows that asymptotically we have

$$\vec{R}(z\to\infty,\varphi)\to\left(a\sqrt{\Delta}z\cos\varphi, a\sqrt{\Delta}z\sin\varphi, z\right), \tag{59}$$

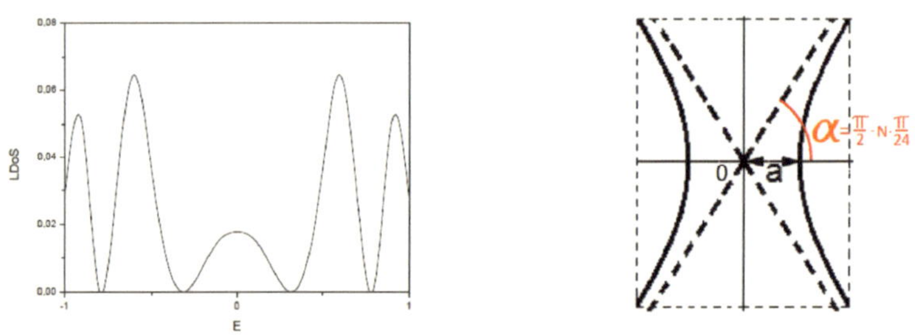

Figure 19. Left: LDOS on the bridge of the graphitic perturbed wormhole. Right: Derivation of the Δ parameter.

from which follows

$$z = x\cdot\tan\alpha = \left(a\sqrt{\Delta}\right)^{-1}x, \tag{60}$$

so

$$\Delta = \frac{1}{a^2 \tan^2 \alpha} = \frac{1}{a^2 \tan^2\left(\frac{\pi}{2} - N \cdot \frac{\pi}{24}\right)}. \tag{61}$$

In **Figure 20**, we see the comparison of LDOS for different kinds of the perturbed wormhole. From the plots it follows that the intensity is rising with the increasing number of the defects, and it is closer and closer approaching the results in **Figure 15**, where the case of 12 defects is shown.

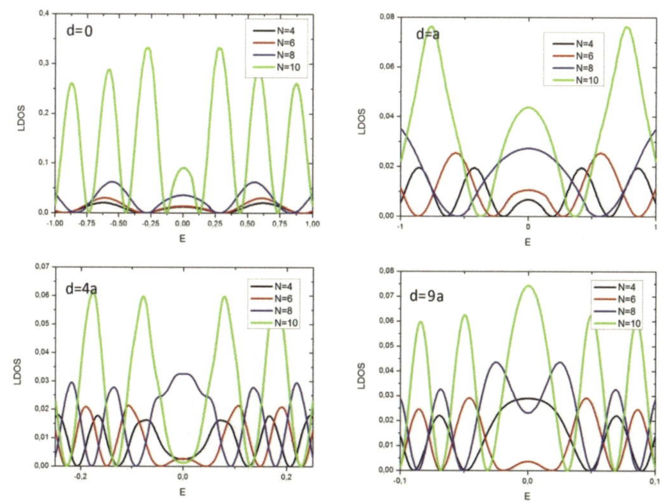

Figure 20. Comparison of LDOS for different numbers of the defects in the perturbed wormhole at different distances d from the wormhole bridge.

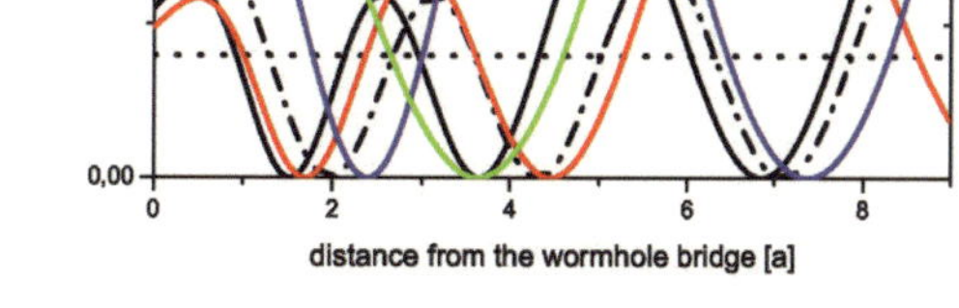

Figure 21. Zero modes of the perturbed wormhole for different numbers of the defects.

In **Figure 21**, LDOS of zero modes is shown for a varying distance from the wormhole bridge in the units of the radius a of the wormhole center. It was also acquired in the numerical way. For the unperturbed case (0 defects), the resulting plot resembles a line. In reference [27], the exponential solution is found for this case but with a very slow increase, so this could be that case. It is also seen from the plot that for the increasing number of the defects, the solution is approaching expressions in Eqs. (54) and (55) for the zero modes of the unperturbed wormhole.

Of course, the massive fermions could also appear in the case of the perturbed wormhole. We will not perform a detailed derivation of the electronic structure for the case of this eventuality, and we only note that the corrections to LDOS would be an analogy of the corrections shown in **Figure 17**.

5. Conclusion

We performed the calculations of the electronic structure for the graphitic nanocone and the graphene wormhole. In the first case, our aim was to find the quadratically integrable solution that includes the boundary effects and considers the real geometry. This goal was partially achieved, but we need to verify the properties of the found solution close to the tip. The precision of the calculations could be improved by the better choice of the corresponding geometry, by the consideration of the discretion of the energetic spectrum coming from the finite size of the nanostructure, and by the inclusion of next effects coming from the overlap of the neighboring atomic orbitals close to the tip [2]. The localization of the electrons shown in **Figures 11** and **12**, especially in the case of three defects, makes the graphitic nanocone a possible candidate for the construction of the scanning probe in atomic force microscopy.

In the second case of the graphene wormhole, we presented the mathematical motivation for our prediction of the effects that should appear close to the wormhole bridge. Our predictions will be verified with the help of the geometric optimizations and ab initio calculations. On this base, the most suitable candidates for the experiments will be chosen.

Author details

Jan Smotlacha[1,2*] and Richard Pincak[3,1]

*Address all correspondence to: smota@centrum.cz

1 Bogoliubov Laboratory of Theoretical Physics, Joint Institute for Nuclear Research, Dubna, Moscow region, Russia

2 Faculty of Nuclear Sciences and Physical Engineering, Czech Technical University, Brehova, Prague, Czech Republic

3 Institute of Experimental Physics, Slovak Academy of Sciences, Kosice, Slovak Republic

References

[1] Pincak R, Smotlacha J, Pudlak M. Spin-orbit interaction in the graphitic nanocone. Eur. Phys. J. B. 2015;88:17–22. DOI: 10.1140/epjb/e2014-50413-9

[2] Pincak R, Smotlacha J, Pudlak M. Calculation of the electronic structure near the tip of a graphitic nanocone. Physica B. 2014;441:58–61. DOI: 10.1016/j.physb.2014.02.012

[3] Smotlacha J, Pincak R. Boundary conditions and Green function approach of the spin-orbit interaction in the graphitic nanocone [Internet]. 2015. Available from: http://arxiv.org/abs/1511.03004. [Accessed: 2015-11-10]

[4] Gonzalez J, Guinea F, Herrero J. Propagating, evanescent, and localized states in carbon nanotube-graphene junctions. Phys. Rev. B. 2009; 79:165434. DOI: 10.1103/PhysRevB.79.165434

[5] Pincak R, Smotlacha J. Analogies in electronic properties of graphene wormhole and perturbed nanocylinder. Eur. Phys. J. B. 2013;86:480–486. DOI: 10.1140/epjb/e2013-40594-0

[6] Pincak R, Smotlacha J. The chiral massive fermions in the graphitic wormhole. Quantum Matter. 2016; 5:114–124. DOI: 10.1166/qm.2016.1262

[7] Wallace PR. The band theory of graphite. Phys. Rev. 1947;71:622–634. DOI: 10.1103/PhysRev.71.622

[8] Slonczewski JC, Weiss PR. Band structure of graphite. Phys. Rev. 1958;109:272–279. DOI: 10.1103/PhysRev.109.272

[9] Pincak R, Smotlacha J, Osipov VA. Electronic states of zigzag graphene nanoribbons with edges. Physica B. 2015;475:61–65. DOI: 10.1016/j.physb.2015.06.025

[10] Wakabayashi K, Sasaki K, Nakanishi T, Enoki T. Electronic states of graphene nanoribbons and analytical solutions. Sci. Technol. 2010;11:054504 DOI: 10.1088/1468-6996/11/5/054504

[11] DiVincenzo DP, Mele EJ. Self-consistent effective-mass theory for intralayer screening in graphite intercalation compounds. Phys. Rev B. 1984;29:1685–1694 DOI: 10.1103/PhysRevB.29.1685

[12] Callaway J. Quantum Theory of the Solid State. New York: Academic Press;1974. 824 p. DOI: 10.1107/S0567739478000662

[13] Kochetov EA, Osipov VA, Pincak R. Electronic properties of disclinated flexible membrane beyond the inextensional limit: application to graphene. J. Phys. Condens. Matter. 2010;22:395502. DOI: 10.1088/0953-8984/22/39/395502

[14] Smotlacha J, Pincak R, Pudlak M. Electronic structure of disclinated graphene in an uniform magnetic field. Eur. Phys. J. B. 2011;84:255–264. DOI: 10.1140/epjb/e2011-20384-6

[15] Sitenko YA, Vlasii ND. Electronic properties of graphene with a topological defect. Nucl. Phys. B. 2007;787:241–259. DOI: 10.1016/j.nuclphysb.2007.06.001

[16] Ando T. Spin-orbit interaction in carbon nanotubes. J. Phys. Soc. Jpn. 2000;69:1757–1763. DOI: 10.1143/JPSJ.69.1757

[17] Gonzalez J, Herrero J. Graphene wormholes: a condensed matter illustration of Dirac fermions in curved space. Nucl. Phys. B. 2010;825:426–443. DOI: 10.1016/j.nuclphysb.2009.09.028

[18] Saito R, Dresselhaus G, Dresselhaus MS. Tunneling conductance of connected carbon nanotubes. Phys. Rev. B. 1996;53:2044–2050. DOI: 10.1103/PhysRevB.53.2044

[19] Rostami H, Asgari R. Electronic ground-state properties of strained graphene. Phys. Rev. B. 2012;86:155435. DOI: 10.1103/PhysRevB.86.155435

[20] Borghi G, Polini M, Asgari R, MacDonald AH. Fermi velocity enhancement in monolayer and bilayer graphene. Solid State Commun. 2009;149:1117–1122. DOI: 10.1016/j.ssc.2009.02.053

[21] Katsnelson MI, Novoselov KS, Geim AK. Chiral tunnelling and the Klein paradox in graphene. Nature Phys. 2006;2:620–625. DOI: 10.1038/nphys384

[22] Alhaidari AD, Jellal A, Choubabi EB, Bahlouli H. Dynamical mass generation via space compactification in graphene. Quantum Matter. 2013;2:140–143. DOI: 10.1166/qm.2013.1039

[23] Greiner W. Relativistic Quantum Mechanics: Wave Equations. 3rd ed. Berlin: Springer; 1994. 424 p. DOI: 10.1007/978-3-662-04275-5

[24] Thaller B. The Dirac Equation. 1st ed. Springer, Berlin: Springer;1992. 357 p. DOI: 10.1007/978-3-662-02753-0

[25] Tan PH et al. The shear mode of multilayer graphene. Nature Mater. 2012;11:294–300. DOI: 10.1038/nmat3245

[26] Alden JS, Tsen AW, Huang PY, Hovden R, Brown L, Park J, Muller DA, McEuen PL. Strain solitons and topological defects in bilayer graphene [Internet]. 2013. Available from: http://arxiv.org/abs/1304.7549. [Accessed: 2013-04-29]

[27] Pincak R, Smotlacha J, Pudlak M. Electronic properties of disclinated nanostructured cylinders. NanoMMTA. 2013; 2:81–95. DOI: 10.2478/nsmmt-2013-0005

Chapter 3

Electronic Structure and Topological Quantum Phase Transitions in Strained Graphene Nanoribbons

Fanyao Qu, Ginetom S. Diniz and Marcos R. Guassi

Additional information is available at the end of the chapter

http://dx.doi.org/10.5772/64493

Abstract

In this chapter, we discuss the new classes of matter, such as the quantum spin Hall (QSH) and quantum anomalous Hall (QAH) states, that have been theoretically predicted and experimentally observed in graphene and beyond graphene systems. We further demonstrate how to manipulate these states using mechanical strain, internal exchange field, and spin-orbit couplings (SOC). Spin-charge transport in strained graphene nanoribbons is also discussed assuming the system in the QAH phase, exploring the prospects of topological devices with dissipationless edge currents. A remarkable zero-field topological quantum phase transition between the time-reversal-symmetry-broken QSH and quantum anomalous Hall states is predicted, which was previously thought to take place only in the presence of external magnetic field. In our proposal, we show as the intrinsic SOC is tuned, how it is possible to two different helicity edge states located in the opposite edges of the graphene nanoribbons exchange their locations. Our results indicate that the strain-induced pseudomagnetic field could be coupled to the spin degrees of freedom through the SOC responsible for the stability of a QSH state. The controllability of this zero-field phase transition with strength and direction of the strain is also explored as additional phase-tuning parameter. Our results present prospect of strain, electric and magnetic manipulation of the QSH, and QAH effect in these novel two-dimensional (2D) materials.

Keywords: Graphene, graphene nanoribbon, quantum spin Hall, quantum anomalous Hall, topological insulator, 2D materials, strain

1. Introduction

Starting from the work by Landau and Peierl's work [1, 2], two-dimensional (2D) materials were regarded as theoretical structures, thermodynamically unstable to be obtained in laboratory. This is because of fusion temperature decreases as function of thickness of thin films, causing the material to segregate in islands or decomposing in typically thicknesses of tens of atomic layers [3, 4]. In 1947, Wallace [5] demonstrated the electronic properties of what became the first theoretical work predicting the one-atom thick of carbon atoms. Past 57 years, his theoretical predictions were experimentally synthesized by Novoselov et al., which now is widely known as graphene. In 2004, Geim and Novoselov [6, 7] created by mechanical exfoliation an one-atom thick layer made of graphite, the so-called graphene. Due to their well-succeeded experiment, many other techniques have been developed to grow graphene on several possible substrate materials such as on hydrogenated silicon carbide, copper, cobalt, and gold [8–16].

Before 2004, graphite systems were also widely studied [5, 17, 18], and their electronic properties used to theoretically describe other materials based on carbon, such as fullerene [19] and carbon nanotubes [20]. These chemical elements have attracted much attention because of their exotic electronic and mechanical properties, such as high tensile strength and, in the case of nanotubes, tunable electronic structure according to chirality, radius, and high thermal conductivity. A new type of derivative of graphene arose after 2004: the graphene nanoribbon [21], in which some electronic properties of the graphene were modified and could be controlled. These properties depends on the type of crop that was carried out on graphene and can be simpler cuts, called zigzag and armchair or being modeled in a specific way, such as triangles to form quantum dots [21–25] or even with Z formats [26].

The interest in two-dimensional materials started from the nineteenth century, mainly for its electronic transport properties after the discovery of the Hall effect. In 1988, Haldane predicted that another type of Hall effect, called anomalous quantum Hall effect, could be observed in a two-dimensional crystal with hexagonal lattice [27]. Recently, the new classes of matter, such as quantum Hall effect (QHE) [28, 29], quantum anomalous Hall (QAH) effect [30–32], and quantum spin Hall (QSH) effect [33, 34], have been discovered or predicted in the graphene, as well as other 2D materials such as topological insulators [35–37], HgTe-CdTe quantum wells [38, 39], silicene [40], two-dimensional germanium [40], and transition metal dichalcogenides [41]. Among these new classes of matter, the QSH and QAH states possess topologically protected edge states at the boundary, where the electron backscattering is forbidden, offering a potential application to electronic devices to transport current without dissipation [24, 42]. However, the QSH state and QAH are very different states of matter. The quantum spin Hall is characterized by a gap completely insulating the bulk, and their edge states are helical with no gap, wherein opposite spins propagate in opposite directions on each side of the sample and are protected by time reversal symmetry (TRS) [27, 33, 36, 38–40, 43, 44]. In the case of quantum anomalous Hall, chiral edge states takes place, also without gap, where one spin channel is suppressed because of the TRS break [35, 37, 45]. Therefore, to observe topological phase transitions (QPT) between quantum spin Hall and quantum anomalous Hall states, it is necessary to apply a condition, which might break the TRS [28]. An external magnetic field is a potential solution but from applicability point of view, an internal exchange field (EX) which takes the main spin band to be completely filled while the minority spin band becomes empty, becomes an more attractive alternative [31, 32, 35, 46]. As it is known, a pseudo-magnetic field induced by strain B_S leads to Landau quantization and edge states that circulate in opposite directions [47, 48], and the strain creates graphene pseudo-magnetic fields. Then, without breaking the TRS, the strain may induce a gap in the bulk and edges without helical gap. Thus, strain, EX and SOC can be used as a versatile tool for control of topological phase transitions [32]. These facts motivated us to propose ways in which the spin-orbit coupling, uniaxial mechanical strain and exchange (instead of an external magnetic field) to be used to carry out phase transitions in graphene nanoribbons [49].

In this chapter, firstly, we make a brief description on tight-binding model. Then, we report energy band structure of the graphene and the individual effects of the intrinsic SOC, the Rashba SOC, and the EX. After that, we present the effects of applied uniaxial strain on both electronic structure and transport property of the graphene. Then, we demonstrate the effects of strain on single-particle energy and quantum transport property of graphene nanoribbons. Finally, we show systematically the strain-engineered QPT from the QSH to QAH states.

2. Electronic structure and transport properties of graphene

2.1. Electronic structure of graphene tight-binding approach

An isolated atom has its own electronic levels ranging and depending on its fundamental characteristics. When two or more atoms are approximate to each other, their electronic levels are recombined to obtain a new structure for the system as a whole. And the periodic clustering of atoms in a structure is meant by the crystal lattice. In the case of an insulating material, superposition of the wave functions of the valence electrons in the crystal lattice atoms is low. In the case of a conductive material, such superposition of the wave function of the electrons is large and acquires great mobility through the solid. Semiconductor materials have an electronic distribution that is not very well located, because there is not a too strong electrical attraction between electrons and protons on the atomic nucleus but low overlap between the valence electrons from neighboring atoms are observed. The tight-binding method is useful in those cases [50]. Thus, one can assume that the lattice Hamiltonian $\hat{H}_{\text{latt}}$ can be approximated in the vicinity of each point of the grid system by Hamiltonian $\hat{H}$ of the atom located at that point. However, this has some disadvantages, because this method does not allow us to include continuous spectra and also does not have good description for levels below the valence states.

In order to apply tight-binding method to graphene, we begin with the wave function of an electron on its lattice as a linear combination of atomic orbitals of A and B sites, the two distinct atoms in graphene unit cell

$$|\Psi\rangle = \alpha(\vec{k})\,|\,\phi_{A},\rangle s + \beta(\vec{k})\,|\,\phi_{B},s'\rangle; \tag{1}$$

where $\alpha(\vec{k})$ and $\beta(\vec{k})$ are coefficients, s is the spin projection, and Bloch functions are

$$|\phi_C\rangle = \frac{1}{\sqrt{N}} \sum_n e^{i\vec{k}\cdot\vec{R}_n} |C_n, s\rangle \tag{2}$$

where $C = A,B$. In these functions, $\vec{R}_n$ is an carbon atom position concerning the origin of coordinate system, the exponential carries the periodicity of the lattice; $\frac{1}{\sqrt{N}}$ is the normalization factor (N is the number of atoms in graphene).

Lattice Hamiltonian are made up of two terms: on-site $\hat{H}_{on}$ and hopping $\hat{H}_h$ energy

$$\hat{H}_{latt} = \hat{H}_{on} + \hat{H}_h \tag{3}$$

The first one, $\hat{H}_{on}$ is

$$\hat{H}_{on} = \epsilon_0 \sum_n^N \{ |A_n\rangle\langle A_n| + |B_n\rangle\langle B_n| \}; \tag{4}$$

with A and B representing sites, nonequivalent types of carbon atoms on the unit cell. This energy can be set as $_0=0$ at Fermi level and the term vanishes. Here, the spin index was suppressed because only identical spins couple. The term $\hat{H}_h$ defines the hopping of electrons between nearest-neighbor atoms, since we use first-neighbors tight-binding approach.

$$\hat{H}_h = -\sum_{\langle i,j \rangle} t_{ij} \left(|A_i\rangle \langle B_j| + |B_j\rangle \langle A_i| \right); \tag{5}$$

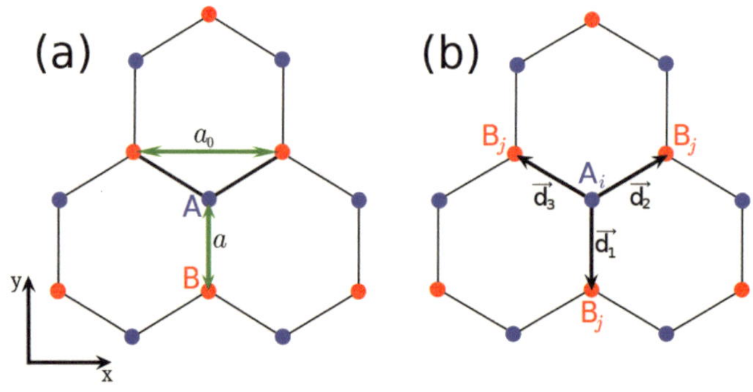

Figure 1. (a) Distance between two carbon atoms a and lattice constant a_0. (b) Nearest-neighbors of an A site. Copyright (2015) by the American Physical Society [49].

http://dx.doi.org/10.5772/64493

where i and j are site index and summation over $\langle i,\ j\rangle$ indicates only nearest neighbor atoms are being considered. The t_{ij} is called hopping factor and gives probability amplitude of an electron on i site hop to a j site. Its value is the same for each nearest neighbor t_{ij}=2.7eV [51, 52]. The distance vectors from an A site to their nearest neighbors are, as one can see in **Figure 1**,

$$\vec{d}_1 = -a\hat{y}, \qquad \vec{d}_2 = +\frac{a_0}{2}\hat{x} + \frac{a}{2}\hat{y} \quad and \quad \vec{d}_3 = -\frac{a_0}{2}\hat{x} + \frac{a}{2}\hat{y} \tag{6}$$

and B-site distance vector can be obtained in a similar way. Here, $\hat{x}$ and $\hat{y}$ are the unitary vectors along x axis and y axis. The distance between two carbon atoms is a=0.142 nm, and a_0=$\sqrt{3}a$ is the lattice constant.

To find energy equations for our system, we solve Schrödinger's equation using Eqs. (1) and (5). This will give

$$E\alpha(\vec{k}) = -t\left[e^{ik_y a} + 2cos\left(\frac{k_x a_0}{2}\right)e^{-ik_y a/2}\right]\beta(\vec{k}), \tag{7}$$

$$E\beta(\vec{k}) = -t\left[e^{-ik_y a} + 2cos\left(\frac{k_x a_0}{2}\right)e^{ik_y a/2}\right]\alpha(\vec{k}); \tag{8}$$

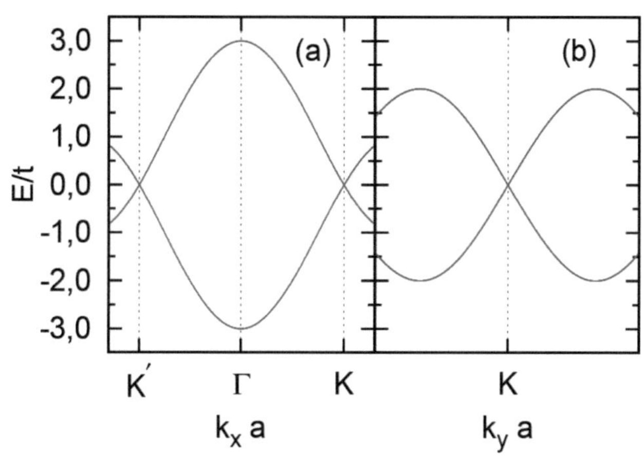

Figure 2. Energy band of graphene (a) along k_y=0 and (b) k_x=$4\pi/3\sqrt{3}$ directions.

In **Figure 2**, valence and conduction bands are shown in profile: with k_y=0 in (a), where symmetric points K, K', and Γ are shown; and k_x=$4\pi/3\sqrt{3}$ in (b), showing K point. In both cases, no bandgap is observed in the graphene.

Quantum anomalous Hall and quantum spin Hall effect can be induced in graphene without a magnetic field if we consider Rashba and intrinsic spin-orbit coupling and also exchange field [53, 54]. Intrinsic spin-orbit coupling is weak on graphene [55–57], however, graphene is easily affected by disturbance at low energies and the effects due to spin-orbit coupling should become relevant at low temperatures [33]. Although it is challenging the experimental envision, this type of coupling can be controlled with graphene deposition on other materials. The exchange interaction that occurs between electron spins can be observed in graphene with stabilization of a ferromagnetic phase, when it has a low doping [53, 58].

The intrinsic spin-orbit interaction evolves the next nearest-neighbors and is written as

$$\hat{H}_{so} = \frac{2i}{\sqrt{3}}\lambda_{so} \sum_{\langle\langle i,j\rangle\rangle} \sum_{ss'} \left\{ | A_i, s\rangle \, \vec{\gamma}\cdot(\vec{d}_{kj}\times\vec{d}_{ik}) \, \langle A_j, s'| + h.c. \right\} \tag{9}$$

being λ_{so} the strength parameter, estimated up to 2.4 K [33]. Here, the s and s' are the z-components of real spin, and the summation $\langle\langle i,j\rangle\rangle$ is over the next nearest-neighbors of a carbon atom, as shown in **Figure 3(a)**, the vectors d_{mn} indicate the distances between an atom on m site and another on n site, as shown in **Figure 3(b)**. To find energy equations for the SOC, we solve Schrödinger's equation using Eqs. (1) and (9). It leads to

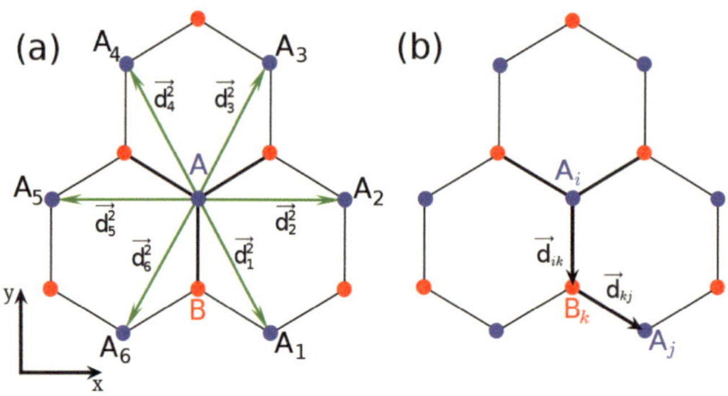

Figure 3. Next nearest-neighbors distance vectors of an A site (a). Definition of $\vec{d}_{kj}$ and $\vec{d}_{ik}$ vectors used in Hamiltonian 9. Figure adapted from Guassi et al. [49].

$$E\alpha(\vec{k}) = -4\lambda_{so}\alpha(\vec{k})\sin\left(\frac{k_x a_0}{2}\right)\left[\cos\left(k_y\frac{3}{2}a\right) - \cos\left(\frac{k_x a_0}{2}\right)\right]s_z \tag{10}$$

and the negative of right side for $\beta(\vec{k})$. Clearly, **Figure 4** shows that an increase in parameter strength λ_{so} increases the gap of graphene, which is associated with a massive term in the Dirac-like Hamiltonian [59].

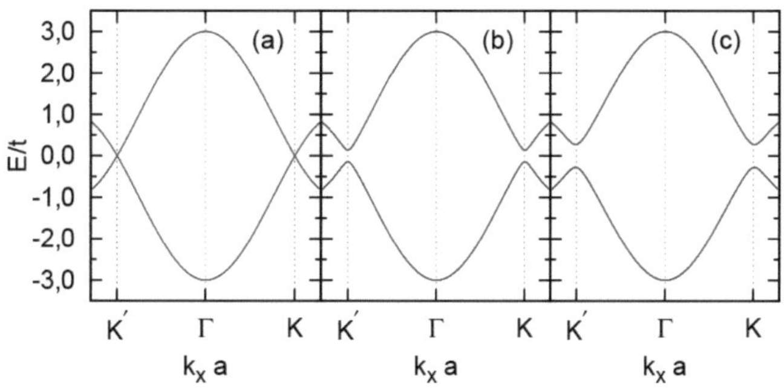

Figure 4. Energy band along $k_y=0$ direction of graphene with (a) $\lambda_{so}=0$, (b) $\lambda_{so}=0.03t$, and (c) $\lambda_{so}=0.06t$.

Rashba spin-orbit coupling can be induced in graphene with application of an external electric field perpendicular to the sheet plane [60], interaction of carbon atoms with a substrate [43] or by curving the sheet [61–63]. Its Hamiltonian reads

$$\hat{H}_R = i\sum_{\langle ij\rangle}\sum_{ss'}\left\{| A_i, s > (\vec{u}_{ij}\cdot\vec{\gamma})\left\langle B_j, s'\right| + h.c.,\right\} \tag{11}$$

where the summation $\langle i,\ j\rangle$ is over the nearest-neighbors of a carbon atom, $\vec{\gamma}$ is the vector whose components are the Pauli spin matrices, $\vec{u}_{ij}$ are defined by [60]

$$\vec{u}_{ij} = \frac{e}{2m^2 a v_F}\vec{E}\times\vec{R}_{ij} = -\frac{\lambda_R}{a}\hat{k}\times\vec{R}_{ij}, \tag{12}$$

with e the electron charge, m the rest mass of the electron, and v_F the Fermi velocity. The vector $\vec{R}_{ij}=\vec{R}_j-\vec{R}_i$ gives difference between two atom positions on the lattice. Observing **Figure 1**, the vector $\vec{R}_{ij}$ can be renamed to $\vec{d}_l$, with (l=1, 2, 3).

Using electron wave function, Eq. (1) and Hamiltonian part 11, we find four spin nondegenerated energy equations:

$$E\alpha(\vec{k},\uparrow) = i\lambda_R\left[e^{ik_y a} + 2\cos\left(\frac{k_x a_0}{2}+\frac{2\pi}{3}\right)e^{-ik_y a/2}\right]\beta(\vec{k},\downarrow) \tag{13}$$

$$E\beta(\vec{k},\uparrow) = i\lambda_R\left[e^{-ik_y a} + 2\cos\left(\frac{k_x a_0}{2}-\frac{2\pi}{3}\right)e^{ik_y a/2}\right]\alpha(\vec{k},\downarrow) \tag{14}$$

$$E\alpha(\vec{k},\downarrow) = i\lambda_R\left[e^{ik_y a} + 2\cos\left(\frac{k_x a_0}{2} - \frac{2\pi}{3}\right)e^{-ik_y a/2}\right]\beta(\vec{k},\uparrow) \tag{15}$$

$$E\beta(\vec{k},\downarrow) = -i\lambda_R\left[e^{-ik_y a} + 2\cos\left(\frac{k_x a_0}{2} + \frac{2\pi}{3}\right)e^{ik_y a/2}\right]\alpha(\vec{k},\uparrow) \tag{16}$$

which, together with energy Eqs. (7) and (8), allow one to plot energy levels in **Figure 5**. As one can note, the Rashba SOC lifts the spin degeneracy, breaking the SU(2) symmetry. However, due to time-reversal symmetry, we still have $E(\vec{k},\uparrow)=E(-\vec{k},\downarrow)$.

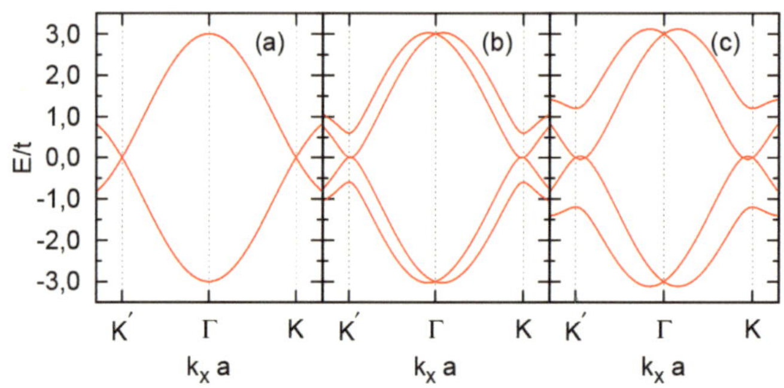

Figure 5. Energy band along $k_y=0$ direction of graphene with (a) $\lambda_R=0$, (b) $\lambda_R=0.20t$, and (c) $\lambda_R=0.40t$.

Calculations of *ab initio* have recently shown that graphene doped with Fe on its surface may have intrinsic ferromagnetism [30]. This interaction arises when the change in the spin of an electron changes the electrostatic repulsion between electrons near it. Its Hamiltonian includes the coupling of orbital motion and the spin of the electrons with the exchange field. In this paper, to simplify the calculations and not lose generality, only the portion of spin will be considered. Now, the Hamiltonian is

$$H_{ex} = M\sum_{i;s}\{|C_i,s>\gamma_z<C_i,s|+h.c.\}, \tag{17}$$

where $C=A,B$; strength parameter M is proportional to $J_{eff}\mu'_z$, where J_{eff} is the exchange interaction and μ'_z is effective magnetic momentum of electron that associates to the exchange field. Thus, H_{ex} describes the magnetic momentum response of spin of an electron to the exchange field, like in Zeeman effect [45]. The γ_z is the Pauli matrix. This part of Hamiltonian gives energies

$$E\alpha(\vec{k},\uparrow) = M\alpha(\vec{k},\uparrow), \tag{18}$$

$$E\alpha(\vec{k},\downarrow) = -M\alpha(\vec{k},\downarrow); \tag{19}$$

the same for $\beta(\vec{k},s)$. Combining with Eqs. (7) and (8), we can obtain the energy dispersion. In **Figure 6**, it clearly shows that the time-reversal symmetry is broken, since $E(\vec{k},\uparrow)\neq E(-\vec{k},\downarrow)$. Initially, one might think that graphene nanoribbon subject to an exchange field should not bear the quantum spin Hall state, as this would be protected by the time-reversal symmetry [33, 43]. But it has been found a similar state, called pseudo-Hall quantum spin state or quantum spin Hall state with broken time-reversal symmetry [45] in which was possible to observe the spin polarized current on the edges of nanoribbon. In addition, the exchange field is critical to control the transition between electronic states of quantum anomalous Hall to quantum spin Hall [49].

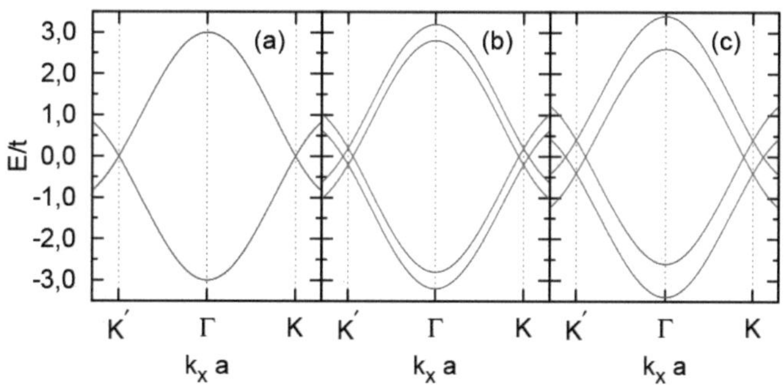

Figure 6. Energy band along $k_y = 0$ direction of graphene with (a) $M = 0$, (b) $M = 0.20t$, and (c) $M = 0.40t$.

2.2. Electronic structure of strained graphene

Deformation can naturally be observed when graphene is grown on top of other materials, because of distinct atomic arrangements between the atoms of graphene and the substrate. The application of an external tension on graphene sheet or nanoribbons can change its electronic properties, as with the nanotubes [64–68]. Some calculations [69] and experiments [70] have shown that these deformations can reach about 20% of the initial interatomic distance without permanently deform the graphene.

Strain is calculated in graphene using the strain matrix defined in [52]:

$$\epsilon = \varepsilon \begin{pmatrix} cos^2\theta - \nu sin^2\theta & (1+\nu)cos\theta sin\theta \\ (1+\nu)cos\theta sin\theta & sin^2\theta - \nu cos^2\theta \end{pmatrix} \tag{20}$$

where ε is strain modulus, θ is the direction of strain, $\theta=0$ being parallel to a zigzag chain and $v=0.165$ is the Poisson ratio [71]. Therefore, matrix elements read,

$$\epsilon_{11} = \varepsilon[cos^2\theta - vsin^2\theta] \tag{21}$$

$$\epsilon_{12} = \varepsilon[(1+v)cos\theta sin\theta] \tag{22}$$

$$\epsilon_{21} = \varepsilon[(1+v)cos\theta sin\theta] \tag{23}$$

$$\epsilon_{22} = \varepsilon[sin^2\theta - vcos^2\theta] \tag{24}$$

The transformation of an atom position from (x, y) to the new position (x', y') after the strain application will be

$$\begin{pmatrix} x' \\ y' \end{pmatrix} = \begin{pmatrix} (1+\varepsilon_{11}) & \varepsilon_{12} \\ \varepsilon_{21} & (1+\varepsilon_{22}) \end{pmatrix} \begin{pmatrix} x \\ y \end{pmatrix}. \tag{25}$$

Or, in vectorial form

$$\vec{d}^s = (I+\epsilon)\vec{d} \tag{26}$$

where $\vec{d}$ and $\vec{d}^s$ are the vectors that defines the nearest-neighbor in the graphene without and with application of the strain, I is identity matrix, and is the matrix 20.

Hopping term t is also modified by strain [52]

$$t_{i,j} = t_i = te^{-3.37(\vec{d}_i^S/a-1)}. \tag{27}$$

With these modifications, energy equations for strained graphene are calculated in the same way as in Section 2.1 and gives

$$E\alpha(\vec{k}) = -\sum_{l}^{3} t_l exp\{i\vec{k}\cdot\vec{d}_l^s\}\alpha(\vec{k}) \tag{28}$$

with summation made over the nearest-neighbors. For $\beta(\vec{k})$, the expression is the complex hermitian of $\alpha(\vec{k})$. **Figure 7** shows the contour plot of valence band subjected to a strain with

http://dx.doi.org/10.5772/64493

modulus of 15% along zigzag chains, i.e., $\theta=0$ (a) and toward the armchair chains, i.e., $\theta=\pi/2$ (b). Notice that no bandgap is opened up in both cases [52]. Nevertheless, the form of the 1st BZ has been deformed by the strain. Therefore, the Dirac points K and K' are the strain shifted. The new positions of the deformed Dirac cones are well determined by the following equation [72]:

$$\vec{K}\cdot\left(\vec{d}_1^s-\vec{d}_2^s\right)=\arccos\left(\frac{t_3^2-t_1^2-t_2^2}{2t_1t_2}\right) \tag{29}$$

where $\vec{K}=(K_x,K_y)$

$$H_{so}=\frac{2i}{\sqrt{3}}\lambda_{so}\sum_{\langle\langle i,j\rangle\rangle}\sum_{ss'}\left\{|A_i,s\rangle\,\vec{\gamma}\cdot(\vec{d}_{kj}^s\times\vec{d}_{ik}^s)\left\langle A_j,s'\right|+h.c.\right\} \tag{30}$$

now, with new distances $\vec{d}_{ik}^s$, modified by strain. This will give us new energies expressions,

$$E\alpha(\vec{k},\uparrow)=i\lambda_{so}\mu\sum_{l}^{3}\sigma_l\left(e^{i\vec{k}\cdot\vec{d}_l^{s2}}-e^{-i\vec{k}\cdot\vec{d}_l^{s2}}\right)\alpha(\vec{k},\uparrow) \tag{31}$$

and nearly the same for other expressions, with

$$\alpha(\vec{k},\uparrow)=-\alpha(\vec{k},\downarrow)=-\beta(\vec{k},\uparrow)=\beta(\vec{k},\uparrow) \tag{32}$$

In Eq. (31), $\vec{d}_l^{s2}$ are the vectors between the next nearest-neighbor sites and

$$\mu=(1+\epsilon_{11})(1+\epsilon_{22})+\epsilon_{12}\epsilon_{21} \tag{33}$$

In **Figure 8**, we show the effects of the combination of intrinsic SOC and uniaxial strain applied along θ=0 direction for different strain modulus ε and λ_R=0.0. The bandgap generated by the intrinsic SOC is still present, although as strain is increased the bandgap reduces, the energy dispersion can also displays different features as the direction of applied strain is varied [52]. The Rashba Hamiltonian can include strain as well, and this will leave us with equations

$$E\alpha(\vec{k},\uparrow)=-i\varphi_{s-}\beta(\vec{k},\downarrow) \tag{34}$$

$$E\alpha(\vec{k},\downarrow)=-i\varphi_{s+}^{*}\beta(\vec{k},\uparrow) \tag{35}$$

$$E\beta(\vec{k},\uparrow) = i\varphi_{s+}\alpha(\vec{k},\downarrow) \tag{36}$$

$$E\beta(\vec{k},\downarrow) = i\varphi_{s-}^{*}\alpha(\vec{k},\uparrow) \tag{37}$$

where

$$\varphi_{s-} = \lambda_R\left\{\rho_1 e^{-3i\xi_1} - \rho_1 cos(\xi_2) + \sqrt{3}\rho_2 sin(\xi_2)\right\}e^{i\xi_1} \tag{38}$$

$$\varphi_{s+} = \lambda_R\left\{\rho_1 e^{3i\xi_1} - \rho_1 cos(\xi_2) - \sqrt{3}\rho_2 sin(\xi_2)\right\}e^{-i\xi_1} \tag{39}$$

and we defined

$$\rho_1 = 1 + \epsilon_{22} + i\epsilon_{12} \tag{40}$$

$$\rho_2 = 1 + \epsilon_{11} - i\epsilon_{21} \tag{41}$$

$$\xi_1 = \frac{k_x a}{2}\epsilon_{12} + \frac{k_y a}{2}(1 + \epsilon_{22}) \tag{42}$$

$$\xi_2 = \frac{k_x a_0}{2}(1 + \epsilon_{11}) + \frac{k_y a_0}{2}\epsilon_{21} \tag{43}$$

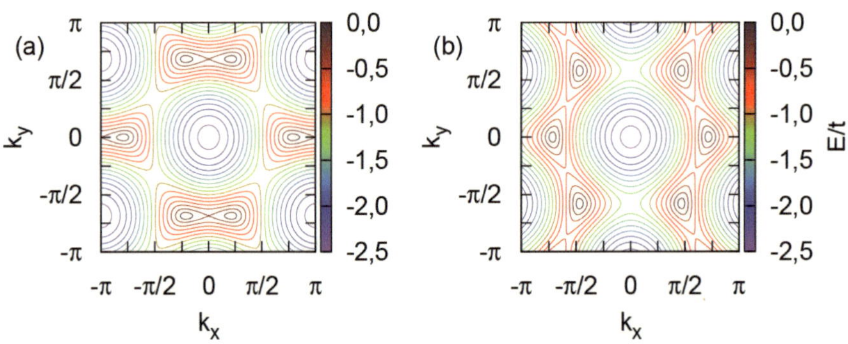

Figure 7. Contour plot of valence band of strained graphene with $\varepsilon = 0.15$ for $\theta = 0$ (a) and $\pi/2$ (b).

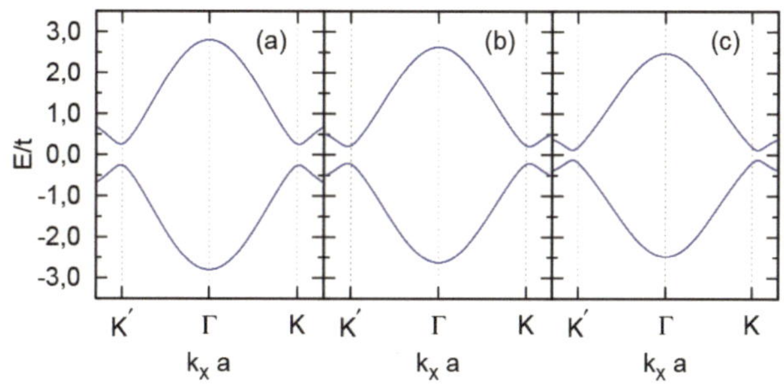

Figure 8. Energy band of strained graphene with $\lambda_s o = 0.06t$ and $\theta = 0$, for $\varepsilon = 0.05$ (a), $\varepsilon = 0.10$ (b), $\varepsilon = 0.15$ (c). Here, we have set $\lambda_R = 0.0$.

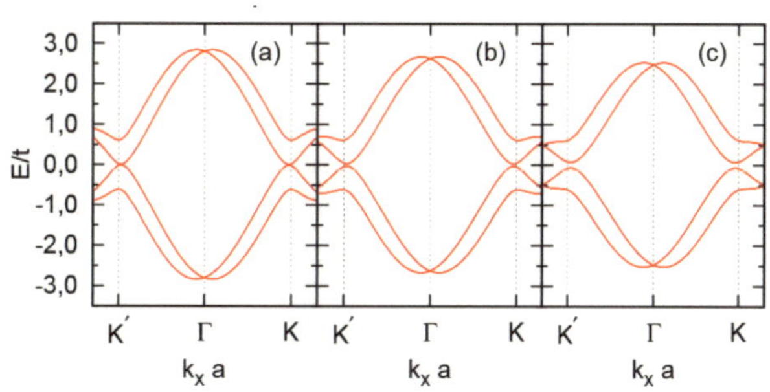

Figure 9. Energy band of strained graphene with $\lambda_R = 0.20t$ and $\theta = 0$, for $\varepsilon = 0.05$ (a), $\varepsilon = 0.10$ (b), $\varepsilon = 0.15$ (c).

In **Figure 9**, we can still observe the broken spin-degeneracy, but compared with the case where the uniaxial strain is absent, one can note that the effect of strain is to renormalize the Rashba SOC and shifts the Dirac point relative to the original one.

2.3. Quantum anomalous Hall effect in strained graphene

In this section, we discuss the prospects of external manipulation of the quantum anomalous Hall effect (QAHE) in graphene by strains [73–76]. We present here our results of the microscopic study of the QAHE in graphene under uniaxial strains [32]. For this purpose, we have theoretically explored the dependence of electronic structure, topological and transport properties upon the orientation and modulus of uniaxial strain, in the presence of Rashba, Intrinsic SO, and an exchange field interaction [32].

To identify the topological properties of the Dirac gap and study the origin of QAHE, we have calculated the Berry curvature of the nth bands $\Omega_{xy}^{n}(k_x,k_y)$ using the Kubo formula:

$$\Omega_{xy}^{n}(k_x,k_y) = -\sum_{n'\neq n} \frac{2Im\langle \Psi_{nk} | v_x | \Psi_{n'k} \rangle \langle \Psi_{n'k} | v_y | \Psi_{nk} \rangle}{(\omega_{n'} - \omega_n)^2} \tag{44}$$

where $\omega_n = E_n/\hbar$ with E_n the energy eigenvalue of the nth band and $v_{x(y)} = \hbar^{-1}\partial H/\partial k_{x(y)}$ is the Fermi velocity operator. When the Fermi level lies within the bulk gap, i.e., in the insulating regime, according to the Kubo formula, the corresponding Hall conductance is quantized as $\sigma_{xy} = Ce^2/h$, where C is defined as the Chern number and can be calculated by [77]

$$C = \frac{1}{2\pi}\sum_n \int_{BZ} d^2k\Omega^n_{xy}, \tag{45}$$

where the summation is taken over all the occupied states below the Fermi level, and the integration is carried out over the whole first Brillouin zone.

Since the Berry curvatures are highly peaked around the Dirac points **K** and **K**$'$ [78], then a low energy approximation can be used in the calculation of the Chern number [30]. This allows us to derive an effective tight-binding Hamiltonian of the strained graphene, by expanding $H(\vec{k})$ at the vicinity of the strain-shifted Dirac points, i.e., $\vec{k} = \eta\mathbf{K} + \vec{q}$, where $\eta = \pm 1$ labels the two valleys, and $\vec{q} = (q_x, q_y)$ is a small crystal momentum around $\eta\mathbf{K}$. The validity of the low energy approximation requires the strain modulus to be upper limited, such that does not go beyond the threshold of an appearance of a band gap, thus the band is still linear and gapless at the strain-shifted Dirac points, in the absence of SOCs and exchange field interactions [79]. This condition is fulfilled by the relation on the strain-dependent hopping parameters $|t_1 - t_2| \leq t_3 \leq |t_1 + t_2|$, where $t_{i=1,2,3}$ is the hopping along each C–C bond [80]. Under this condition, we calculate the Chern number using the following equation:

$$C = \frac{1}{2\pi}\sum_{K,K'}\sum_{n=1,2}\int_{-\infty}^{\infty} dq_x dq_y \Omega^n_{xy}(q_x, q_y) \tag{46}$$

It is interesting to mention that in the above integral, a momentum cutoff is set around each valley for which the Chern number calculation is guaranteed to converge.

As known, intrinsic spin-orbit (ISO) interaction respects the crystal symmetries and does not couple states of opposite spins. But it opens up a topologically nontrivial bulk band gap at zero magnetic field [43]. This bulk band gap hosts two counter-propagating edge modes per edge in the graphene nanoribbon, with opposite spins: this topological phase is known as the QSH phase and may be regarded as two opposite QH phases (i.e., each spin performs the QH effect, with opposite chirality) [27]. Therefore, the Chern number must vanish in a system with TRS. In contrast, the Rashba term explicitly violates the $z \to -z$ mirror symmetry. Moreover, it mixes different spin components and depresses the ISO induced band gap [81]. When the exchange field is applied and only ISO is turned on, the combination of the ISO coupling and exchange field leads to the breaking of the TRS which is preserved in the QSH phase. However, due to the absence of spin-flip terms in the Hamiltonian, the helical edge-state structure persists. Thus, both the Chern number and the conductance are equal to zero. Unlikely, when

Rashba SOC is considered, in addition to ISO and exchange field, the system can be in a regime, which depends on λ_R, λ_{so}, and M parameters, that may result in a phase transition from zero conductance to finite conductance [32].

Let us now calculate the Hall conductivity of the strained graphene considering both Rashba SOC and ISO. **Figure 10(a)** and **(b)** shows the Hall conductance for $\lambda_R(\varepsilon=0)=0.1t$ and $\lambda_{so}(\varepsilon=0)=0.06t$ along $\theta=0$ and $\theta=\pi/2$, respectively. One can clearly note the two distinct phases: Insulating (I) characterized by $\mathcal{C}=0$ and the QAHE phase with $\mathcal{C}=2$, where $\mathcal{C}=\mathcal{C}_K+\mathcal{C}_{K'}$. The two different phases can be accessed by appropriately tuning the exchange field M and the strain modulus ε. **Figure 10(d)** shows the dependence of the Hall conductance $\sigma_{xy}=\mathcal{C}e^2/h$ on the exchange field and the strain parameters with $\theta=\pi/2$ for $\lambda_R(\varepsilon=0)$=0.1t and $\lambda_{SO}(\varepsilon=0)$ =0.06t. We find that a finite ISO drives a phase transition from QAHE to an insulator phase [32]. We also notice that for M being smaller than 0.24t, the conductance σ_{xy} of unstrained graphene is equal to zero, corresponding to an insulator phase in the graphene, also called a time-reversal-symmetry-broken quantum spin-Hall phase [34]. At M_c=0.24t, an abrupt change from 0 to $2e^2/h$ takes place, which indicates a quantization of the Hall conductance and an occurring of a phase transition at $M=M_c$. After that, it remains $2e^2/h$, in which the unstrained graphene stays in the phase of QAHE. Furthermore, the applied strain drives Hall conductance curve forward to the right-hand side for strained graphene. Consequently, as the strain modulus increases from zero, the critical exchange field M_c becomes larger, such as for ε=0.2, M_c=0.275t with a relative change of M_c being approximately +14.5%. Astonishingly, in the case of $\theta=0$, as demonstrated in **Figure 10(c)**, there is an increase in the exchange field with similar behavior for the Hall conductance. However, beyond an specific value of strain modulus, indicated by the vertical dashed line in **Figure 10(e)**, the system presents an opposite strain-strength dependence, i.e., an increase in the strain parameter shifts the Hall conductance curve to the left-hand side. For instance, in the case of ε=0.2, we have obtained M_c=0.1t with a relative change of M_c being equal to -58.3%.

The distinct behaviors observed along different strain directions for the QAHE phase transition can be explained by the competition of the Rashba SOC and ISO in the bulk band gap-closing phenomena for a given critical exchange field M_c [31, 45]. In the case of $\theta=\pi/2$, an increase in the strain modulus leads to an approximately linear enhancement in the ISO parameter as can be observed in **Figure 10(e)**, which results in an smaller bulk band gap in the presence of an exchange field. On the other hand, the Rashba SOC is not very sensitive to the variation of strain strength. Therefore, the variation of Hall conductance mainly reflects the dependence of ISO on the strain strength. In contrast, for values of strain modulus larger than ε=0.078 in the case of θ=0, there is drastic reduction in the effective ISO interaction, hence Rashba becomes dominant and the critical exchange field for the phase transition becomes smaller as one can note in **Figure 10(c)** with a critical M_c=0.1t for ε=0.2 for the QAHE phase transition [32].

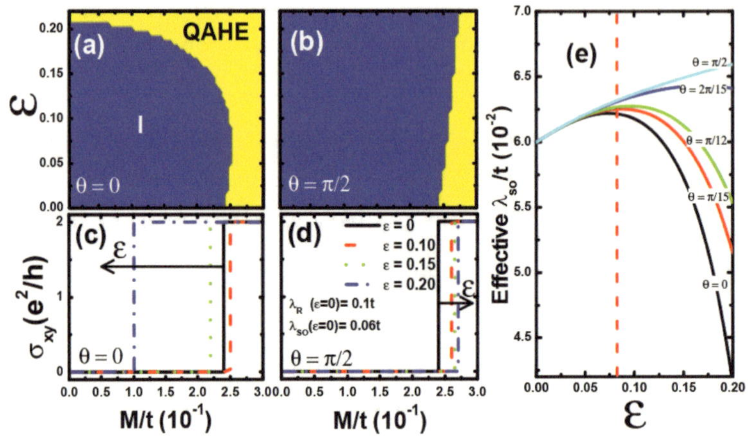

Figure 10. Phase diagram of the QAHE for strained graphene along two distinct directions: (a) along $\theta=0$ and (b) along $\theta=\pi/2$. The Hall conductance as a function of the exchange interaction M, for uniaxial strain direction along $\theta=0$ and $\theta=\pi/2$ with four different strain strengths is shown in (c) and (d), respectively. The arrows in panels (c) and (d) indicate the direction for which the strength is increased from ε=0 to ε=0.2. The parameters $\lambda_R(\varepsilon=0)=0.1t$ and $\lambda_{so}(\varepsilon=0)=0.06t$ have been used in panels (a)–(d). (e) Effective λ_{so} as function of strain strength along different directions θ. The vertical dashed line in panel (e) indicates the limiting strain modulus, for which the effective ISO parameter changes its behavior according to the direction and modulus of strain. Reproduced with permission from Diniz et al. [32]. Copyright (2013), AIP Publishing LLC.

3. Electronic structure and transport properties of graphene nanoribbon

A graphene nanoribbon is defined as a graphene sheet in which one of its dimensions is narrow and the other approximately infinite. The unique properties arising due to the reduced dimensions become very important because shape of the edges and width of nanoribbon defines its electronic structure. The main nanoribbons classification is based on the edge design, which can be armchair, zigzag, chiral, and bearded nanoribbons depending on the edge terminations [82]. We will focus on the electronic dispersion of only two types: armchair and zigzag.

3.1. Electronic structure of graphene nanoribbon

The electron wave function in a armchair nanoribbon is

$$|\phi_{A,n}\rangle=\frac{1}{\sqrt{M}}\sum_{m,n}^{N}e^{i\vec{k}\cdot\vec{R}_m}\,|A_m,n\rangle \tag{47}$$

where

$$\vec{R}_m = m\vec{a}_0 \tag{48}$$

is a vector between the atom and its neighbors in the next unit cell with same type of site. In armchair nanoribbons, the unit cell m and its width is defined as showed in **Figure 11(a)**. One can write the vectors that separate the nearest neighbors for a A site, **Figure 11(b)**:

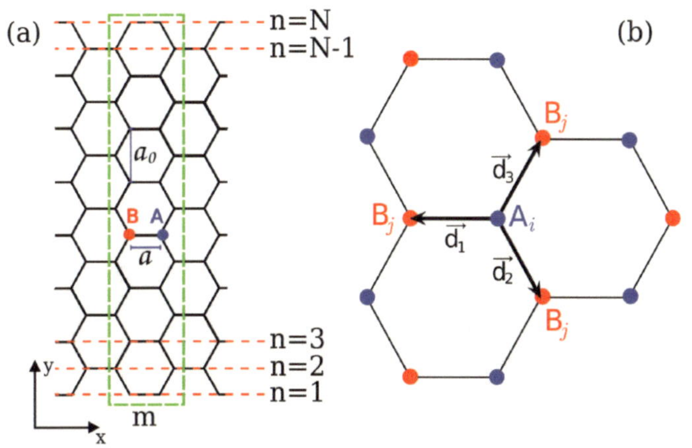

Figure 11. (a) Armchair unit cell m showing the sites A and B, distance between carbon atom a and lattice constant a_0 is also shown. (b) Distance vectors between two nearest neighbors sites and site index.

$$\vec{d}_1 = -a\hat{\imath} \tag{49}$$

$$\vec{d}_2 = \frac{a}{2}\hat{\imath} - \frac{a_0}{2}\hat{\jmath} \tag{50}$$

$$\vec{d}_3 = \frac{a}{2}\hat{\imath} + \frac{a_0}{2}\hat{\jmath} \tag{51}$$

And similar vectors could be find for B sites. Then, Hamiltonian of armchair nanoribbon is

$$\begin{aligned}\hat{H}_h = -\sum_{\langle m,n\rangle}^{N} \{ & t_1 \,|\, A_m, n\rangle\langle B_{m-1/2}, n \,|\, + \\ & + t_2 \,|\, A_m, n\rangle\langle B_m, n-1 \,|\, + t_3 \,|\, A_m, n\rangle\langle B_m, n+1 \,|\, + h.c\}\end{aligned} \tag{52}$$

where the summation $\langle m,n\rangle$ is over nearest neighbors; the product $|\, a_m, n\rangle\langle B_{m-1/2}, n \,|$ is the hopping of electrons between an atom in the A site at position m and n and an atom in neighboring site B which is at position $m-\frac{1}{2}$ and n. Solving Schrödinger's equation will give energy expressions

$$E\alpha(\vec{k},n) = -t\left[\beta(\vec{k},n)e^{-ik_x a_0/2} + \beta(\vec{k},n+1) + \beta(\vec{k},n-1)\right] \tag{53}$$

$$E\beta(\vec{k},n) = -t\left[\alpha(\vec{k},n)e^{ik_x a_0/2} + \alpha(\vec{k},n+1) + \alpha(\vec{k},n-1)\right] \tag{54}$$

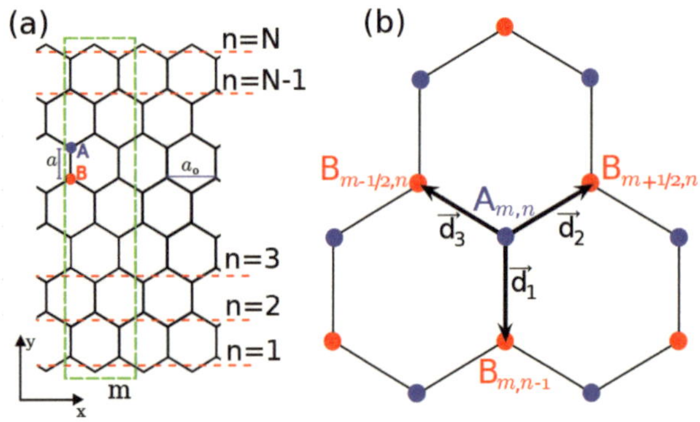

Figure 12. (a) mth unit cell of zigzag graphene ribbon, and (b) distance vectors between two nearest neighbor sites and index of them. In (a), A and B indicate two sites, a is interatomic distance, and a_0 represents lattice constant.

Zigzag nanoribbon has unit cell m and width defined as showed in **Figure 12(a)**. The vectors that separate the nearest neighbors for a A site, **Figure 12(b)**, are the same as the graphene case. The Hamiltonian is

$$\begin{aligned} H = -\sum_{m,n}^{N}\{&t_1 \mid A_m, n\rangle\langle B_m, n-1\mid + \\ &+t_2 \mid A_m, n\rangle\langle B_{m+1/2}, n\mid + t_3 \mid A_m, n\rangle\langle B_{m-1/2}, n\mid + h.c\} \end{aligned} \tag{55}$$

Now, the energies are

$$E\alpha(\vec{k},n) = -t\left[\beta(\vec{k},n)2cos\left(\frac{k_x a_0}{2}\right) + \beta(\vec{k},n-1)\right] \tag{56}$$

$$E\beta(\vec{k},n) = -t\left[\alpha(\vec{k},n)2cos\left(\frac{k_x a_0}{2}\right) + \alpha(\vec{k},n+1)\right] \tag{57}$$

http://dx.doi.org/10.5772/64493

and energy dispersion for a $N = 24$ zigzag nanoribbon can be seen in **Figure 13**, where the edge states are marked blue and gray lines denotes bulk states. In **Figure 14**, intrinsic spin-orbit is applied with $\lambda_{so} = 0.05t$. The electronic density is very localized near the edges of the nanoribbon, **Figure 14(b)**.

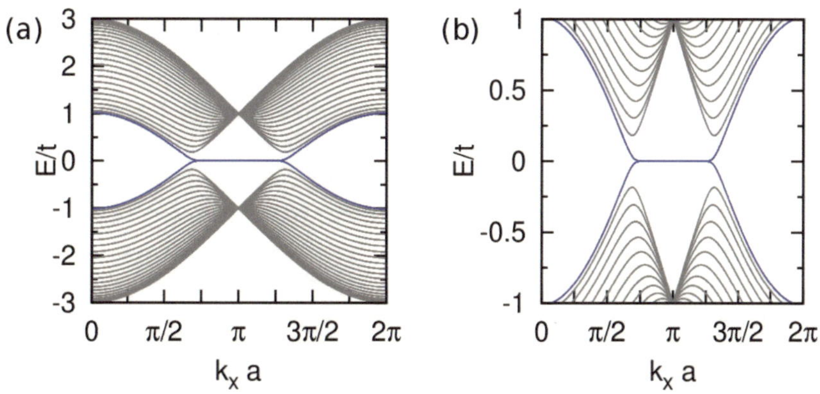

Figure 13. Energy band of bulk- (gray lines) and edge- (blue lines) states of zigzag nanoribbon with width $N = 24$.

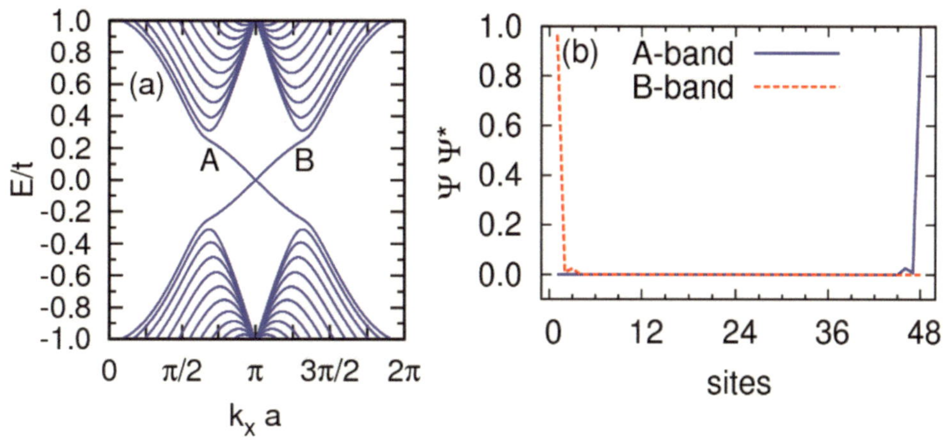

Figure 14. (a) Energy band of zigzag nanoribbons with $N = 24$ and $\lambda_{so} = 0.05t$, and (b) electronic probabilities of A- and B-edge states defined by cross points between edge states and Fermi level $E = 0.05t$, as indicated in (a).

Electronic structure of strained graphene nanoribbon-For the case of strained graphene nanoribbons, we need to replace the strain-invariant hopping integrals by the strain-dependent ones [52, 83] as described in Section 2.2. Many interesting properties are observed in the optical conductivity [72] and electronic transport [84, 85] when the uniaxial strain is considering in the graphene nanoribbons. In the next section, we discuss in detail the effects of uniaxial strain in the spin-charge electronic transport and QAH effect.

3.2. Transport properties of strained graphene nanoribbon

Here, we aim to analyze the electronic transport control in GNR with different terminations in the QAH phase by means of uniform strain deformations [86]. The electronic transport can be performed using a two-terminal device akin to a field electron transistor (FET). QAH phase can be determined experimentally, by spin-resolved density of states, that can be accessed by spatially scanning tunneling microscope (STM) or by scanning tunneling spectroscopy (STS) [87–89]. To calculate the spin-resolved conductance, we have implemented the standard surface Green's function approach [90, 91]. The GNR device is divided into three regions: left lead, central conductor, and right lead. The uniaxial strain is applied to either the longitudinal ($\theta=0$) or the transversal ($\theta=\pi/2$). The central conductor is the only region under the influence of SOC effects and exchange field; it is also connected to semi-infinite leads by nearest-neighbor hopping. To avoid surface mismatch in the case of strained GNR [85], we have considered that the leads are also strained. Therefore, a perfect atomic matching at the interface leads/central conductor is achieved. The Green's function of the device (omitting the spin indices) is then calculated by

$$\mathcal{G}_C^{a/r}(E) = \left(E \pm i\eta - H_C - \Sigma_L - \Sigma_R\right)^{-1}, \tag{58}$$

where a/r denotes the advanced/retarded Green's function, E is the energy ($\eta \to 0$) of the injected electron (the Fermi energy at a given doping). H_C stands for the Hamiltonian in the central region, and $\Sigma_{L/R}$ are the self-energies that describe the influence of the left/right leads, $\Sigma_l = H_{lC}^{\dagger} g_l H_{lC}$, where g_l is the Green's function for the $l=L,R$ semi-infinite lead obtained through an iterative procedure of the tight-binding Hamiltonian [90], and H_{lC} couples each lead to the central region. The spin resolved conductance through the system is given by [86],

$$G_{\sigma\sigma'} = G_0 Tr\left[\Gamma_{\sigma}^{L} \mathcal{G}_{C,\sigma\sigma'}^{r} \Gamma_{\sigma'}^{R} \mathcal{G}_{C,\sigma'\sigma}^{a}\right], \tag{59}$$

where the trace runs through the lattice sites at the central conductor, $G_0 = e^2/h$ is the quantum of conductance per spin, and Γ_{σ}^{l} are the couplings for the leads, related to the spin-diagonal self-energies by $\Gamma^l = i[\Sigma_l^r - \Sigma_l^a]$ [90].

To study the conductance characteristics in the presence of both Rashba SOC and exchange field [86], we set the parameters $\lambda_R=0.1t_0$, $M=0.2t_0$, and $\lambda_{so}=0t_0$. Notice that with these parameters, the system is in the QAH phase [31]. Nevertheless, if the ISO parameter is different from zero, there is an upper-limited value of λ_{so} [31, 32], beyond which a new phase characterized by a vanishing Chern Number C=0 can take place; this phase is the so-called TRS-broken QSH phase [31, 32, 34]. The spin-resolved conductance $G_{\sigma\sigma'}$ is shown in **Figure 15**: for (a) unstrained, (b) strained along $\theta=0$, (c) strained along $\theta=\pi/2$, and (d) the total conductance $\sum_{\sigma\sigma'} G_{\sigma\sigma'}$ of a ZGNR. Notice that there is a suppression for both the spin conserving and the spin-flip conductance components for either unstrained or strained ZGNR in the energy range considered. However, these backscattering (transmission dips) at certain precise energies at the first plateau are different depending on the strain configuration, and a close inspection shows that conducting channels for non-spin flip and spin-flip conductances oscillate. Depending on the Fermi energy and set parameters, certain conductance components can even be completely suppressed. This suppression is attributed to the appearance of quasi-localized states in the device, which may produce sharp scattering resonances, also known as resonant backscattering which is a general behavior of quasi-1D quantum systems [92]. For higher energies, however, the large number of conducting channels leads to a nonvanishing transmission, as the channels get mixed along the device and results in the appearance of an interchannel backscattering leaded by interference effects. Therefore, in the QSH phase protected by the TRS, nonmagnetic impurities do not cause backscattering on each boundary, and the spin transport in the edge states is dissipationless at zero temperature.

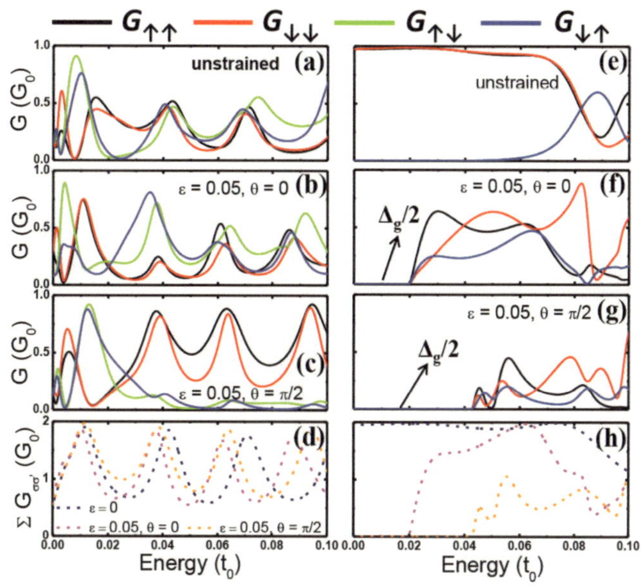

Figure 15. Effects of strain on spin-resolved conductance $G_{\sigma\sigma'}$ of ZGNR with $N_Z=26$ (a)–(c) and AGNR with N_A=47 (e)–(g), respectively. Panels (d) and (h) show the total conductance. The parameters used in all panels are λ_R=0.1t_0, λ_{so} =0, and M=0.2t_0. Reproduced with permission from Diniz et al. [86]. Copyright (2014), AIP Publishing LLC.

In the QAH phase, however, there is a weak scattering between forward and backward movers, leading to a low-dissipation spin transport. At low energy, this interesting strain-controllable behavior of conducting channel suppression might be efficiently used to filter electrical current of desired spins, in spin filtering devices [86]. In **Figure 15(d)**, we show the total conductance, which is nearly robust against strains, specially close to the charge neutrality point, where the deviations due to strain are quite small. In contrast, the conductance of AGNR shows a drastic modification as one can notice in **Figure 15(e)–(h)**, with the development of a transport gap, which is insensitive to the electron spin that is injected and collected in the device. However, this induced transport gap is dependent upon the direction of the applied strain, with a larger conduction suppression along $\theta=0$ (red dashed line) with Δ_g= 0.04 t_0, and Δ_g= 0.086 t_0 while along $\theta=\pi/2$, that can be observed in **Figure 15(h)**. Also, the total conductance exhibits different plateaus: around $2G_0$ and approximately G_0 in AGNR without and with strain, respectively, which is one less quantum of conductance available for the electron to be transmitted along the strained device for energies beyond the transport gap.

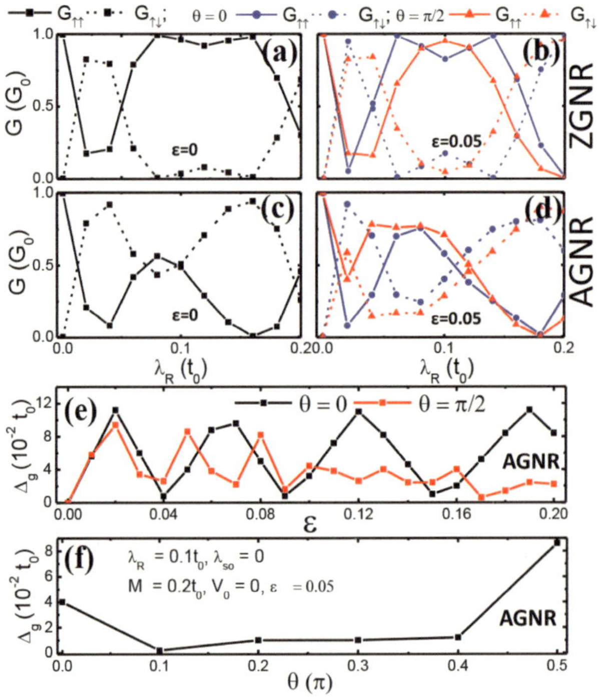

Figure 16. (a) Conductance profiles $G_{\uparrow\uparrow}$ and $G_{\uparrow\downarrow}$ for 26-ZGNR (a)–(b) and 47-AGNR (c)–(d) as function of λ_R subjected to different configurations of strain. Panels (e) shows the band gap Δ_g of an AGNR as function of strain parameter ε for $\theta=0$ and $\theta=\pi/2$, respectively. (f) Δ_g of an AGNR as function of the direction of strain for fixed ε=0.05. Reproduced with permission from Diniz et al. [86]. Copyright [2014], AIP Publishing LLC.

Another remarkable phenomenon is the oscillatory dependence of the spin components of $G_{\sigma\sigma'}$ on the value of λ_R [86], which is shown in **Figure 16(a)–(d)**, where the curves correspond to different topological GNRs and strain setups for $E=0.05t_0$. The same parameters are used as the **Figure 15**, except for M. To reveal the effects of Rashba SOC, we set M=0 in the calculation. Then, the system is time-reversal invariant and the conductance components $G_{\uparrow\uparrow}=G_{\downarrow\downarrow}$ and $G_{\uparrow\downarrow}=G_{\downarrow\uparrow}$. This oscillatory behavior is reminiscent of the spin field effect transistor (FET) and has a similar source [93], as the spin *precesses* as it propagates in the presence of the Rashba field, acquiring a net phase that is proportional to $\lambda_R L$, where L is the length of the device.

Further inspecting the strain-induced band gap in 47-AGNR in the presence of SOC and exchange field interactions, one notices that in **Figure 16(e)** a similar band gap oscillation characteristic as reported in a earlier work [84]. In the regime of small strain, the band gap shows approximately linear response, with increasing values of strain, however, it starts to oscillate. Further investigation shows that the amplitude and period of the gap oscillation are tuned by direction of the strain, as shown in **Figure 16(e)**. A specific dependence of transport gap on the angle of the strain is clearly depicted in **Figure 16(f)**. Notice that the transport gap is indeed strongly tuned by strain direction. It equals approximately zero at 0.1π, while it reaches 0.086 t_0 at 0.5π.

3.3. Quantum phase transitions in strained graphene nanoribbon

Quantum spin Hall and quantum anomalous Hall (QAH) states have topologically protected edge states, where the electron back scattering is forbidden, making these systems good candidates for electronic devices with dissipationless electronic transport [33, 35, 38, 41]. The potential possibility to explore the different Quantum Hall phases in strained graphene has motivated us to study the strain-related physics at zero magnetic field in graphene nanoribbons [49].

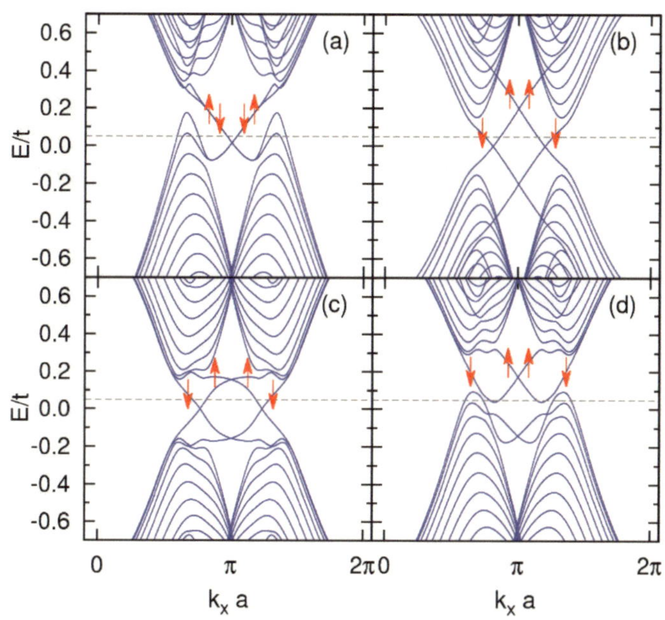

Figure 17. Energy band of ZGNR with intrinsic- and Rashba-SOC terms (a), intrinsic SOC and EX (b), Rashba SOC and EX (c), and intrinsic- and Rashba-SOCs and EX (d). The Fermi level is assumed to be above zero, as indicated by the dashed horizontal line, and thus has four intersections with the conduction bands. This gives rise to four edge currents on the ribbon edges. The following parameters are used: (a) $\lambda_{so}=0.06t$, $\lambda_R=0.20t$; (b) $\lambda_{so}=0.06t$, $M=0.20t$; (c) $\lambda_R=0.20t$, $M=0.20t$; (d) $\lambda_{so}=0.06t$, $\lambda_R=0.20t$, and $M=0.20t$ for the ZGNR with width $W=48$. The arrows represent the major components of spin. Copyright (2015) by the American Physical Society [49].

If the mirror symmetry about the graphene plane is preserved, then the intrinsic SOC which opens gaps around Dirac points is the only allowed spin-dependent term in the Hamiltonian. Otherwise, if the mirror symmetry is broken, then a Rashba term is allowed, which mixes spin-up and spin-down states around the band crossing points. Besides, Rashba SOC pushes the valence band up and the conduction band down, reducing the bulk gap. Following Reference [45], we present our results for the ZGNR in **Figure 17**, which shows the effects of intrinsic- and Rashba-SOCs and EX upon the energy band of the ZGNR [49]. Notice in **Figure 17(a)** that the interplay between intrinsic- and Rashba-SOCs partially lifts the degeneracies of both bulk- and edge-state, breaks particle-hole symmetry and pushes the valence band up. In turn, the presence of the EX breaks the TRS and lifts the Kramer's degeneracy of electron spin, pushing the spin-up (spin-down) bands upward (downward), as shown in **Figure 17(b)**. In strong contrast with **Figure 17(b)**, the presence of Rashba SOC and EX induces coupling between edge and bulk states, which significantly modifies the group velocity of edge states, as shown in **Figure 17(c)**. The combined effects of intrinsic, Rashba SOCs and EX are shown in **Figure 17(d)**, which are in agreement with results reported in Reference [45] (see for instance **Figure 2**). Notice that the Fermi level enters into the valence band and the energies of some edge modes are smaller than the valence band maximum.

The intrinsic SOC can be strongly enhanced by impurity (adatom) coverage on the surface of graphene, which produces strong lattice distortions [59]. In this context, one may ask how the quantum phase transition in a graphene ribbon changes as the intrinsic SOC is tuned [49]. Following the discussion of Reference [45], the effects of strain fields are shown in **Figure 18** (with a similar representation to the one introduced in Reference [45]) with parameters W=48, $\lambda_R=0.20t$, $M=0.20t$, and uniaxial strain $\varepsilon=0.10$ along $\theta=0$. The left panel of **Figure 18** shows the effects of intrinsic SOC on the energy spectrum of a ZGNR. The Fermi level is set at $E_F=0.05t$. The corresponding edge state probability distributions across the width of the nanoribbon, for each of the four edge states indicated by A, B, C, and D, are shown in the middle panel. Schematic diagrams of charge current distributions on the edges of ZGNR are illustrated in the right panel. To determine the edge current direction, $I=-|e|v_x$ (indicated by the arrow), the electron group velocity $v_x=\partial E(k)/\partial k_x$ has been calculated [45]. In the case of weak intrinsic SOC, at the ribbon boundaries, the edge states pair A and D would form a single handed loop (in the sense that the turning point is at infinity along the ribbon length), meanwhile there is the formation of another loop with opposite handedness, which is formed by the edge states pair B and C. Both edge states A and B, consequently I_A and I_B, are located at the same edge, as indicated in **Figure 18(c)**. Thus, the chirality of the current loop due to the A and D edge states would produce a Chern number of $(C_1=\pm1)$, which is the same as that of current loop owing to B and C edge states. Since the system is akin to two integer quantum Hall subsystems, its Chern number((C)) is equal to $(C_1=+1)\oplus(C_2=+1)$, i.e., $C=(+1)+(+1)=2$ or $(C_1=-1)\oplus(C_2=-1)$, with $C=(-1)+(-1)=-2$. Therefore, the ZGNR with a weak intrinsic SOC is in the QAH phase. For a ZGNR with strong intrinsic SOC, however, one can notice that the edge states pair A and C are located on the same edge, whereas the B and D edge states are in the opposite edge, as shown in **Figure 18(f)**. Due to the handedness of the current loop of edge states A and D, the Chern number would give a contribution of $(\mathcal{C}_1=-1)$, and the pair B and C, which has an opposite handedness, produces a Chern number of $(\mathcal{C}_2=+1)$. Since the ZGNR is composed of these two integer quantum Hall subsystems, its Chern number $(\mathcal{C})$ is obtained by $(\mathcal{C}_1=+1)\oplus(\mathcal{C}_2=-1)$, i.e., $\mathcal{C}=(+1)+(-1)=0$. Therefore, the GNR is in the TRS broken QSH phase.

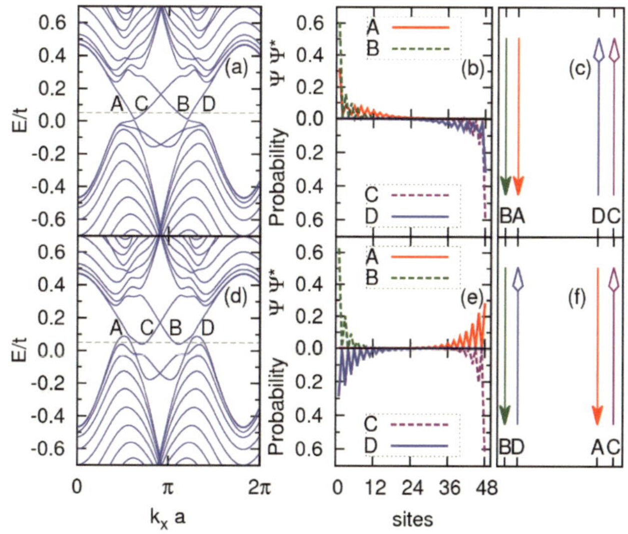

Figure 18. Energy spectrum of ZGNR with $W=48$, $\lambda_R=0.20t$, $M=0.20t$, $\varepsilon=0.10$, and $\theta=0$, for (a) $\lambda_{so}=0.035t$ and (d) $\lambda_{so}=0.055t$, respectively. The Fermi level $E=0.05t$ corresponds to four different edge states, as indicated by A, B, C, and D. The corresponding probability distributions $|\psi|^2$ across the width of the ribbon, and diagrams of charge current distributions are shown in the middle (b)–(e) and right panels (c)–(f), respectively. The arrows indicate the current flux. Copyright (2015) by the American Physical Society [49].

To understand the QPT and show intuitively how it takes place [49], we follow Reference [45] and introduce the average value of the position $\langle y\rangle_n$, as a parameter to label the angular momentum of the current [49]. It is defined as: $\langle y\rangle_n=\sum_i y_i\,|\varphi_n(y_i)|^2$, where n represents the edge states at the Fermi level and i is the site index along the width of ribbon. We chose the origin of y axis at the lower boundary of the ribbon. **Figure 19(a)** shows the average values $\langle y\rangle_n$ of edge states as a function of λ_{so} in the ribbon with the width $W=48$, $\lambda_R=0.20t$, $M=0.20t$, $\varepsilon=0.10$, and $\theta=0$, where $n=A,B,C$, and D, respectively. The direction and magnitude of a group velocity are indicated by the direction and length of an arrow, respectively. When the intrinsic SOC is vanishing, the Rashba SOC and EX are dominant, A and B are on the same boundary of the ribbon, and thus both $\langle y\rangle_A$ and $\langle y\rangle_B\rightarrow 0$. The edge states C and D also follow the same behavior, but are localized at the other edge of the ribbon, thus $\langle y\rangle_C$ and $\langle y\rangle_D\rightarrow W$. The system is in the QAH phase. When the λ_{so} increases, however, three different topological phases are found. In the regime of small λ_{so} $(0.03t<\lambda_{so}<0.04t)$, the positions of the edge states are only very slightly shifted. With increasing λ_{so}, the states A and D become delocalized, swiftly moving to the center of the ribbon from different boundaries owing to the edge- and bulk-states coupling. In the regime of large λ_{so} (λ_{so}>0.05), the locations of state A and D have been exchanged. Since the group velocity of state A is opposite to D, the exchange of their locations results in a change of chirality. Therefore, the system is in the QSH phase. It is worthy to point out that owing to the finite-size (finite-width) effect, the edge states are not exactly localized at the two boundaries. Remarkably, a similar behavior is also presented in **Figure 19(b)** in which $\langle y \rangle$ versus strain is plotted. At first glance, it seems to be hard to understand this exotic behavior. But, recalling the discussion of phase transition in bulk graphene, one can logically speculate that this is a manifestation of strain-induced QPT between QSH and QAH states in the ZGNR. This strain-induced QSH state shares many emergent properties similar to the usual zero-strain QSH effect. We notice that with realistic values for uniaxial strain the critical value for the spin-orbit coupling is reduced by a factor between 10% and 20%. Thus, the combination of strain and appropriate substrates shows a promising direction to realize the phase transition in current settings.

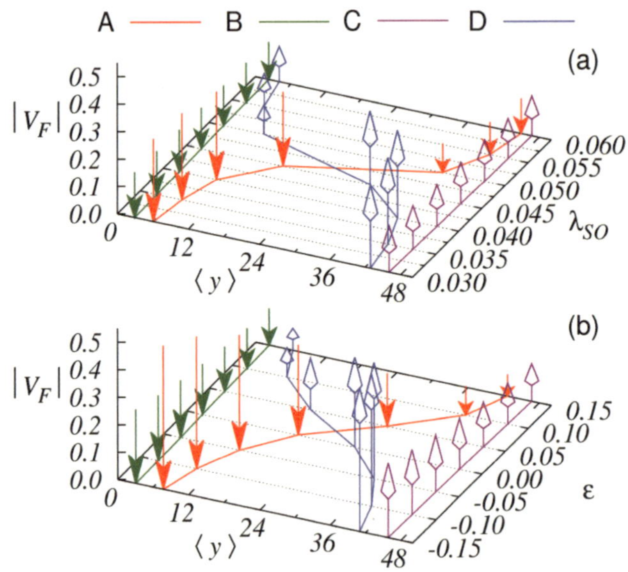

Figure 19. (a) Average values $\langle y \rangle$ of edge states versus λ_{so} in ZGNR, subjected to a strain with $\varepsilon = 0.10$ and $\theta = 0$. (b) $\langle y \rangle$ as a function of strain with $\theta = 0$ for $\lambda_{so} = 0.05t$. $W = 48$, $\lambda_R = 0.20t$, and $M = 0.20t$ are used in the computations. Vertical axis is the Fermi velocity V_F modulus. The arrows point in the directions of band velocities and their lengths present the magnitudes of V_F. Copyright (2015) by the American Physical Society [49].

To seek the controllable topological QPTs induced either by strain (EX), or intrinsic SOC, or any of their combinations [49], the phase diagrams in which the phase is characterized by the difference in the average value of position $\langle y \rangle_C$ and $\langle y \rangle_A$, defined as $\langle y \rangle_{AC} = \langle y \rangle_C - \langle y \rangle_A$, are constructed, as shown in **Figure 20**. **Figure 20(a)** and **(b)** plots the phase diagrams of ε versus

λ_{so} for $\theta = 0$ and $\theta = \pi/2$, respectively. It is trivial to notice that if $\langle y \rangle_{AC} \cong 0$, the edge states A and C are localized at the same boundary, corresponding to a QSH phase, as indicated by blue. Otherwise, if $\langle y \rangle_{AC} \cong W$, the system is in the QAH phase, as marked by red. The other values of $\langle y \rangle_{AC}$ correspond to delocalized state A. Notice that both strength and direction of the strain change considerably the phase diagram. In the regime of small intrinsic SOC, the GNR lies in the QAH state. The critical λ_{so}^c at which topological QPT depends strongly on both the strength and direction of the strain. The larger the strain, the smaller the λ_{so}^c is required to reach the QSH state. In addition, the strain drives the GNR from the QAH into QSH states for a given λ_{so}^c. It is also noted that in the case of $\theta = \pi/2$, when the λ_{so}^c changes in the boundary between the QSH and QAH states, the correspondent critical value of ε varies faster than that for $\theta = 0$.

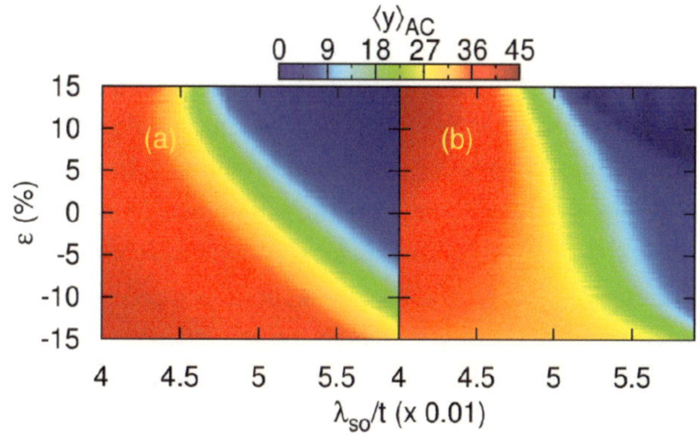

Figure 20. Phase diagrams (strain vs. intrinsic SOC) of a ribbon with $W = 48$, $\lambda_R = 0.20t$, and $M = 0.20t$, characterized by a difference in the average value of position between mode A and C, defined as $\langle y \rangle_{AC} = \langle y \rangle_C - \langle y \rangle_A$, for $\theta = 0$ (a) and $\theta = \pi/2$ (b), respectively. Copyright (2015) by the American Physical Society [49].

The underlying physics of the strain tuned phase diagram is as follows. It is well established that uniaxial mechanical strain does not break the sublattice symmetry, but rather deforms the Brillouin zone, such as, the Dirac cones located in graphene at points K (K') being shifted in the opposite directions [52, 83]. This is reminiscent of the effect of pseudomagnetic field B_S induced by the strain on charge carriers, i.e., accumulating charge in place where the B_S is maximum. Because the B_S does not break TRS, the strain will not have any direct effect on the spin degrees of freedom of the electrons, even though it couples with sublattice pseudospin. Therefore, at first glance, it seems that the strain only induces a renormalization of the energy scales. Actually, *this is not true* for graphene with SOC. Since SOC couples the spin and the momentum degrees of freedom of the carriers, B_S could affect real spin of an electron through the SOC. Therefore, a strong pseudomagnetic field should lead to Landau quantization and a QSH state due to opposite signs of B_S for electrons in valleys K (K'). In this context, the strain enhances the carrier localization and pushes the edge states much closer to the boundaries of the ribbon. Hence, the QSH state could be stabilized by the strain. Finally, it is worthwhile to argue that since inter-valley scattering requires a large momentum transfer [94], it is strongly suppressed in wide ZGNRs in which we are interested.

4. Conclusion

In summary, we have performed a systematic investigation of the effects of uniaxial strains, exchange field, staggered sublattice potential, and SOC on the electronic and transport properties of graphene and graphene nanoribbons. We have employed the tight-binding approximation, and Green's function formalism in order to fully describe the electronic and transport properties of these interesting nanostructures.

Using an effective low energy approximation, we were able to describe the Berry curvature and the associated Chern numbers for different orientation and uniaxial strain strength, as function of exchange field interaction. The QSH–QAH phase transition associated to the tunability of Chern number for the bulk graphene displays an interesting behavior according to specific directions of strains: an increase in the critical exchange field M_c for the QAHE phase transition for $\theta=\pi/2$ as the strain modulus is enhanced, in contrast to the $\theta=0$, which shows a reduction (above a limiting strain modulus of approximately $\varepsilon=0.078$) in the critical exchange field M_c for the QAHE phase [32]. The investigated spin-resolved electronic transport and LDOS of GNR devices have demonstrate that it is possible to achieve a total electron transmission suppression of specific spin specie, which can be further tailored by uniaxial tensile strain on specific directions [86]. In addition, we have implemented a formalism to describe the zero-field topological QPT between QSH and QAH states in GNRs in the presence of internal EX, uniaxial strain, and intrinsic and Rashba SOCs [49].

Our results demonstrated in this Chapter offer the prospect to efficiently manipulate the electronic structure, transport properties, and consequently the QAHE by strain engineering of the graphene. We also envision that our work can be extended to other layered materials (for instance, transition metal dichalcogenides), with a great potential application on novel electronic devices with the focus on dissipationless charge current.

Author details

Fanyao Qu[1*], Ginetom S. Diniz[2,3] and Marcos R. Guassi[1]

*Address all correspondence to: fanyao@unb.br

1 Institute of Physics, University of Brasília, Brasília, DF, Brazil

2 Institute of Physics, Federal University of Uberlândia, Uberlândia, MG, Brazil

3 Institute of Science and Technology, Federal University of Goiás, Jataí, GO, Brazil

References

[1] R. E. Peierls. Quelques proprietes typiques des corpses solides. *Ann. I. H. Poincare*, 5:177, 1935.

[2] L. D. Landau. Zur theorie der phasenumwandlungen ii. *Phys. Z. Sowjetunion*, 11:26, 1937.

[3] J. A. Venables, G. D. T. Spiller, and M. Hanbucken. Nucleation and growth of thin films. *Rep. Prog. Phys.*, 47:399, 1984.

[4] J. W. Evans, P. A. Thiel, and M. C. Bartelt. Morphological evolution during epitaxial thin film growth: Formation of 2D islands and 3D mounds. *Sur. Sci. Rep.*, 61:1, 2006.

[5] P. R.Wallace. The band theory of graphite. *Phys. Rev.*, 71:622–634, 1947.

[6] K. S. Novoselov, A. K. Geim, S. V. Morozov, D. Jiang, Y. Zhang, S. V. Dubonos, I. V. Grigorieva, and A. A. Firsov. Electric field effect in atomically thin carbon films. *Science*, 306(5696):666–669, 2004.

[7] A. K. Geim and K. S. Novoselov. The rise of graphene. *Nature materials*, 6(3):183–191, 2007.

[8] J. L. Tedesco, B. L. VanMil, R. L. Myers-Ward, J. M. McCrate, S. A. Kitt, P. M. Campbell, G. G. Jernigan, J. C. Culbertson, C. R. Eddy Jr, and D. K. Gaskill. Hall effect mobility of epitaxial graphene grown on silicon carbide. *Appl. Phys. Lett.*, 95(12):122102, 2009.

[9] T. Shen, J. J. Gu, M. Xu, Y. Q. Wu, M. L. Bolen, M. A. Capano, L. W. Engel, and P. D. Ye. Observation of quantum-hall effect in gated epitaxial graphene grown on SiC (0001). *Appl. Phys. Lett.*, 95(17):172105, 2009.

[10] J. Hass, F. Varchon, J. Millan-Otoya, M. Sprinkle, N. Sharma, W. A. de Heer, C. Berger, P. N. First, L. Magaud, and E. H. Conrad. Why multilayer graphene on 4 h-sic (000 1^{-}) behaves like a single sheet of graphene. *Phys. Rev. Lett.*, 100(12):125504, 2008.

[11] G. M. Rutter, J. N. Crain, N. P. Guisinger, T. Li, P. N. First, and J. A. Stroscio. Scattering and interference in epitaxial graphene. *Science*, 317(5835):219–222, 2007.

[12] M. Sprinkle, D. Siegel, Y. Hu, J. Hicks, A. Tejeda, A. Taleb-Ibrahimi, P. Le Fevre, F. Bertran, S. Vizzini, H. Enriquez et al. First direct observation of a nearly ideal graphene band structure. *Phys. Rev. Lett.*, 103(22):226803, 2009.

[13] X. Li, W. Cai, J. An, S. Kim, J. Nah, D. Yang, R. Piner, A. Velamakanni, I. Jung, E. Tutuc et al. Large-area synthesis of high-quality and uniform graphene films on copper foils. *Science*, 324(5932):1312–1314, 2009.

[14] S. Nie, M. W. Joseph, N. C. Bartelt, O. D. Dubon, and K. F. McCarty. Origin of the mosaicity in graphene grown on cu (111). *Phys. Rev. B*, 84(15):155425, 2011.

[15] A. Varykhalov and O. Rader. Graphene grown on co (0001) films and islands: Electronic structure and its precise magnetization dependence. *Phys. Rev. B*, 80(3):035437, 2009.

[16] T. Oznuluer, E. Pince, E. O. Polat, O. Balci, O. Salihoglu, and C. Kocabas. Synthesis of graphene on gold. *Appl. Phys. Lett.*, 98(18):183101, 2011.

[17] J. W. McClure. Diamagnetism of graphite. *Phys. Rev.*, 104:666, 1956.

[18] J. C. Slonczewski and P. R.Weiss. Band structure of graphite. *Phys. Rev.*, 109:272, 1958.

[19] R. E. Smalley. Discovering the fullerenes. *Rev. Mod. Phys.*, 69(3):723–730, 1997.

[20] S. Iijima et al. Helical microtubules of graphitic carbon. *nature*, 354(6348):56–58, 1991.

[21] L. Brey and H. A. Fertig. Electronic states of graphene nanoribbons studied with the dirac equation. *Phys. Rev. B*, 73:235411, 2006.

[22] W. Häusler and R. Egger. Artificial atoms in interacting graphene quantum dots. *Phys. Rev. B*, 80(16):161402, 2009.

[23] Z. Z. Zhang, K. Chang, and F. M. Peeters. Tuning of energy levels and optical properties of graphene quantum dots. *Phys. Rev. B*, 77:235411, 2008.

[24] K. Wakabayashi, Y. Takane, M. Yamamoto, and M. Sigrist. Electronic transport properties of graphene nanoribbons. *New J. Phys.*, 11(9):095016, 2009.

[25] Motohiko Ezawa. Peculiar width dependence of the electronic properties of carbon nanoribbons. *Phys. Rev. B*, 73:045432, 2006.

[26] Adam Rycerz, Electron transport and quantum-dot energy levels in Z-shaped graphene nanoconstriction with zigzag edges, Acta Phys. Polon. A 118, 238–243 (2010).

[27] F. D. M. Haldane. Model for a quantum hall effect without landau levels: Condensed-matter realization of the "parity anomaly". *Phys. Rev. Lett.*, 61:2015–2018, 1988.

[28] Y. Zhang et al. Experimental observation of the quantum hall effect and berry's phase in graphene. *Nature*, 438:201, 2005.

[29] V. P. Gusynin and S. G. Sharapov. Unconventional integer quantum hall effect in graphene. *Phys. Rev. Lett.*, 95:146801, 2005.

[30] Zhenhua Qiao, Shengyuan A. Yang, Wanxiang Feng, Wang-Kong Tse, Jun Ding, Yugui Yao, JianWang, and Qian Niu. Quantum anomalous hall effect in graphene from rashba and exchange effects. *Phys. Rev. B*, 82:161414, 2010.

[31] Z. Qiao et al. Microscopic theory of quantum anomalous hall effect in graphene. *Phys. Rev. B*, 85:115439, 2012.

[32] G. S. Diniz, M. R. Guassi, and F. Qu. Engineering the quantum anomalous hall effect in graphene with uniaxial strains. *J. Appl. Phys.*, 114(24), 2013.

[33] C. L. Kane and E. J. Mele. Quantum spin hall effect in graphene. *Phys. Rev. Lett.*, 95(22): 226801, 2005.

[34] Yunyou Yang, Zhong Xu, L. Sheng, Baigeng Wang, D. Y. Xing, and D. N. Sheng. Time-reversal-symmetry-broken quantum spin hall effect. *Phys. Rev. Lett.*, 107:066602, 2011.

[35] C. Chang, J. Zhang, X. Feng et al. Experimental observation of the quantum anomalous hall effect in a magnetic topological insulator. *Science*, 340:167, 2013.

[36] X. Qi and S. Zhang. Topological insulators and superconductors. *Rev. Mod. Phys.*, 83:1057–1110, 2011.

[37] R. Yu, W. Zhang, H. Zhang, S. Zhang, X. Dai, and Z. Fang. Quantized anomalous hall effect in magnetic topological insulators. *Science*, 329:61–4, 2010.

[38] B. A. Bernevig, T. L. Hughes, and S. C. Zhang. Quantum spin hall effect and topological phase transition in hgte quantum wells. *Science*, 314(5806):1757–1761, 2006.

[39] M. König, S. Wiedmann, C. Brüne, A. Roth, H. Buhmann, L. W. Molenkamp, Xiao-Liang. Qi, and Shou-Cheng Zhang. Quantum spin hall insulator state in hgte quantum wells. *Science*, 318(5851):766–770, 2007.

[40] C. Liu, W. Feng, and Y. Yao. Quantum spin hall effect in silicene and two-dimensional germanium. *Phys. Rev. Lett.*, 107:076802, 2011.

[41] X. Qian, J. Liu, L. Fu, and J. Li. Quantum spin hall effect in two-dimensional transition metal dichalcogenides. *Science*, 346:1344, 2014.

[42] Ferry, D. K., Goodnick, S. M., Bird, J. (2009). *Transport in Nanostructures*. Cambridge University Press. DOI: 10.1017/CBO9780511840463.

[43] C. L. Kane and E. J. Mele. z_2 topological order and the quantum spin hall effect. *Phys. Rev. Lett.*, 95(14):146802, 2005.

[44] C. Liu, T. L. Hughes, X. Qi, K.Wang, and S. Zhang. Quantum spin hall effect in inverted type-ii semiconductors. *Phys. Rev. Lett.*, 100:236601, 2008.

[45] T. Chen, Z. Xiao, D. Chiou, and G. Guo. High chern number quantum anomalous hall phases in single-layer grapheme with Haldane orbital coupling. *Phys. Rev. B*, 84(16): 165453, 2011.

[46] S. Oh. The complete quantum hall trio. *Science*, 340(6129):153–154, 2013.

[47] F. Guinea, M. I. Katsnelson, and A. K. Geim. Energy gaps and a zero-field quantum hall effect in graphene by strain engineering. *Nat. Phys.*, 6(1):30–33, 2009.

[48] M. C. Rechtsman, J. M. Zeuner, A. Tünnermann, S. Nolte, M. Segev, and A. Szameit. Strain-induced pseudomagnetic field and photonic landau levels in dielectric structures. *Nature Photon.*, 3:91, 2009.

[49] M. R. Guassi, G. S. Diniz, N. Sandler, and F. Qu. Zero-field and time-reserval-symmetry-broken topological phase transitions in graphene. *Phys. Rev.* B, 92:075426, 2015.

[50] N. W. Ashcroft and N. D. Mermin. *Solid State Physics*. New York: Holt, Rinehart and Winston 1976.

[51] A. H. Castro Neto, F. Guinea, N. M. R. Peres, K. S. Novoselov, and A. K. Geim. The electronic properties of graphene. *Rev. Mod. Phys.*, 81:109–162, 2009.

[52] V. M. Pereira, A. H. Castro Neto, and N. M. R. Peres. Tight-binding approach to uniaxial strain in graphene. *Phys. Rev. B*, 80(4):045401, 2009.

[53] Zhenhua Qiao, Wei Ren, Hua Chen, L. Bellaiche, Zhenyu Zhang, A. H. MacDonald, and Qian Niu. Quantum anomalous hall effect in graphene proximity coupled to an antiferromagnetic insulator. *Phys. Rev. Lett.*, 112:116404, 2014.

[54] Adrian G. Swartz, Patrick M. Odenthal, Yufeng Hao, Rodney S. Ruoff, and Roland K. Kawakami. Integration of the ferromagnetic insulator euo onto graphene. *ACS Nano*, 6(11):10063–10069, 2012.

[55] Yugui Yao, Fei Ye, Xiao-Liang Qi, Shou-Cheng Zhang, and Zhong Fang. Spin-orbit gap of graphene: First-principles calculations. *Phys. Rev.* B, 75(4):041401, 2007.

[56] J. C. Boettger and S. B. Trickey. First-principles calculation of the spin-orbit splitting in graphene. *Phys. Rev. B*, 75:121402, 2007.

[57] H. Min, J. E. Hill, N. A. Sinitsyn, B. R. Sahu, L. Kleinman, and A. H. MacDonald. Intrinsic and rashba spin-orbit interactions in graphene sheets. *Phys. Rev. B*, 74(16):165310, 2006.

[58] N. M. R. Peres, F. Guinea, and A. H. Castro Neto. Coulomb interactions and ferromagnetism in pure and doped graphene. *Phys. Rev. B*, 72(17):174406, 2005.

[59] A. H. Castro Neto and F. Guinea. Impurity-induced spin-orbit coupling in graphene. *Phys. Rev. Lett.*, 103:026804, 2009.

[60] M. Zarea and N. Sandler. Rashba spin-orbit interaction in graphene and zigzag nanoribbons. *Phys. Rev. B*, 79(16):165442, 2009.

[61] Daniel Huertas-Hernando, F. Guinea, and Arne Brataas. Spin-orbit coupling in curved graphene, fullerenes, nanotubes, and nanotube caps. *Phys. Rev. B*, 74:155426, 2006.

[62] M. P. López-Sancho and M. C. Muñoz. Intrinsic spin-orbit interactions in flat and curved graphene nanoribbons. *Phys. Rev. B*, 83:075406, 2011.

[63] Yu. S. Dedkov, M. Fonin, U. Rüdiger, and C. Laubschat. Rashba effect in the graphene/ni(111) system. *Phys. Rev. Lett.*, 100:107602, 2008.

[64] A. P. M. Barboza, A. P. Gomes, B. S. Archanjo, P. T. Araujo, A. Jorio, A. S. Ferlauto, M. S. C. Mazzoni, H. Chacham, and B. R. A. Neves. Deformation induced semiconductor-metal transition in single wall carbon nanotubes probed by electric force microscopy. *Phys. Rev. Lett.*, 100:256804, 2008.

[65] Mingyuan Huang, Yang Wu, Bhupesh Chandra, Hugen Yan, Yuyao Shan, Tony F. Heinz, and James Hone. Direct measurement of strain-induced changes in the band structure of carbon nanotubes. *Phys. Rev. Lett.*, 100:136803, 2008.

[66] A. G. Souza Filho, N. Kobayashi, J. Jiang, A. Grüneis, R. Saito, S. B. Cronin, J. Mendes Filho, Ge. G. Samsonidze, G. Dresselhaus, and M. S. Dresselhaus. Strain-induced interference effects on the resonance raman cross section of carbon nanotubes. *Phys. Rev. Lett.*, 95:217403, 2005.

[67] Liu Yang and Jie Han. Electronic structure of deformed carbon nanotubes. *Phys. Rev. Lett.*, 85:154–157, 2000.

[68] Jien Cao, Qian Wang, and Hongjie Dai. Electromechanical properties of metallic, quasimetallic, and semiconducting carbon nanotubes under stretching. *Phys. Rev. Lett.*, 90:157601, 2003.

[69] Fang Liu, Pingbing Ming, and Ju Li. *Ab initio* calculation of ideal strength and phonon instability of graphene under tension. *Phys. Rev. B*, 76:064120, 2007.

[70] K. S. Kim, Y. Zhao, H. Jang, S. Y. Lee, J. M. Kim, K. S. Kim, J. H. Ahn, P. Kim, J. Choi, and B. H. Hong. Large-scale pattern growth of graphene films for stretchable transparent electrodes. *Nature (London)*, 457:706, 2009.

[71] L. Blakslee, D. G. Proctor, E. J. Seldin, G. B. Stence, and T. Wen. Elastic constants of compression-annealed pyrolytic graphite. *J. Appl. Phys*, 41:3373, 1970.

[72] F. M. D. Pellegrino, G. G. N. Angilella, and R. Pucci. Strain effect on the optical conductivity of graphene. *Phys. Rev. B*, 81:035411, 2010.

[73] H. Zhao, K. Min, and N. R. Aluru. Size and chirality dependent elastic properties of graphene nanoribbons under uniaxial tension. *Nano Letters*, 9(8):3012–3015, 2009.

[74] Bitan Roy and Igor F. Herbut. Topological insulators in strained graphene at weak interaction. *Phys. Rev. B*, 88:045425, 2013.

[75] Igor F. Herbut. Pseudomagnetic catalysis of the time-reversal symmetry breaking in graphene. *Phys. Rev. B*, 78:205433, 2008.

[76] D. A. Abanin and D. A. Pesin. Interaction-induced topological insulator states in strained graphene. *Phys. Rev. Lett.*, 109:066802, 2012.

[77] D. J. Thouless, M. Kohmoto, M. P. Nightingale, and M. den Nijs. Quantized hall conductance in a two-dimensional periodic potential. *Phys. Rev. Lett.*, 49:405–408, 1982.

[78] Jeil Jung, Zhenhua Qiao, Qian Niu, and Allan H. MacDonald. Transport properties of graphene nanoroads in boron nitride sheets. *Nano Letters*, 12(6):2936–2940, 2012.

[79] Vitor M. Pereira, A. H. Castro Neto, and N. M. R. Peres. Tight-binding approach to uniaxial strain in graphene. *Phys. Rev. B*, 80:045401, 2009.

[80] Yasumasa Hasegawa, Rikio Konno, Hiroki Nakano, and Mahito Kohmoto. Zero modes of tight-binding electrons on the honeycomb lattice. *Phys. Rev. B*, 74:033413, 2006.

[81] D. Bercioux and A. De Martino. Spin-resolved scattering through spin-orbit nanostructures in graphene. *Phys. Rev. B*, 81:165410, 2010.

[82] Kyoko Nakada, Mitsutaka Fujita, Gene Dresselhaus, and Mildred S. Dresselhaus. Edge state in graphene ribbons: Nanometer size effect and edge shape dependence. *Phys. Rev. B*, 54:17954, 1996.

[83] M. Oliva-Leyva and G. G. Naumis. Understanding electron behavior in strained graphene as a reciprocal space distortion. *Phys. Rev. B*, 88:085430, 2013.

[84] Y. Lu and J. Guo. Band gap of strained graphene nanoribbons. *Nano Research*, 3(3):189–199, 2010.

[85] D. A. Bahamon and Vitor M. Pereira. Conductance across strain junctions in graphene nanoribbons. *Phys. Rev. B*, 88:195416, 2013.

[86] Ginetom S. Diniz, Marcos R. Guassi, and Fanyao Qu. Controllable spin-charge transport in strained graphene nanoribbon devices. *Journal of Applied Physics*, 116(11), 113705: 1–7, 2014.

[87] Markus Morgenstern. Scanning tunneling microscopy and spectroscopy of graphene on insulating substrates. *physica status solidi (b)*, 248(11):2423–2434, 2011.

[88] Elena Stolyarova, Kwang Taeg Rim, Sunmin Ryu, Janina Maultzsch, Philip Kim, Louis E. Brus, Tony F. Heinz, Mark S. Hybertsen, and George W. Flynn. High-resolution scanning tunneling microscopy imaging of mesoscopic graphene sheets on an insulating surface. *PNAS*, 104(22):9209–9212, 2007.

[89] Guohong Li, Adina Luican-Mayer, Dmitry Abanin, Leonid Levitov, and Eva Y. Andrei. Evolution of landau levels into edge states in graphene. *Nat. Commun.*, 4:1744, 2013.

[90] M. B. Nardelli. Electronic transport in extended systems: Application to carbon nanotubes. *Phys. Rev. B*, 60(11):7828, 1999.

[91] M. P. Lopez Sancho, J. M. Lopez Sancho, and J. Rubio. Quick iterative scheme for the calculation of transfer matrices: application to mo (100). *Journal of Physics F: Metal Physics*, 14(5):1205, 1984.

[92] R. Gómez-Medina, P. San José, A. García-Martín, M. Lester, M. Nieto-Vesperinas, and J. J. Sáenz. Resonant radiation pressure on neutral particles in a waveguide. *Phys. Rev. Lett.*, 86:4275–4277, 2001.

[93] S. Datta and B. Das. Electronic analog of the electro-optic modulator. *Appl. Phys. Lett.*, 56(7):665–667, 1990.

[94] J. Wurm, M. Wimmer, and K. Richter. Symmetries and the conductance of graphene nanoribbons with long-range disorder. *Phys. Rev. B*, 85:245418, 2012.

Chapter 4

Nonstandard Transition GUE-GOE for Random Matrices and Spectral Statistics of Graphene Nanoflakes

Adam Rycerz

Additional information is available at the end of the chapter

http://dx.doi.org/10.5772/64240

Abstract

Spectral statistics of weakly disordered triangular graphene flakes with zigzag edges are revisited. Earlier, we have found numerically that such systems may show spectral fluctuations of Gaussian unitary ensemble (GUE), signaling the time-reversal symmetry (TRS) breaking at zero magnetic field, accompanied by approximate twofold valley degeneracy of each energy level. Atomic-scale disorder induces the scattering of charge carriers between the valleys and restores the spectral fluctuations of Gaussian orthogonal ensemble (GOE). A simplified description of such a nonstandard GUE–GOE transition, employing the mixed ensemble of 4 × 4 real symmetric matrices was also proposed. Here, we complement our previous study by analyzing numerically the spectral fluctuations of large matrices belonging the same mixed ensemble. Resulting scaling laws relate the ensemble parameter to physical size and the number of atomic-scale defects in graphene flake. A phase diagram, indicating the regions in which the signatures of GUE may by observable in the size-doping parameter plane, is presented.

Keywords: graphene, quantum chaos, random matrix, time-reversal symmetry, gaussian ensemble

1. Introduction

The notion of emergent phenomena was coined out by Anderson in his milestone science paper of 1979 [1]. In brief, emergence occurs when a complex system shows qualitatively different properties then its building blocks. Numerous examples of emergent systems studied in contemporary condensed matter physics, including high-temperature superconductors and heavy-fermion compounds [2], are regarded as systems with spontaneous symmetry breaking [3]. A link between emergence and spontaneous symmetry breaking, however, does not seem to have a permanent character. In a wide class of electronic systems, such as semiconducting heterostructures containing a two-dimensional electron gas (2DEG), physical properties of itinerant electrons are substantially different than properties of free electrons (or electrons in atoms composing the system), and are also highly-tunable upon variation of external electromagnetic fields [4]. To give some illustration of this tunability, we only mention that electrons in GaAs heterostructures can be usually described by a standard Schrödinger equation of quantum mechanics with the effective mass $m_{\rm eff}$=0.067$m_{\rm e}$ (where $m_{\rm e}$ is the free electron mass), whereas in extreme cases of quantum states formed in quantum Hall systems, effective quasiparticles may not even show the Fermi–Dirac statistics [5, 6].

It is rather rarely noticed that graphene, a two-dimensional form of carbon just one atom tick [7], also belongs to the second class of emergent systems (i.e., without an apparent spontaneous symmetry breaking) described briefly above. In a monolayer graphene, effective Hamiltonian for low-energy excitations has a Dirac–Weyl form, namely

$$\mathcal{H}_{\rm eff} = v_F[\boldsymbol{p} + e\boldsymbol{A}(\boldsymbol{r},t)] \cdot \boldsymbol{\sigma} + U(\boldsymbol{r},t), \tag{1}$$

where $v_{\rm F}=10^6$ m/s is the energy-independent Fermi velocity, $\sigma=(\sigma_x, \sigma_y)$ with the Pauli matrices σ_x and σ_y, $p=-i\hbar(\partial_x, \partial_y)$ is the in-plane momentum operator, the electron charge is $-e$, and the external electromagnetic field is defined via scalar and vector potentials, $U(r, t)$ and $A(r, t)$, with the in-plane position $r=(x, y)$ and the time t.[1] In other words, the system build of nonrelativistic elements (carbon atoms at normal conditions) turns out to host ultrarelativistic quasiparticles, providing a beautiful example of an emergent phenomenon, which binds together two rather distant areas of relativistic quantum mechanics and condensed matter physics [8]. This observation applies generically to bilayer or multilayer graphenes [9], as well as to HgTe/CdTe quantum wells [10], although microscopic models describing such *other Dirac systems* are slightly different. It is also worth to mention so-called artificial graphenes, in which waves (of different kinds) obey their effective Dirac equations [11–13].

A peculiar nature of Dirac fermions in graphene originates from the chiral structure of the Hamiltonian $\mathcal{H}_{\rm eff}$, accompanied by the fact that coupling to the external electromagnetic field is described by additive terms, which are linear in both scalar and vector potentials. A remarkable consequence of these facts is the quantization of the visible light absorption [14], an unexpected macroscopic quantum effect recently found to have analogs in other Dirac systems [15, 16], and even in a familiar graphite [17]. Another intriguing effect of this kind appears for dc conductivity of ballistic graphene [18]. In the so-called pseudodiffusive transport regime, the conductance of a rectangular sample (with the width W and the length L) scales as $G=\sigma_0 \times W/L$ for $W \gg L$, where $\sigma_0=(4/\pi)e^2/h$ is the universal quantum value of the conductivity [19, 20], whereas the shot-noise power and all the other charge-transfer characteristics are indistinguishable from those of a classical diffusive conductor regardless

[1] Strictly speaking, $\mathcal{H}_{\rm eff}$ Eq. (1) applies to quasiparticles near the K valley in the dispersion relation. To obtain the effective Hamiltonian for other valley (K') it is sufficient to substitute $\sigma_y \to -\sigma_y$.

the sample shape [21]. At high magnetic fields, the pseudo diffusive charge transport is predicted theoretically to reappear for resonances with Landau levels in both monolayer [22] and bilayer graphene [23]. In the presence of disorder, a fundamental property of the Hamiltonian—the time reversal symmetry (TRS)—starts to play a decisive role. In particular, effective TRS in a single valley may be broken even in the absence of magnetic fields, leading to observable (and having the universal character) consequences for the conductance and spectral fluctuations [24, 25], as well as for the peculiar scaling behavior predicted for the conductivity [26, 27].

Although the interest in graphene and other Dirac systems primarily focus on their potential applications [28, 29], quite often linked to the nonstandard quantum description [8], we believe that the fundamental perspective sketched in the above also deserves some attention. In the remaining part of this article, we first overview basic experimental, theoretical and numerical findings concerning signatures of quantum chaos in graphene and its nanostructures (Section 2). Next, we present our new numerical results concerning the additive random matrix model originally proposed in Ref. [25] to describe a nonstandard GUE–GOE transition, accompanied by lifting out the valley degeneracy (Section 3). The consequences of these findings for prospective experiments on graphene nanoflakes, together with the phase diagram depicting the relevant matrix ensembles in the system size-doping plane, are described in Section 4. The concluding remarks are given in Section 5.

2. Gauge fields, fluctuations and chaos in nanoscale graphene structures

Dirac fermions confined in graphene quantum dots [30] have provided yet another surprising situation, in which a piece of handbook knowledge needed a careful revision [31].

Quantum chaotic behavior appears generically for systems, whose classical dynamics are chaotic, and manifest itself via the fact that energy levels show statistical fluctuations following those of Gaussian ensembles of random matrices [32]. In particular, if such a system posses the time-reversal symmetry (TRS), its spectral statistics follows the Gaussian orthogonal ensemble (GOE). A system with TRS and half-integer spin has the symplectic symmetry and, in turn, shows spectral fluctuations of the Gaussian symplectic ensemble (GSE). If TRS is broken, as in the presence of nontrivial gauge fields, and the system has no other antiunitary symmetry [33], spectral statistics follow the Gaussian unitary ensemble (GUE). For a particular case of massless spin-1/2 particles, it was pointed out by Berry and Mondragon [34], that the confinement may break TRS in a persistent manner (i.e., even in the absence of gauge fields), leading to the spectral fluctuations of GUE.

When applying the above symmetry classification to graphene nanosystems [24, 25], one needs, however, to take into account that Dirac fermions in graphene appear in the two valleys, K and $K^{'}$, coupled by TRS. (In particular, real magnetic field breaks TRS and has the same sign in the two valleys, whereas the strain-induced gauge field preserves TRS and has opposite signs in the two valleys.) If the valley pseudospin is conserved, a special (symplectic) time-reversal symmetry (STRS) becomes relevant, playing a role of an effective TRS in a single valley

[24]. Both real magnetic and strain-induced gauge fields may break STRS leading to the spectral fluctuations of GUE [35]. As demonstrated numerically in Ref. [25], such fluctuations also appear for particular closed nanosystems in graphene in the presence of random scalar potentials slowly varying on the scale of atomic separation. Such nanosystems include equilateral triangles with zigzag or Klein edges, i.e., with terminal atoms belonging to one sublattice. Generic graphene nanoflakes with irregular edges show spectral fluctuations of GOE [24], as strong intervalley scattering restores TRS in the absence of gauge fields (see **Figure 1**). In contrast, the boundary effects are suppressed in open graphene systems, for which signatures of the symplectic symmetry class were reported [36].

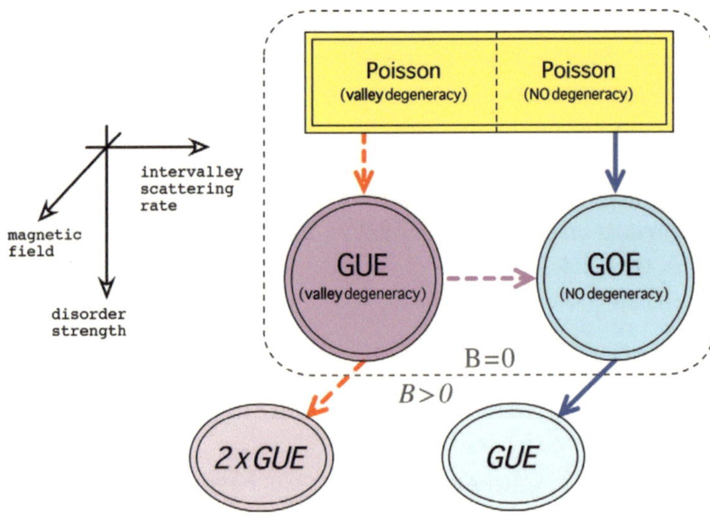

Figure 1. Transitions between symmetry classes and random matrix ensembles relevant for *closed* nanosystems in graphene characterized by the disorder strength, the intervalley scattering rate, and (optionally) placed in the weak magnetic field B. (Reprinted with permission from Ref. [25].)

It is worth mention here, that triangular graphene flakes, similar to studied theoretically in Ref. [25], have been recently fabricated [37, 38]. However, due to the hybridization with metallic substrates, quantum-dot energy levels in such systems are significantly broaden, making it rather difficult to determine the symmetry class via spectral statistics.

3. Transition GUE–GOE for real symmetric matrices

3.1. Additive random-matrix models: brief overview

Additive random-matrix models are capable of reproducing the evolutions of spectral statistics in many cases when a complex system undergoes transition to quantum chaos or transition between symmetry classes [32, 39]. The discussion usually focus on the auxiliary random Hamiltonian of the form

$$H(\lambda) = \frac{H_0+\lambda V}{\sqrt{1+\lambda^2}}, \tag{2}$$

where $H_0=(H_0)^{\dagger}$ and $V=V^{\dagger}$ are members of different Gaussian ensembles,[2] and the parameter $\lambda \in [0, \infty]$.

For instance, if elements of H_0 are real numbers chosen to follow a Gaussian distribution with zero mean and the variance $\langle (H_0)_{ij}^2 \rangle = (1+\delta_{ij})/N$, where δ_{ij} is the Kronecker delta and N is the matrix size, while elements of V are complex numbers in which real and imaginary parts are generated independently according to Gaussian distribution with zero mean and the variance $\langle (\mathrm{Re}V_{ij})^2 \rangle = (1+\delta_{ij})/2N$, $\langle (\mathrm{Im}V_{ij})^2 \rangle = (1-\delta_{ij})/2N$ (respectively), the Hamiltonian $H(\lambda)$ (2) refers to transition GOE–GUE. For $N=2$, statistical distribution of the spacing between energy levels $S=|E_1-E_2|$ can be found analytically [40], and reads

$$P_{\mathrm{GOE-GUE}}(\lambda; S) = \sqrt{\frac{2+\lambda^2}{2}}\, S c^2(\lambda) \exp\left[-\frac{S^2 c^2(\lambda)}{2}\right] \mathrm{erf}\left[\frac{S c(\lambda)}{\lambda}\right], \tag{3}$$

where erf(x) is the error function, i.e., $\mathrm{erf}(x) = (2/\sqrt{\pi})\int_0^x \exp(-t^2)dt$, and

$$c(\lambda) = \sqrt{\frac{\pi(2+\lambda^2)}{4}}\left[1 - \frac{2}{\pi}\left(\arctan\left(\frac{\lambda}{\sqrt{2}}\right) - \frac{\sqrt{2}\,\lambda}{2+\lambda^2}\right)\right]. \tag{4}$$

The above follows from the normalization condition

$$\langle S \rangle = \int_0^\infty S\, P_{\mathrm{GOE-GUE}}(\lambda; S)dS = 1 \quad \text{for} \quad 0 < \lambda < \infty. \tag{5}$$

The limiting forms of the spacing distribution given by Eqs. (3) and (4) are

$$P_{\mathrm{GOE-GUE}}(\lambda \to 0; S) = \frac{\pi}{2} S \exp\left(-\frac{\pi S^2}{4}\right) \equiv P_{\mathrm{GOE}}(S), \tag{6}$$

$$P_{\mathrm{GOE-GUE}}(\lambda \to \infty; S) = \frac{32}{\pi^2} S^2 \exp\left(-\frac{4S^2}{\pi}\right) \equiv P_{\mathrm{GUE}}(S), \tag{7}$$

coinciding with well-known Wigner surmises for GOE and GUE, respectively [32]. For $N \gg 1$, it was also shown that actual spacing distributions obtained numerically can be

[2] To describe transition to quantum chaos rather then transition between symmetry classes in a chaotic system, one can choose H_0 to be a diagonal random matrix, elements of which follow a Gaussian distribution with zero mean and the variance $\langle (H_0)_{ij}^2 \rangle = \delta_{ij}$.

approximated (with an astonishing accuracy) by $P_{\mathrm{GOE\text{-}GUE}}(\lambda_{\mathrm{fit}}; S)$, where the empirical parameter $\lambda_{\mathrm{fit}} \propto \lambda\sqrt{N}$ [39]. Similar scaling laws apply generically to all transitions between basic symmetry classes.

Relatively recently, spectra of models employing self-dual random matrices have attracted some attention [41]. In such models, the matrix H_0 in Eq. (2) is equivalent (up to a unitary transformation) to the matrix having a block structure

$$\widetilde{H}_0 = \begin{pmatrix} C & 0 \\ 0 & C^T \end{pmatrix}, \tag{8}$$

where random matrix C is an $N \times N$ member of one of Gaussian ensembles, C^{T} denotes the transposition of C. The matrix V in Eq. (2) is a generic $2N \times 2N$ member of the other ensemble (hereinafter, we redefine the $H(a)$ size as $2N$). In turn, for $\lambda=0$, each eigenvalue is doubly degenerate. For $\lambda \neq 0$, we have the degeneracy splitting accompanied by transition between selected symmetry classes. Even in the simplest case of $N=2$, closed-form analyticexpressions for level-spacing distributions corresponding to arbitrary $0<\lambda<\infty$ are missing. The approach presented in Ref. [41] employs the relevant expressions for joint probability densities for eigenvalues [42], allowing one to express level-spacings distribution in terms of two-dimensional integrals to be evaluated numerically.

In the remaining part of section, we focus on the transition between self-dual GUE to GOE, show that the corresponding Hamiltonian $H(\lambda)$, and can be represented as real-symmetric random matrix, and present our empirical expressions approximating spacing distributions obtained numerically.

3.2. Self-dual GUE to GOE via 4×4 real-symmetric matrices

We focus now on the situation, when the matrix C in Eq. (8) is chosen to be an $N \times N$ member of GUE, whereas V in Eq. (2) is a $2N \times 2N$ member of GOE.

For $N=2$, the matrix $\widetilde{H}_0$ can be written as

$$\widetilde{H}_0^{4\times 4} = \begin{pmatrix} a & c+id & 0 & 0 \\ c-id & b & 0 & 0 \\ 0 & 0 & a & c-id \\ 0 & 0 & c+id & b \end{pmatrix}, \tag{9}$$

where a and b are real random numbers following Gaussian distribution with zero mean and the variance $\langle a^2\rangle=\langle b^2\rangle=1/2$, whereas c and d are real random numbers following Gaussian distribution with zero mean and the variance $\langle c^2\rangle=\langle d^2\rangle=1/4$. Exchanging the second row with the third row, as well as the second column with the third column, we find the matrix $\widetilde{H}_0^{4\times 4}$ is equivalent, up to an orthogonal transformation, to

http://dx.doi.org/10.5772/64240

$$\widetilde{H}_0^{4\times 4} \overset{0}{\leftrightarrow} \begin{pmatrix} a & 0 & c+id & 0 \\ 0 & a & 0 & c-id \\ c-id & 0 & b & 0 \\ 0 & c+id & 0 & b \end{pmatrix}. \tag{10}$$

The matrix on the right-hand side of Eq. (10) is self-dual, and can be further transformed as

$$U \begin{pmatrix} a & 0 & c+id & 0 \\ 0 & a & 0 & c-id \\ c-id & 0 & b & 0 \\ 0 & c+id & 0 & b \end{pmatrix} U^\dagger = \begin{pmatrix} a & 0 & c & d \\ 0 & a & -d & c \\ c & -d & b & 0 \\ d & c & 0 & b \end{pmatrix}, \tag{11}$$

where

$$U = \frac{1}{\sqrt{2}} \mathbb{I}_{2\times 2} \otimes \begin{pmatrix} 1 & 1 \\ i & -i \end{pmatrix} = \frac{1}{\sqrt{2}} \begin{pmatrix} 1 & 1 & 0 & 0 \\ i & -i & 0 & 0 \\ 0 & 0 & 1 & 1 \\ 0 & 0 & i & -i \end{pmatrix}. \tag{12}$$

Exchanging the second with the third row and column in the rightmost matrix in Eq. (11) we arrive to

$$H_0^{4\times 4} = \begin{pmatrix} a & c & 0 & d \\ c & b & -d & 0 \\ 0 & -d & a & c \\ d & 0 & c & b \end{pmatrix} = \begin{pmatrix} A & B \\ -B & A \end{pmatrix}, \tag{13}$$

where the blocks A and B are real-symmetric ($A^{\mathrm{T}}=A$) and skew-symmetric ($B^{\mathrm{T}}=-B$) random matrices.

Spectral statistics of the Hamiltonian $H(\lambda)=\left(H_0^{4\times 4}+\lambda V^{4\times 4}\right)/\sqrt{1+\lambda^2}$, with $V^{4\times 4}$ being a 4×4 GOE matrix, were thoroughly studied before [43]. Here, we revisit our findings, before discussing spectra of larger matrices in next subsection.

The nearest-neighbor spacings distribution can be approximated by

$$P(\alpha,\kappa;S) = \frac{\alpha P_{\mathrm{GOE}}(\alpha S)+\beta(\alpha)P_{\mathrm{GOE-GUE}}(\kappa;\beta(\alpha)S)}{2}, \tag{14}$$

with $\beta(\alpha)=\alpha/(2\alpha-1)$, $P_{\mathrm{GOE}}(S)$ given by Eq. (6), $P_{\mathrm{GOE-GUE}}(\kappa;S)$ given by Eq. (3), and the parameters α and κ which can be approximated by empirical functions

$$\alpha \approx \overline{\alpha}_{4\times 4}(\lambda) = 1.118 \times \left[\sqrt[3]{1+(0.60/\lambda)^3}\right]^{0.98}, \tag{15}$$

and

$$\kappa \approx \overline{\kappa}_{4\times 4}(\lambda) = \sqrt{\left(\frac{1+\lambda^{-2}}{1+(0.33)^{-2}}\right)^{0.29}} - 1. \tag{16}$$

Eqs. (15) and (16) represent simplified versions of the corresponding formulas given in Ref. [43]. A comparison with the numerical will be given later in this section.

3.3. Self-dual GUE to GOE via $2N \times 2N$ real-symmetric matrices

We consider now the case of large random matrices ($N \gg 1$). A generalization of the reasoning presented in previous subsection brought as to the unperturbed Hamiltonian H_0 with the block structure as given by the last equality in Eq. (13), but $A = A^{\mathrm{T}}$ and $B = -B^{\mathrm{T}}$ are now $N \times N$ random matrices. The elements of each block are independently generated according to a Gaussian distribution with zero mean and the variance $\mathrm{Var}(A_{ij}) = (1+\delta_{ij})/2N$ and $\mathrm{Var}(B_{ij}) = (1-\delta_{ij})/2N$, respectively. In turn, H_0 can be unitary mapped onto the matrix $\widetilde{H}_0$ given by Eq. (8) with C being an $N \times N$ member of GUE. The additive random-matrix model $H(\lambda)$ is complemented with the perturbation V being a $2N \times 2N$ member of GOE.

Ensembles of large pseudo-random Hamiltonians $H(\lambda)$ were generated and diagonalized numerically, to check whether the standard scaling law $\lambda_{\mathrm{fit}} \simeq (2N)^{1/2}\lambda$ [44] applies to spacings distribution of such matrices. Our presentation is limited to the matrix sizes $2N = 200$, 400, and 1000; the statistical ensemble consists of the total amount of 10^6, 10^5, or 10^4 matrices (respectively), same for each considered value of the parameter λ. To avoid the boundary effects, we limit our numerical study to about 30% of the energy levels such that $|E| \leq 0.5$. Selected examples are presented in **Figure 2**.

We find that nearest-neighbor level spacings of large matrix $H(\lambda)$ follow the empirical distribution having the general form as given by Eq. (14).

$$P_{\overline{\alpha},\overline{\kappa}}(\tilde{\lambda}, S) = P(\overline{\alpha}_{N\gg 1}(\tilde{\lambda}), \overline{\kappa}_{N\gg 1}(\tilde{\lambda}); S), \tag{17}$$

with the empirical relations of Eqs. (15) and (16) [see blue solid lines in **Figure 3**] now replaced by

$$\overline{\alpha}_{N\gg 1}(\tilde{\lambda}) = 1.114 \times \left[\sqrt[3]{1 + (0.60/\tilde{\lambda})^3}\right]^{0.98}, \tag{18}$$

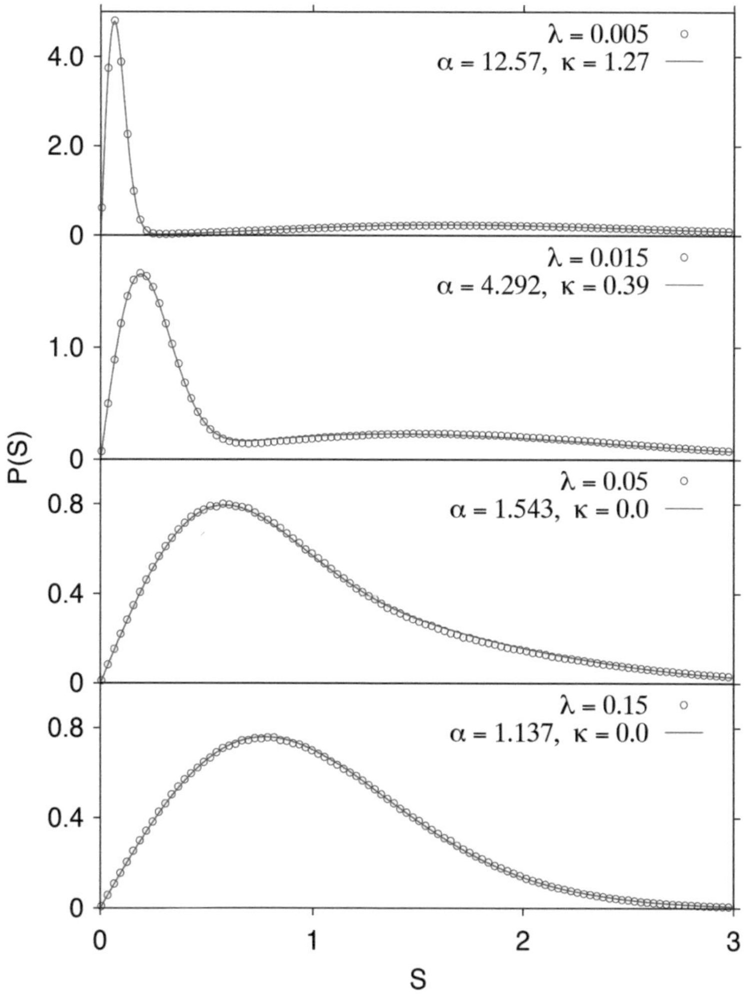

Figure 2. Level-spacing distributions for 10^5 randomly-generated Hamiltonians $H(\lambda)$ with the size $2N = 400$ (data-points). The scaling parameters λ is varied between the panels. The least-squares fitted functions $P(\alpha, \kappa; S)$ defined by Eq. (14) are also shown (solid lines).

and

$$\overline{\kappa}_{N\gg 1}(\tilde{\lambda}) = \begin{cases} \sqrt{\left[\frac{1+\tilde{\lambda}^{-2}}{1+(0.27)^{-2}}\right]^{0.29} - 1} & \text{if } \tilde{\lambda} < 0.27, \\ 0 & \text{if } \tilde{\lambda} \geq 0.27. \end{cases} \tag{19}$$

The above formulae are marked in **Figure 3** with red dashed lines. We also find that the scaling law $\bar{\lambda} = \lambda_{\mathrm{fit}} \simeq (2N)^{1/2}\lambda$ [with λ being the original parameter of $H(\lambda)$] is satisfied for the matrices considered with a surprising accuracy (see **Figure 4**).

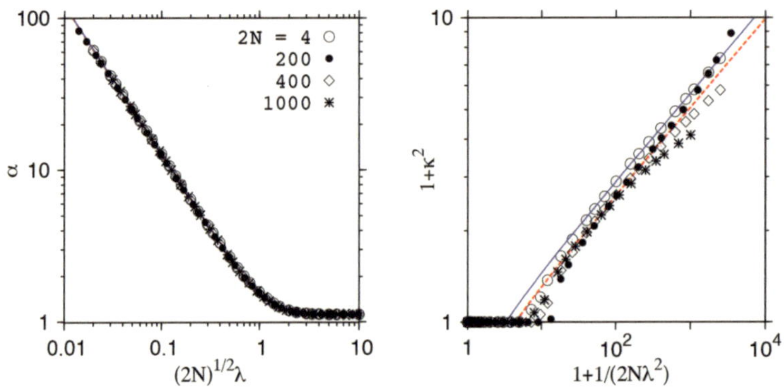

Figure 3. Least-squares fitted parameters of $P(\alpha,\ \kappa; S)$ Eq. (14) for different values of $2N$ as functions of the scaled model parameter $(2N)^{1/2}\lambda$ (datapoints). The empirical relations $\bar{\alpha}_{4\times4}(\lambda)$ Eq. (15) and $\bar{\kappa}_{4\times4}(\lambda)$ Eq. (16) valid for $2N = 4$ are shown with blue solid lines; the relations $\bar{\alpha}_{N\gg1}(\lambda)$ Eq. (18) and $\bar{\kappa}_{N\gg1}(\lambda)$ Eq. (19) for large matrices are shown with red dashed lines.

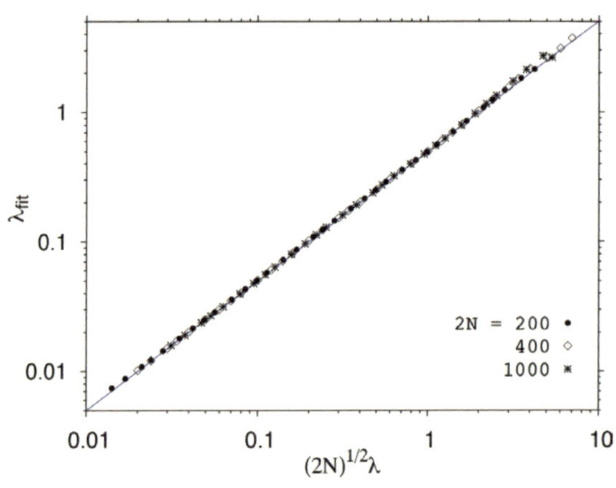

Figure 4. Scaling law for the best fitted parameters $\bar{\lambda} = \lambda_{\mathrm{fit}}$ in the distribution $P_{\bar{\alpha},\bar{\kappa}}(\bar{\lambda},\ S)$ Eq. (17) approximating $P(S)$ obtained numerically for random Hamiltonians $H(\lambda)$ with $2N = 200$, 400, and 1000. [See the main text for details.] Blue solid line marks $\lambda_{\mathrm{fit}} = \lambda$.

4. Consequences for graphene nanoflakes

4.1. Level-spacing distributions revisited

In this section, the empirical distribution $P_{\tilde{\alpha},\tilde{\kappa}}(\lambda, S)$ (17) with least-square fitted $\lambda = \lambda_{\rm fit}$ is utilized to rationalize level-spacing distributions for triangular graphene nanoflakes with zigzag edges.

At zero magnetic field, the tight-binding Hamiltonian for weakly-disordered graphene can be written as

$$\mathcal{H}_{\rm TBA} = \Sigma_{\langle ij \rangle}\left[t_{ij}\,|i\rangle\langle j| + {\rm h.c.}\right] + \Sigma_i\left[M_V(\boldsymbol{r}_i) + U_{\rm imp}(\boldsymbol{r}_i)\right] |i\rangle\langle i|, \tag{20}$$

where $t_{ij} = -t$ if the orbitals $|i$ and $|j$ are nearest neighbors on the honeycomb lattice (with $t = \frac{2}{3}\sqrt{3}\hbar v_{\rm F}/a \approx 3$ eV, and $a = 0.246$ nm being the lattice spacing), otherwise $t_{ij} = 0$. (The symbol $\Sigma_{\langle ij \rangle}$ denotes that each pair $\langle ij \rangle$ is counted only once.) The terms $M_{\rm V}(r_i)$ and $U_{\rm imp}(r_i)$ represent the potentials abruptly and slowly varying on the scale of atomic separation (respectively). Here, we put $M_V(r_i) = 0.7\,t$ if r_i is the outermost atom position at zigzag edge, otherwise $M_{\rm V}(r_i) = 0$. The random contribution $U_{\rm imp}(r_i)$ is generated in as follows: first, we randomly choose $N_{\rm imp}$ lattice sites R_n ($n = 1, \ldots, N_{\rm imp}$) out of $N_{\rm tot}$. Next, the amplitudes $U_n \in (-\delta, \delta)$ are randomly generated. Finally, the potential is smoothed over a distance $\xi = \sqrt{3}\,a$ by convolution with a Gaussian, namely

$$U_{\rm imp}(\boldsymbol{r}) = \Sigma_{n=1}^{N_{\rm imp}} U_n \exp\left(-\frac{|\boldsymbol{r}-\boldsymbol{R}_n|^2}{2\xi^2}\right). \tag{21}$$

A model of substrate-induced disorder, constituted by Eqs. (20) and (21), was widely used to reproduce numerically several transport properties of disordered graphene samples [45–48]. Here, we revisit the spectra of closed graphene flakes considered in Ref. [25], within a simplified empirical model $P_{\tilde{\alpha},\tilde{\kappa}}(\lambda_{\rm fit}, S)$ (17), in order discuss the consequences for prospective experimental observation of the zero-field time-reversal symmetry breaking in such systems.

A compact measure of the disorder strength is given by the dimensionless correlator

$$K_0 = \frac{\mathcal{A}}{(\hbar v_F)^2}\frac{1}{N_{tot}^2}\Sigma_{i=1}^{N_{\rm tot}}\Sigma_{j=1}^{N_{\rm tot}}\langle U_{\rm imp}(\boldsymbol{r}_i) U_{\rm imp}(\boldsymbol{r}_j)\rangle, \tag{22}$$

where the system area $\mathcal{A} = \frac{1}{4}\sqrt{3}N_{\rm tot}\,a^2$, and the averaging takes place over possible realizations of the disorder in Eq. (21). For $\xi \gg a$, Eq. (22) leads to

$$K_0 = \frac{64\pi^2\sqrt{3}}{27}\frac{N_{imp}}{N_{tot}}\left(\frac{\delta}{t}\right)^2\left(\frac{\xi}{a}\right)^4. \tag{23}$$

For $\xi = \sqrt{3}a$, used for numerical demonstration in the remaining of this article, Eq. (23) still provides a good approximation of the actual value of K_0 and can be rewritten as

$$K_0 \approx 364.7 \times \frac{N_{imp}}{N_{tot}} \left(\frac{\delta}{t}\right)^2. \tag{24}$$

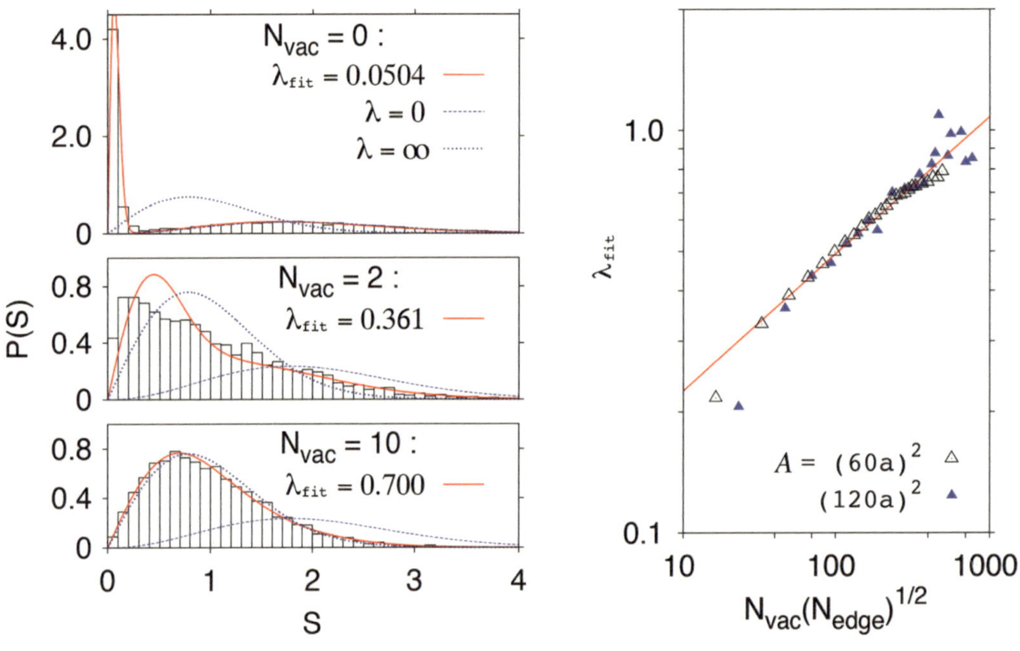

Figure 5. *Left:* Level-spacing distributions $P(S)$ for triangular graphene nanoflakes with zigzag edges. The flake area is $\mathcal{A} \approx (120\,a)^2$, the disorder strength is $K_0 \approx 0.125$, the number of edge vacancies $N_{\rm vac}$ is varied betweenthe panels. Numerical results (replotted with permission from Ref. [25]) are shown with black solid lines. The other lines correspond to empirical distributions $P_{\tilde{\alpha},\tilde{\kappa}}(\lambda,\ S)$ Eq. (17) with $\lambda = \lambda_{\rm fit}$ (red solid line), $\lambda = 0$ (blue dashed line) or $\lambda = \infty$ (blue dotted line). *Right:* Least-squares fitted parameters for different numbers of edge vacancies $1 \leq N_{\rm vac} \leq 30$ and the flake areas $\mathcal{A} \approx (60\,a)^2$ (open symbols) and $\mathcal{A} \approx (120\,a)^2$ (closed symbols), corresponding to the total number of terminal atoms $N_{\rm edge} = 270$ and 540 (respectively). Solid line depicts the approximating power-law relation given by Eq. (25).

The numerical results are presented in **Figure 5**, where we have fixed the remaining disorder parameters at $\delta / t = 0.1$ and $N_{\rm imp} / N_{\rm tot} = 0.034$ leading to $K_0 = 0.125$.[3] Level-spacing distributions $P(S)$ obtained numerically for triangular nanoflakes with zigzag edges [see left panels

[3] The disorder parameters are actually same as in Figures 8 and 9 of Ref. [25], where we have mistakenly omitted the factor π in the numerical evaluation of K_0.

in **Figure 5**, black solid lines] are replotted with permission from Ref. [25], where we used approximately 1500 energy levels with energies $0.1 \leq |E|/t \leq 0.5$ out of the total number of $N_{tot}(N_{vac})=32758-N_{vac}$ (corresponding the flake area $\mathcal{A} = (120\,a)^2$), with N_{vac} being the number of vacancies, randomly distributed along the system boundary. Typically, best-fitted parameters $\lambda=\lambda_{fit}$ of the simplified distribution $P_{\bar{\alpha},erline\kappa}(\lambda, S)$ (17) coincide with given in Ref. [25] up to a second decimal place. New values of λ_{fit} for $1 \leq N_{vac} \leq 30$ and two flake sizes $N_{tot}(0)=8278$ and $N_{tot}(0)=32,758$ are displayed in the right panel of **Figure 5**. The dependence of λ_{fit} on N_{vac} and N_{tot} can be rationalize within a power-law

$$\lambda_{fit} \approx 0.103 \times \left(N_{vac}\sqrt{N_{edge}}\right)^{0.34}, \tag{25}$$

where the total number of terminal sites

$$N_{edge} = 3\sqrt{N_{tot} + N_{vac} + 3} - 3. \tag{26}$$

4.2. Phase diagram for triangular flakes with zigzag edges

Eq. (25) is now employed to estimate the maximal system size N_{tot}, and the maximal number of edge vacancies N_{vac}, for which signatures of TRS breaking still can be identified in the spectrum. This is possible as long as $\lambda_{fit} < \lambda_\star = 0.27$ (see Eq. (19)), as for any $\lambda_{fit} \geq \lambda_\star$ we have $\bar{\kappa}(\lambda_{fit})=0$ and level-spacing distribution simply evolves from that characterizing GOE matrix with approximate twofold degeneracy of each level toward GOE without such a degeneracy. For instance, we obtain

$$N_{tot} \lesssim 9500 \quad \text{for} \quad N_{vac} = 1, \tag{27}$$

$$N_{tot} \lesssim 630 \quad \text{for} \quad N_{vac} = 2, \tag{28}$$

$$N_{tot} \lesssim 130 \quad \text{for} \quad N_{vac} = 3. \tag{29}$$

On the other hand, system size and the number of energy levels taken into account must be large enough to distinguish between spectral fluctuations of GUE and spectral fluctuations of other ensembles.

Density of states (per one direction of spin) for bulk graphene reads

$$\rho_{bulk}(E) = \frac{\mathcal{A}}{\pi(\hbar v_F)^2}|E| = \frac{N_{tot}}{\sqrt{3}\,\pi t^2}|E|. \tag{30}$$

The number of energy levels N_{lev} in the interval $(0, E_{max})$ can be approximated by

$$N_{\text{lev}} \approx \int_0^{E_{\max}} \rho_{\text{bulk}}(E)dE \approx 0.0919 \times N_{\text{tot}} \left(\frac{E_{max}}{t}\right)^2. \tag{31}$$

Physically, occupying N_{lev} electronic levels above the Dirac point one produces the electric charge $Q=-2eN_{\text{lev}}$, resulting in a typical experimental limit of $E_{\max}$=0.2−0.3 eV for graphene nanostructures on SiO_2-based substrates [49].

Level-spacing distributions $P(S)$ are normalized such that $\langle 1\rangle = \langle S\rangle = 1$. In turn, the variance $\text{Var}\{S\} \equiv \sigma^2 = \langle S^2\rangle - 1$ raises as the lowest moment allowing one to distinguish between different distributions. In particular, we have

$$\text{Var}\{S\}_{\text{GOE}} = \frac{4}{\pi} - 1 \approx 0.273, \quad \text{Var}\{S\}_{\text{GUE}} = \frac{3\pi}{8} - 1 \approx 0.178, \tag{32}$$

$$\text{Var}\{S\}_{\lambda\to 0} = \frac{3\pi}{4} - 1 \approx 1.356, \tag{33}$$

where Eq. (33) refers to the empirical distribution $P_{\tilde{\alpha},\tilde{\kappa}}(\lambda, S)$ given by Eq. (17) with $\lambda \to 0$. Similar calculation for arbitrary λ is straightforward, but the resulting formula is too lengthy to be presented.[4] When Var$\{S\}$ is calculates for a large but finite collection of spacings $N_{\text{spc}}=N_{\text{lev}}-1$, it becomes a random variable itself, with a variance which can be approximated by

$$\text{Var}\{\text{Var}\{S\}\} \approx \frac{1}{N_{\text{lev}}}(\mu_4 - \sigma^2) = \frac{1}{N_{\text{lev}}}(\langle S^4\rangle - 4\langle S^3\rangle + 8\langle S^2\rangle - \langle S^2\rangle^2 - 4), \tag{34}$$

where $\mu_4=\langle(\langle S\rangle-S)\rangle^4$ denotes the fourth central moment and we have used the normalization $\langle S\rangle=1$. In turn, for spacings following the distribution $P_{\tilde{\alpha},\tilde{\kappa}}(\lambda, S)$ (17) one can find they do not follow GOE if

$$N_{\text{lev}} \gtrsim \left[\frac{3\sqrt{(\mu_4-\sigma^2)_\lambda}}{\text{Var}\{S\}_\lambda - \text{Var}\{S\}_{\text{GOE}}}\right]^2, \tag{35}$$

where the factor 3 in the nominator corresponds to the 3σ level of significance. Substituting Eq. (31) one can rewrite the above as

$$E_{\max}\sqrt{N_{\text{tot}}} \gtrsim 9.9\, t \times \frac{\sqrt{(\mu_4-\sigma^2)_\lambda}}{\text{Var}\{S\}_\lambda - 0.273}. \tag{36}$$

---

[4] We use the property of m-th cumulant of the distribution $P(S)=\frac{1}{2}[aP_1(aS)+bP_2(bS)]$, which is equal to $\langle S^m\rangle_P=1/2(a^{(-m)}\langle S^m\rangle_{(P_1)}+b^{(-m)}\langle S^m\rangle_{(P_2)})$.. For $P_1=P_{\text{GOE}}$ and $P_2=P_{\text{GOE-GUE}}$, see Eqs. (6) and (3), necessary integrals for $m=2,\ 3,\ 4$ can be calculated analytically.

For $\lambda \to 0$, we have $\mu_4 - \sigma^2 \to \frac{21}{16}\pi^2 - 2\pi - 4 \approx 2.671$, leading to

$$N_{tot} \gtrsim 223 \times \left(\frac{t}{E_{max}}\right)^2 \quad \text{for} \quad N_{vac} = 0. \tag{37}$$

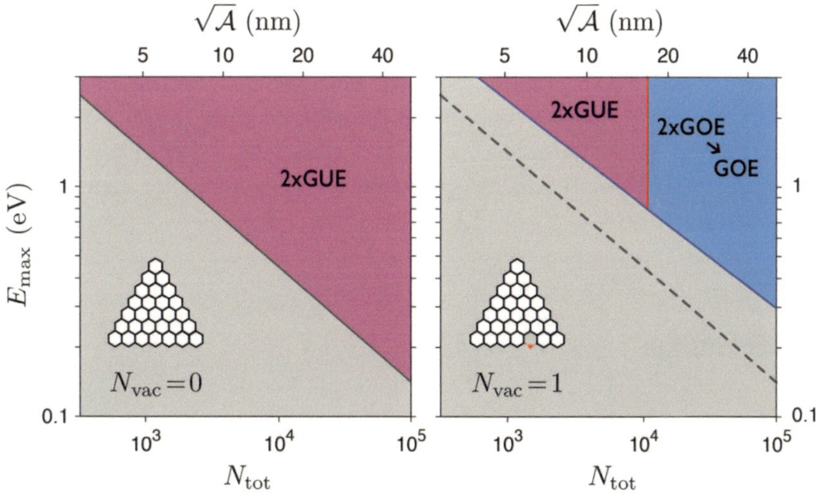

Figure 6. Phase diagram for triangular graphene nanoflakes with zigzag edges. Grey solid line in left panel (replotted as a dashed line in right panel) corresponding to Eq. (37) for $N_{vac}=0$ splits the region where number of available energy levels is insufficient to determine the class of spectral fluctuation (below the line) and the region where one should be able to identify the unitary class with approximate twofold degeneracy (2 × GUE). Blue solid line in right panel is same as solid line in left panel, but for $N_{vac}=1$, calculated numerically from Eq. (13) for $\lambda = \lambda_{fit}(N_{tot})$ (see Eq. (25)). Vertical red line in right panel marks the limit given by Eq. (27), above which the orthogonal class with gradual degeneracy splitting (2 × GOE → GOE) appears.

Limiting values of N_{tot} and E_{max}, following from Eqs. (27), (36), and (37) are in depicted **Figure 6**, presenting the central results of this work. In the absence of edge vacancies (N_{vac}=0), the attainable Fermi energy E_{max}=0.25 eV should make it possible to detect TRS breaking in nanoflakes containing $N_{tot} \gtrsim 3\ 10^4$ carbon atoms, corresponding to the physical diameter of $\sqrt{\mathcal{A}} \approx 15$ nm. For N_{vac}=1, the limit of $N_{tot} \lesssim 9500$ (see Eq. (27)) implies $E_{max} \gtrsim 0.8$ eV is required, slightly exceeding current experimental limits for graphene nanostructures.

5. Concluding remarks

We have revisited level-spacing statistics of triangular graphene nanoflakes with zigzag edges, subjected to weak substrate-induced disorder. Our previous study of the system is complemented by comparing the spectral fluctuations with these of large random matrices belonging to a mixed ensemble interpolating between GUE with self-dual symmetry and generic GOE.

The results show that for a fixed value of maximal Fermi energy E_{max} (in typical experiment, the Fermi energy is tuned in the range $-E_{max} < E < E_{max}$ by top gate electrode), the system size required to detect signatures of the time-reversal symmetry breaking at zero magnetic field is bounded from the bottom by the condition for minimal number of quantum-dot energy levels allowing one to distinguish between different classes of spectral fluctuations. A finite number of vacancies at the system boundary may lead to intervalley scattering restoring TRS, resulting in additional, upper limit for the system size.

In conclusion, we expect that triangular graphene flakes with *perfect* zigzag edges may show signatures of TRS breaking starting from physical sizes exceeding 15 nm. For a finite number of atomic-scale defects (starting from a single edge vacancy), one should search for signatures of the unitary symmetry class in artificial graphene-like systems rather then in real graphene nanoflakes.

Acknowledgements

The author thanks to Huang Liang for the correspondence. The work was supported by the National Science Centre of Poland (NCN) via Grant no. 2014/14/E/ST3/00256. Computations were partly performed using the PL-Grid infrastructure.

Author details

Adam Rycerz

Address all correspondence to: rycerz@th.if.uj.edu.pl

Marian Smoluchowski Institute of Physics, Jagiellonian University, Kraków, Poland

References

[1] Anderson PW. More is different. Science. 1972;177(4047):393–396. DOI: 10.1126/science. 177.4047.393.

[2] Spałek J. Correlated fermions: a new paradigm in physics on the example of solid state physics. European Journal of Physics. 2000;21(6):511–534. DOI: 10.1088/0143-0807/21/6/303.

[3] Anderson PW. Basic Notions of Condensed Matter Physics. 2nd ed. Reading: Westview Press/Addison-Wesley; 1997. 549 p. DOI: 978-0201328301.

[4] Beenakker CWJ, van Houten H. Quantum transport in semiconductor nanostructures. Solid State Physics. 1991;44(1):1–228. DOI: 10.1016/S0081-1947(08)60091-0; e-print arXiv:condmat/0412664.

[5] Stern A. Non-Abelian states of matter. Nature (London). 2010;464:187–193. DOI: 10.1038/nature08915.

[6] Willett RL, Nayak C, Pfeiffer LN, West KW. Aharonov-Bohm oscillations and evidence for non-Abelian anyons at $\nu = 5/2$. Physical Review Letters. 2013;111:186401. DOI: 10.1103/PhysRevLett.111.186401.

[7] Katsnelson MI. Graphene: Carbon in Two Dimensions. 1st ed. Cambridge: Cambridge University Press; 2012. 363 p. DOI: 9780521195409.

[8] Abergel DSL, Apalkov V, Berashevich J, Ziegler K, Chakraborty T. Properties of graphene: a theoretical perspective. Advances in Physics. 2010;59(4):261–482. DOI: 10.1080/00018732.2010.487978.

[9] McCann E and Koshino M. The electronic properties of bilayer graphene. Reports on Progress in Physics. 2013;76(056503):1–28. DOI: 10.1088/0034-4885/76/5/056503.

[10] Bernevig BA, Hughes TL, Zhang SC. Quantum spin Hall effect and topological phase transition in HgTe quantum wells. Science. 2006;314(5806):1757–1761. DOI: 10.1126/science.1133734.

[11] Singha A, Gibertini M, Karmakar B, Yuan S, Polini M, Vignale G, Katsnelson MI, Pinczuk A, Pfeiffer LN, West KW, Pellegrini V. Two-Dimensional Mott–Hubbard electrons in an artificial honeycomb lattice,. Science. 2011;332(6034):1176–1179. DOI: 10.1126/science.1204333.

[12] Torrent D and Sánchez-Dehesa J. Acoustic analogue of graphene: observation of Dirac cones in acoustic surface waves. Physical Review Letters. 2012;108(17):174301. DOI: 10.1103/PhysRevLett.108.174301.

[13] Gomes KK, Mar W, Ko W, Guinea F, Manoharan H C. Designer Dirac fermions and topological phases in molecular graphene. Nature (London). 2012;483(7389):306–310. DOI: 10.1038/nature10941.

[14] Nair RR, Blake P, Grigorenko AN, Novoselov KS, Booth TJ, Stauber T, Peres NMR, Geim AK. Fine structure constant defines visual transparency of graphene. Science. 2008;320(5881):1308. DOI: 10.1126/science.1156965; For the complete derivation of absorbed power, see e-print arXiv:0803.3718.

[15] Maciejko J, Qi X-L, Drew HD, Zhang S-C. Topological quantization in units of the fine structure constant. Physical Review Letters. 2010;105(16):166803. DOI: 10.1103/PhysRevLett.105.166803.

[16] Shuvaev AM, Astakhov GV, Tkachov G, Brüne C, Buhmann H, Molenkamp LW, Pimenov A. Terahertz quantum Hall effect of Dirac fermions in a topological insulator. Physical Review B. 2013;87(12):121104(R). DOI: 10.1103/PhysRevB.87.121104.

[17] Skulason HS, Gaskell PE, Szkopek T. Optical reflection and transmission properties of exfoliated graphite from a graphene monolayer to several hundred graphene layers. Nanotechnology. 2010;21(29):295709. DOI: 10.1088/0957-4484/21/29/295709.

[18] Das Sarma S, Adam Sh, Hwang EH, Rossi E. Electronic transport in twodimensional graphene. Reviews of Modern Physics. 2011;83(2):407–470. DOI: 10.1103/RevModPhys.83.407.

[19] Tworzydło J, Trauzettel B, Titov M, Rycerz A, Beenakker CWJ. Sub-Poissonian shot noise in graphene. Physical Review Letters. 2006;96(24):246802. DOI: 10.1103/PhysRevLett.96.246802.

[20] Danneau R, Wu F, Craciun MF, Russo S, Tomi MY, Salmilehto J, Morpurgo AF, Hakonen PJ. Shot noise in ballistic graphene. Physical Review Letters. 2008;100(19):196802. DOI: 10.1103/PhysRevLett.100.196802.

[21] Rycerz A, Recher P, and Wimmer M. Conformal mapping and shot noise in graphene. Physical Review B. 2009;80(12):125417. DOI: 10.1103/PhysRevB.80.125417.

[22] Prada E, San-Jose P, Wunsch B, Guinea F. Pseudodiffusive magneto transport in graphene. Physical Review B. 2007;75(11):113407. DOI: 10.1103/PhysRevB.75.113407.

[23] Rut G, Rycerz A. Pseudodiffusive conductance, quantum-limited shot noise, and Landau-level hierarchy in biased graphene bilayer. Physical Review B. 2014;89(4):045421. DOI: 10.1103/PhysRevB.89.045421.

[24] Wurm J, Rycerz A, Adagideli I, Wimmer M, Richter K, Baranger HU. Symmetry classes in graphene quantum dots: universal spectral statistics, weak localization, and conductance fluctuations. Physical Review Letters. 2009;102(5):056806. DOI: 10.1103/PhysRevLett.102.056806.

[25] Rycerz A. Random matrices and quantum chaos in weakly disordered graphene nanoflakes. Physical Review B. 2012;85(24):245424. DOI: 10.1103/PhysRevB.85.245424.

[26] Bardarson JH, Tworzydło J, Brouwer PW, Beenakker CWJ. One-parameter scaling at the Dirac point in graphene. Physical Review Letters. 2007;99(10):106801. DOI: 10.1103/PhysRevLett.99.106801.

[27] Ostrovsky PM, Gornyi IV, and Mirlin AD. Interaction-induced criticality in Z2 topological insulators. Physical Review Letters. 2010;105(3):036803. DOI: 10.1103/PhysRevLett.105.036803.

[28] Geim AK, Kim P. Carbon wonderland. Scientific American. April 2008;299:90–97. DOI: 10.1038/scientificamerican0408-90.

[29] Geim AK. Graphene: status and prospects. Science. 2009;324(5934):1530–1534. DOI: 10.1126/science.1158877.

[30] Ponomarenko LA, Schedin F, Katsnelson MI, Yang R., Hill EW, Novoselov KS, Geim AK. Chaotic Dirac billiard in graphene quantum dots. Science. 2008;320(5874):356–358. DOI: 10.1126/science.1154663.

[31] For a recent review of the topic, see: Huang L, Xu H-Y, Lai Y-C, Grebogi C. Level spacing statistics for two-dimensional massless Dirac billiards. Chinese Physics B. 2014;23(7): 070507. DOI: 10.1088/1674-1056/23/7/070507.

[32] Haake F. Quantum Signatures of Chaos. 3rd ed. Berlin: Springer; 2010. 570 p. DOI: 9783642054273.

[33] Robnik M, Berry MV. False time-reversal violation and energy level statistics: the role of anti-unitary symmetry. Journal of Physics A: Mathematical and General. 1986;19(5): 669–682. DOI: 10.1088/0305-4470/19/5/020.

[34] Berry MV, Mondragon RJ. Neutrino billiards: time-reversal symmetry-breaking without magnetic fields. Proceedings of the Royal Society A. 1987;412(1842):53–74. DOI: 10.1098/rspa.1987.0080.

[35] Rycerz A. Strain-induced transitions to quantum chaos and effective time-reversal symmetry breaking in triangular graphene nanoflakes. Physical Review B. 2013;87(19): 195431. DOI: 10.1103/PhysRevB.87.195431.

[36] Pal AN, Kochat V, Ghosh A. Direct observation of valley hybridization and universal symmetry of graphene with mesoscopic conductance fluctuations. Physical Review Letters. 2012;109(19):196601. DOI: 10.1103/PhysRevLett.109.196601.

[37] Hämäläinen SK, Sun Z, Boneschanscher MP, Uppstu A, Ijäs M, Harju A, Vanmaekelbergh D, Liljeroth P. Quantum-confined electronic states in atomically well-defined graphene nanostructures. Physical Review Letters. 2011;107(23):236803. DOI: 10.1103/PhysRevLett.107.236803.

[38] Olle M, Ceballos G, Serrate D, Gambardella P. Yield and shape selection of graphene nanoislands grown on Ni(111). Nano Letters. 2012;12(9):4431. DOI: 10.1021/nl300897m.

[39] Życzkowski K. Parametric dynamics of quantum systems and transitions between ensembles of random matrices. Acta Physica Polonica B. 1993;24(5):967–1025. http://www.actaphys.uj.edu.pl/fulltext?series=Reg&vol=24&page=967.

[40] Lenz G, Życzkowski K. Time-reversal symmetry breaking and the statistical properties of quantum systems. Journal of Physics A: Mathematical and General. 1992;25(21): 5539–5551. DOI: 10.1088/0305-4470/25/21/013.

[41] Schierenberg S, Bruckmann F, Wettig T. Wigner surmise for mixed symmetry classes in random matrix theory. Physical Review E. 2012;85(6):061130. DOI: 10.1103/PhysRevE.85.061130.

[42] Mehta ML. Random Matrices. 3rd ed. Amsterdam: Elsevier/Academic Press; 2004. 706 p. DOI: 9780120884094.

[43] See Appendix D in Rycerz A. Random matrices and quantum chaos in weakly disordered graphene nanoflakes. Physical Review B. 2012;85(24):245424. DOI: 10.1103/PhysRevB.85.245424.

[44] Lenz G. Życzkowski K, Saher D. Scaling laws of the additive random-matrix model. Physical Review A. 1991;44(12):8043. DOI: 10.1103/PhysRevA.44.8043.

[45] Rycerz A, Tworzydło J, Beenakker CWJ. Anomalously large conductance fluctuations in weakly disordered graphene. Europhysics Letters. 2007;79(5):57003. DOI: 10.1209/0295-5075/79/57003.

[46] Lewenkopf CH, Mucciolo ER, Castro Neto AH. Numerical studies of conductivity and Fano factor in disordered graphene. Physical Review B. 2008;77(8):081410(R). DOI: 10.1103/PhysRevB.77.081410.

[47] Horsell DW, Savchenko AK, Tikhonenko FV, Kechedzhi K, Lerner IV, Fal'ko VI. Mesoscopic conductance fluctuations in graphene. Solid State Communications. 2009;149(27–28):1041–1045. DOI: 10.1016/j.ssc.2009.02.058.

[48] Wurm J, Wimmer M, Richter K. Symmetries and the conductance of graphene nanoribbons with long-range disorder. Physical Review B. 2012;85(24):245418. DOI: 10.1103/PhysRevB.85.245418.

[49] Huard B, Sulpizio JA, Stander N, Todd K, Yang B, Goldhaber-Gordon D. Transport measurements across a tunable potential barrier in graphene. Physical Review Letters. 2007;98(23):236803. DOI: 10.1103/PhysRevLett.98.236803.

Section 2

Graphene Synthesis

Chapter 5

Optimization of the Synthesis Procedures of Graphene and Graphite Oxide

María del Prado Lavín López,
José Luis Valverde Palomino,
María Luz Sánchez Silva and
Amaya Romero Izquierdo

Additional information is available at the end of the chapter

http://dx.doi.org/10.5772/63752

Abstract

The optimization of both the *chemical vapor deposition (CVD)* synthesis method to prepare graphene and the *Improved Hummers method* to prepare graphite oxide is reported. Copper and nickel were used as catalysts in the *CVD*-graphene synthesis, CH_4 and H_2being used as precursor gases. Synthesis variables were optimized according to a *thickness value*, calculated using a homemade Excel-VBA application. In the case of copper, the maximum *thickness value* was obtained for those samples synthesized at 1050°C, a CH_4/H_2 flow rate ratio of 0.07 v/v, a total flow of 60 Nml/min, and a time on stream of 10 min. In the case of nickel, a reaction temperature of 980°C, a CH_4/H_2 flow rate ratio of 0.07 v/v, a total flow of 80 Nml/min, and a time on stream of 1 min were required to obtain a high *thickness value*. On the other hand, the *Improved Hummers method* used in the synthesis of graphite oxide was optimized. The resultant product was similar to that reported in literature in terms of quality and characteristics but both time and cost of the synthesis procedure were considerably decreased.

Keywords: graphene, graphite oxide, CVD, Improved Hummers method, thickness value

1. Introduction

Carbon (C) is a chemical element with atomic number 6 and solid at room temperature. Depending on the synthesis conditions during its growth, carbon can be found in nature with different allotrope forms [1]. Among them, the softer and the harder materials known in nature are included: graphite (**Figure 1a**) and diamond (**Figure 1b**), respectively. Recently, new carbon allotropes have been discovered such as fullerenes (**Figure 1c**), carbon nanotubes (**Figure 1d**), carbon nanofibers (**Figure 1e**), and carbon nanospheres. To date, the last carbon allotrope that has been appended is graphene (**Figure 1f**). It consists of a two-dimensional (2D) carbon atom network with sp^2 hybridization and only one atom thick [2]. Each atom is bonded by a covalent bond to other three carbon atoms. These carbon atoms are densely packaged in a honeycomb-shape crystal lattice [3] comprising, in turn, of two superimposed triangular subnets [4]. Although graphene has been known since 1960, it was not until 2004 when Andre Geim and Konstantin Novoselov achieved to obtain an isolated graphene sheet using the Scotch® tape method [5].

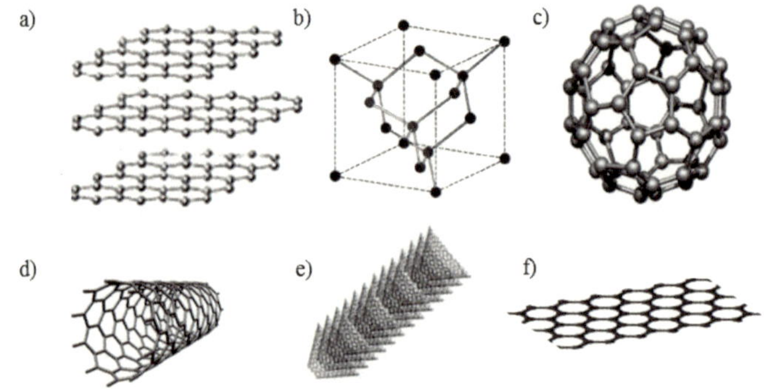

Figure 1. Structure of (a) graphite, (b) diamond, (c) fullerene, (d) carbon nanotube (CNT), (e) carbon nanofiber (CNF), and (f) graphene [6].

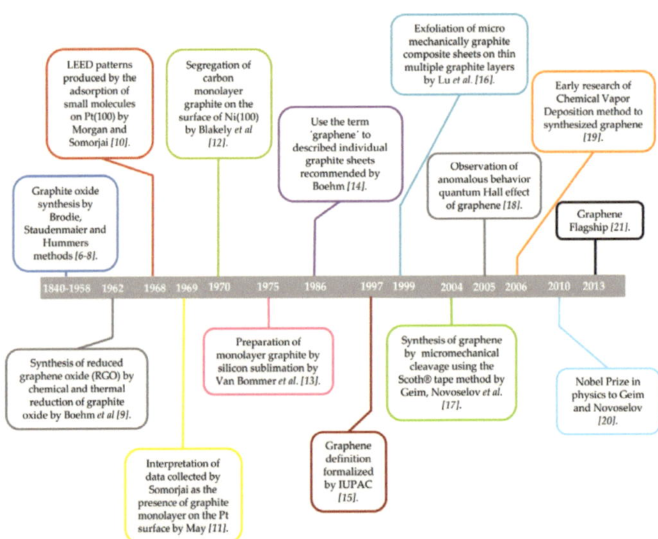

Figure 2. Roadmap of graphene [5, 7–21].

1.1. Graphene chronology

Graphene is one of the most extensively researched materials and is currently regarded as a fascinating material [2]. **Figure 2** shows the chronology of graphene, from 1840 to nowadays.

1.2. Properties and applications

Since 2004, many researchers have been focused on the synthesis of high-yield and high-quality graphene as well as on the search of an easily scalable process to manufacture it [6]. **Table 1** shows the extraordinary properties of graphene related to the applications that can be derived from them.

Property	Application	Ref.
• High-speed electron mobility	Transistors, lasers, photo detectors	[3, 22]
• Large specific surface area • Conductance	Sensors	[2, 23]
• Linear band structure (Dirac spectrum for mass less fermions)	Field effect transistors	[5]
• High electrical conductivity • High-speed electron mobility • High optical transmittance	Transparent conductive film	[23]
• High theoretical surface area • Electron transfer along 2d surface	Clean energy devices	[23]
• Anomalous quantum hall effect	Ballistic transistors	[24]
• Irrelevant spin-orbit coupling	Spin-Valve Devices	[22]
• High conductivity	Conductive materials, electrical batteries, super capacitors	[2]
• Easy absorption of gases	Contamination control	[22]
• Transparency (>99%) • High electronic conductivity	Displays, touch screens	[25]
• Impermeability	Coatings	[26]
• High mechanical stress (hardness)	Construction	[3]

Table 1. Graphene properties and applications.

1.3. Graphene synthesis

Two different routes can be followed to synthesize graphene: ***Bottom Up*** and ***Top Down*** (**Figure 3**). *Bottom-Up* route comprised those methods, which use a carbonaceous gas source to produce graphene. The most relevant ones are *epitaxial growth on Silicon Carbide (SiC)* [27] and *chemical vapor deposition (CVD)* [3]. *Top-Down* route is based on the attack of graphite (used as raw material) to break its layers forming graphene sheets [28]. Methods such as *micromechanical cleavage* [5, 22], *exfoliation of graphite intercalation compounds (GICs)* [29], *arc discharge* [30, 31], *unzipping carbon nanotubes (CNTs)* [32, 33], *graphene oxide exfoliation* [27] and *solvent-base exfoliation* [34–37] comprise the *Top-Down* route.

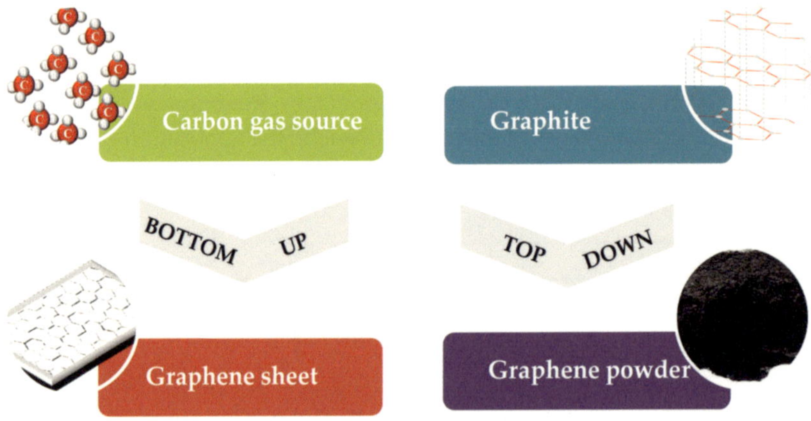

Figure 3. *Bottom-Up* and *Top-Down* routes to synthesize graphene.

Among the different *Bottom-Up* synthesis methods, *CVD* is considered the most extensively one used to synthesize large amounts of high-quality graphene sheets. This method is simple and easily scalable [38]. It is important to highlight that the quality and the type of graphene (monolayer, bilayer, few layer, or multilayer) can be varied as a function of the catalytic metal used [3, 39, 40].

On the other hand, the simultaneous reduction and exfoliation of graphene oxide can be considered, among the different *Top-Down* synthesis methods, the easiest one to synthesize graphene-based powder materials. However, the synthesis of graphite oxide (GrO) is first required since it is the intermediate product leading to graphene oxide from graphite. Graphite oxide synthesis is an exothermic process that involves the use of strong acid solutions. In addition, it can be considered as a tedious procedure because many steps are required before the ultimate product is obtained.

Next, the most relevant results obtained in the *CVD* synthesis of graphene are summarized using nickel and copper as catalytic metals, with particular emphasis on the optimization of the main operating variables (synthesis time, temperature, and amount of gases involved during the synthesis). Similarly, the synthesis of graphite oxide is also described. In the latter case, the optimization study here reported pursued the reduction of the time of preparation and the amount of chemical oxidants used during the synthesis of this intermediate product.

2. Chemical vapor deposition of graphene layers

Chemical vapor deposition (*CVD*) is a *Bottom-Up* technique, which allows to synthesize wafer-scale graphene [41, 42]. In the *CVD* procedure, a metal substrate, which is used as the catalyst, is placed into a furnace and heated to high temperatures. The heat anneals the metal, increasing its domain size [43]. Nitrogen, a carbon source (such as methane), and hydrogen are flowed through the furnace during the graphene synthesis. Hydrogen catalyzes the reaction produced between the carbon source and the metal substrate resulting in carbon atoms coming from the carbon source decomposition, which are deposited onto the metal surface through chemical absorption [3]. Hydrogen activates the carbon bounds of the metal surface and controls the size and morphology of graphene domains [44]. After the reaction, the furnace is cooled to keep the deposited carbon atom layer from aggregating into bulk graphite, which crystallizes into a contiguous graphene layer on the metal surface [45]. This method has the advantage of producing large size and high-quality graphene layers, and the ability to synthesize graphene at wafer scale [41]. Moreover, it is considered a low-cost method leading to a high yield if compared to other growth methods. However, the *CVD* graphene tends to wrinkle due to the difference in thermal expansion between graphene and metal layer. This fact could be decreased via proper annealing [43]. Nickel and copper are commonly used in the *CVD* method as the metal substrate material for graphene synthesis [46]. Other transition metals such as Ru, Co, and Pt are alternative transition metals, although they are used less frequently [47, 48].

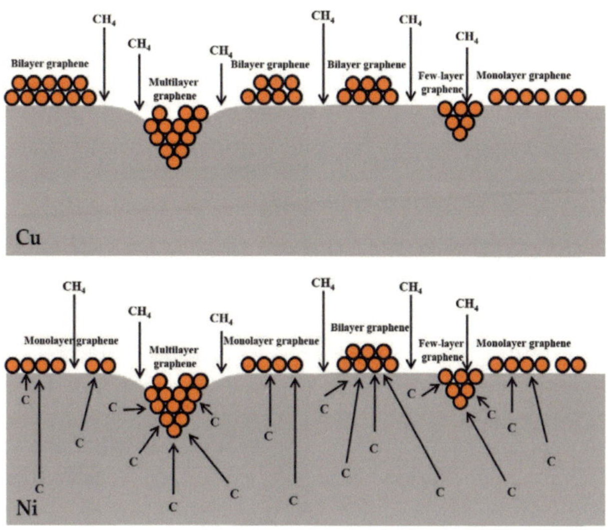

Figure 4. *CVD* growth-mechanism graphene on copper and nickel sheets [50].

Depending on the metal used, two different mechanisms can be differentiated in the *CVD*-graphene synthesis (**Figure 4**). The first one, called *Carbon atoms surface deposition*, which is the growth mechanism observed over copper sheets, is described like the direct deposition of carbon atoms on the catalyst surface. In the second one, called *Carbon atoms surface segregation*, which is the growth mechanism observed when nickel is used as the catalyst, carbon atoms decomposed from the carbonaceous source are diffused onto the catalyst bulk during the annealing step at high temperatures. Then, they precipitate on the catalyst during the cooling period [49].

2.1. Optimization of the CVD operational parameters

Several studies have established a close correlation between the *CVD* growth parameters, the quality of the graphene obtained, and the number of graphene layers, leading to the formation of different types of graphene (named as monolayer, bilayer, few layer, and multilayer) on the metal substrate [51–55].

Zhang et al. [51] and Nie et al. [52] found that the graphene quality improved at higher temperatures of reaction. On the contrary, lower temperatures gave rise to graphene with a number of defects. Rybin et al. reported that the larger the temperature of reaction, the higher the amount of atoms dissolved into the metal layer, leading to the production of more and more graphene layers [53].

Recent studies have showed that the concentration of hydrogen, which is obviously related with the CH_4/H_2 flow rate ratio and the total flow rate of CH_4 and H_2 during the reaction step, also plays an important role in providing quality to *CVD* graphene. Gao et al. found for an atmospheric pressure *CVD* system that high hydrogen concentrations contributes to the degradation of the graphene quality as a result of the occurrence of defects and wrinkles [54]. In a similar way, Vlassiouk et al. detected the presence of a critical value of the partial hydrogen that determines the occurrence of graphene growth (<2 Torr with 30 ppm of CH_4). No graphene nucleation was observed below this value, whereas higher hydrogen concentrations caused degradation in graphene quality [44]. Finally, several groups have shown that the growth of bilayer and few-layer graphene depends on the concentration of active carbon species [55], which was in turn related with the CH_4/H_2 flow ratio and the total flow rate of CH_4 and H_2 used during the reaction step.

On the other hand, monocrystalline metals favor the formation of superficial and uniform monolayer, and bilayer graphene, being hindered the formation of multilayer graphene due to graphene, is grown over smooth and free defect surfaces (**Figure 4**). However, the industrial production of graphene strongly recommends to use polycrystalline metals, since it is much lower than that of monocrystalline one [56].

Figure 5a shows the experimental installation used for *CVD*-graphene synthesis over polycrystalline metals (Ni and Cu). **Figure 5b** shows the stages followed during the graphene synthesis as well the duration, temperature, and gases used in each of them.

Methane and hydrogen were actually used as precursor gases. Graphene samples were grown by *CVD* at atmospheric pressure on 25-μm-thick polycrystalline metal foils in a 40-inch quartz tube heated in a furnace. Firstly, the reduction step was carried out by heating the furnace to 900°C, passing through the tube a flow of N_2 (400 sccm) and H_2 (100 sccm) to prevent metal sheet oxidation. The annealing step was carried out by maintaining the furnace at this temperature for 45 min. Later, the reaction step was started and carried out at different operational conditions in order to improve the quality of the obtained graphene by decreasing the amount of multilayer graphene formed over the metal. The temperature set point was increased and varied in the range of 900–1050°C. A ratio of methane to hydrogen in the range 0.4–0.07 v/v was introduced into the reactor for different times (15 min to 30 s) to complete the

http://dx.doi.org/10.5772/63752

reaction step. The total flow of gases involved during the reaction step ranged from 80 to 100 Nml/min. Finally, the system was cooled down (10°C min^{-1}) by flowing 400 sccm of nitrogen.

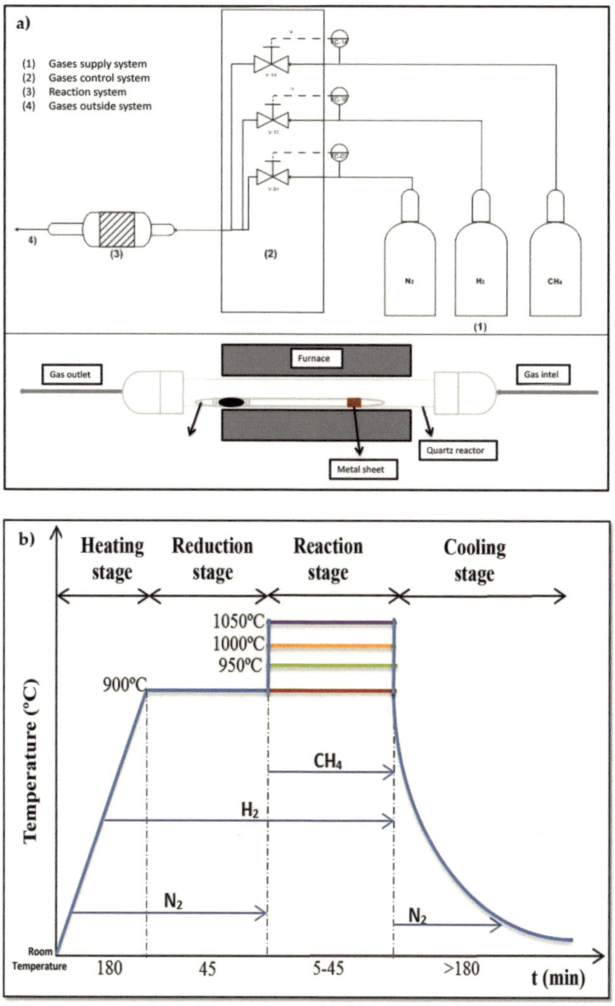

Figure 5. (a) Experimental installation for *CVD* graphene synthesis. (b) Summary of the *CVD*-graphene synthesis steps and conditions.

To control the graphene thickness and determine the percentage of each type of graphene (monolayer, bilayer, few layer, and multilayer) deposited over the polycrystalline metal foils, a homemade Excel-VBA application was designed. This software used the different colors present in a digitalized optical microscope image to evaluate the percentage of the different types of graphene deposited over the metal sheets. By using Raman spectroscopy, the relationship between the different colors present in optical images has been demonstrated with each type of graphene [3]. For this purpose, a logarithmic scale was considered in the Excel-VBA software design. Thus, *thickness values* 1, 10, 100, and 1000 were assigned when the 100% of the sample was covered by multilayer, few-layer, bilayer, and monolayer graphene, respectively. The *thickness value* was calculated as an average of the percentage obtained for each type of graphene calculated by the Excel-VBA application.

Ferrari et al. [57] demonstrated that using the second-order 2D feature obtained in the Raman spectra, it was possible to know the number of graphene layers. Based on that study, Malard et al. [58] investigated the theoretical background associated with the double-resonance Raman-scattering mechanism that gives rise to the main feature in the Raman spectra of the different types of graphene. Thus, the deconvolution of the 2D peak, corresponding to each type of graphene, showed that in the case of monolayer graphene the 2D peak could be fitted with a symmetric single peak only; in the case of bilayer graphene, the 2D peak could be deconvoluted in four different contributions; in the case of few-layer and multilayer graphene, the 2D peak could be deconvoluted in two contributions, which is characteristic of graphite (material consisting of many layers of graphene) (**Figure 6**).

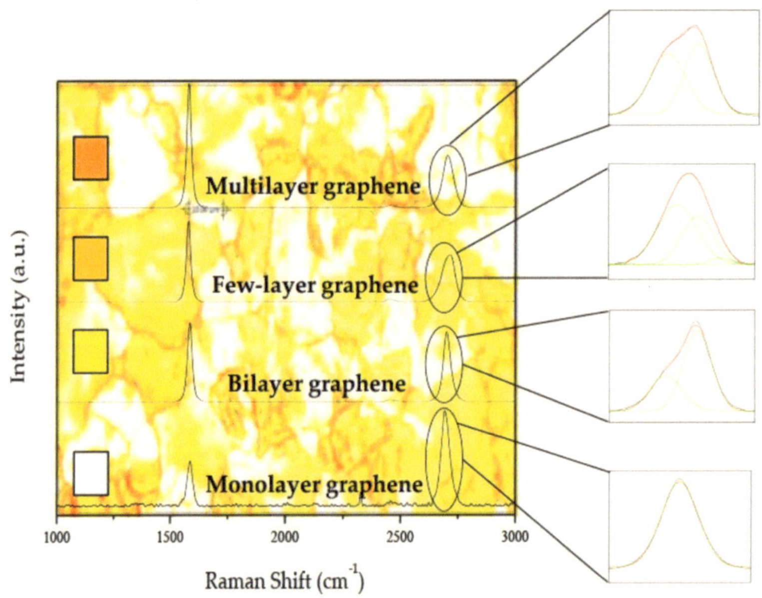

Figure 6. Relationship between optical microscope image and typical Raman spectra of monolayer [48, 59–61], bilayer [47, 59, 60, 62], few layer [63–65], and multilayer [63, 64] graphene.

In this study, polycrystalline copper and nickel were chosen as metal catalyst in the synthesis of *CVD* graphene.

Regarding polycrystalline Cu, 1050°C was required to maximize the amount of monolayer graphene over the metal, whereas 980°C was selected as the optimum reaction temperature in the case of using polycrystalline nickel. In the former case, the *thickness value* was found to be

http://dx.doi.org/10.5772/63752

4.2, the proportion of each type of graphene being the following one: 81% multilayer graphene, 17% few-layer graphene, and 2% bilayer graphene; the presence of monolayer graphene was considered negligible. In the latter case, the *thickness value* was found to be 397, in turn the proportion of each type of graphene being the following one: 0.9% multilayer graphene, 40% few-layer graphene, 22% bilayer grapheme, and 37% monolayer graphene.

Regarding the CH_4/H_2 flow rate ratio, an optimal value of 0.07 v/v was obtained when both metals were used as catalysts. A *thickness value* of 34.7 was obtained using Cu (20% multilayer graphene, 20% few-layer graphene, and 51% bilayer graphene), whereas a *thickness value* of 536 was obtained using Ni (0.5% multilayer graphene, 27% few-layer graphene, 20% bilayer graphene, and 52% monolayer graphene).

Finally, regarding the study of the influence of the total flow of gases (CH_4+ H_2) and reaction time, it could be concluded that the best results in the case of using Cu as the catalyst were obtained for a total gas flow of 60 Nml/min and a reaction time of 10 min, leading to an increased *thickness value* of 60, 56 and 11% of the resulting sample being covered by bilayer graphene and multilayer graphene, respectively. In the case of using Ni as the catalyst, the best results were obtained for a total gas flow of 80 Nml/min and a reaction time of 1 min (*thickness value* of 810). At these conditions, just 0.3% of the sample was covered by multilayer graphene, 11% by few-layer graphene, 9% by bilayer graphene, and 80% by monolayer graphene.

Variable	**Copper**	**Nickel**
Reaction temperature (°C)	1050	980
CH_4/H_2 flow rate ratio (v/v)	0.07	0.07
Total gas flow (CH_4+H_2) during reaction step	60	80
Reaction time (min)	10	1
Graphene type	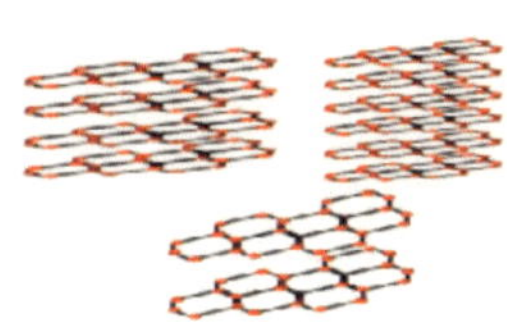	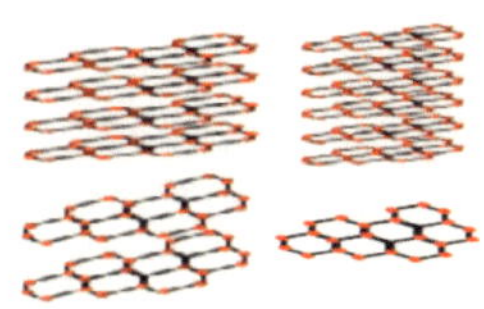
Thickness value	**59**	**810**

Table 2. Optimum synthesis conditions.

Summarizing, the *thickness value* was increasing at each stage of the optimization study regardless of the metal used. However, it was not possible to detect monolayer graphene on polycrystalline copper foil. In the case of using polycrystalline nickel, monolayer graphene covered 80% of the foil for the optimal conditions of synthesis.

Table 2 shows the optimum operating conditions for each metal resulting from this study.

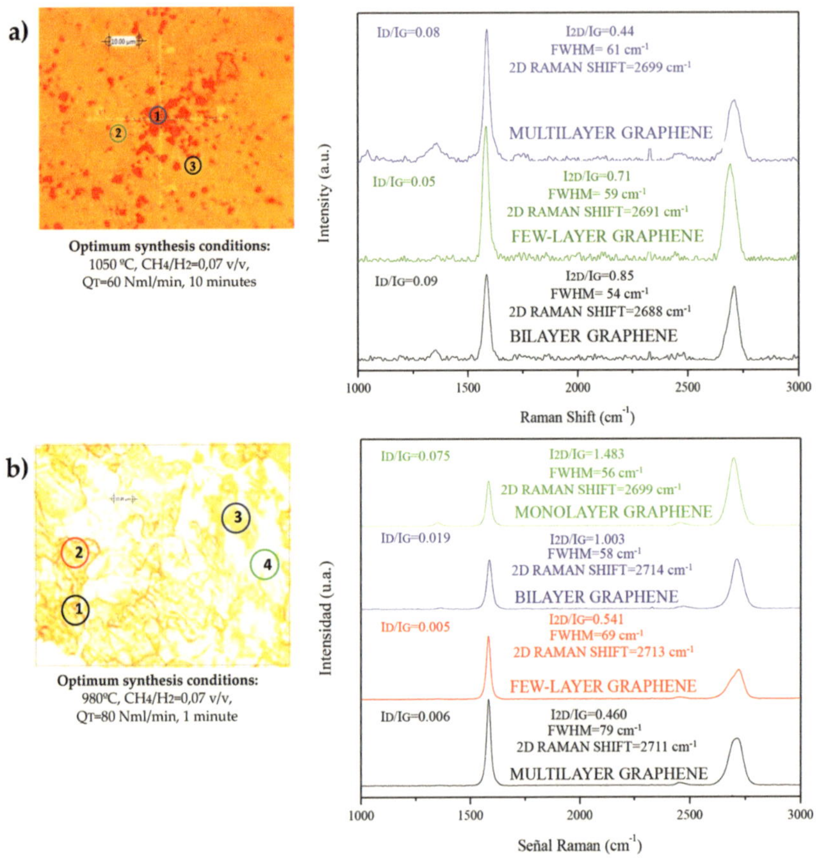

Figure 7. Raman spectra corresponding to graphene samples synthesized using (a) copper and (b) nickel.

Figure 7 shows the Raman spectra corresponding to the graphene samples obtained at the optimal conditions. D peak (1350 cm^{-1}) is related to the presence of defects in graphitic materials [66]. G peak (~1560 cm^{-1}) denotes the symmetry of graphite band and is a way of checking the vibration of sp^{2}-hybridized carbon atoms in the same plane. Finally, 2D peak, visible around 2700 cm^{-1}, is the hallmark of graphene layers [67]. The relationship between the intensity of D and G peak (I_D/I_G) is a way to check the amount of defects present in the graphene sheet. The number of graphene layer is also related to the ratio between the intensity of 2D and G peak (I_{2D}/I_G). 2D peak position in graphene sample should be displaced to lower Raman shift values in comparison with that of graphite. Finally, full width at hall maximum (FWHM) is related to the lifetime of the excited states and is calculated as the Raman shift difference to the half average height of the 2D peak [39]. For both metal catalysts, the I_D/I_G ratio values were low, demonstrating that graphene samples had low amount of defects, whereas the I_{2D}/I_G ratio values increased, as expected, from multilayer to monolayer graphene [68]. The contrary effect could be observed for the FWHM parameter, which decreased from multilayer to monolayer graphene. Finally, 2D peak position was, in all cases, located at around 2700 cm^{-1}, which is characteristic of graphene materials [3, 39].

http://dx.doi.org/10.5772/63752

3. Graphite oxide

Graphite oxide (GrO) can be defined as a set of functionalized graphene sheets that are mainly composed of carbon, oxygen, and hydrogen atoms. This material is considered a precursor of graphene itself [69]. The structure of graphite oxide is similar to that of graphite differing only in the oxygenated groups present in GrO, which give rise to a greater distance between the graphene layers [69]. GrO consists of a hexagonal network of sp^2- and sp^3-hybridized carbon atoms that bear hydroxyl and epoxide functional groups on the 'basal' plane and carboxyl and carbonyl groups at the edges [70].

3.1. Graphite oxide synthesis

Graphite oxide can be synthesized by the Brodie [7], Staudenmaier [8], or Hummers and Offeman [9] methods or by variations of the latter, namely *Modified Hummers method* or *Improved Hummers method* [71]. The main differences between the abovementioned methods are summarized in **Table 3**, with particular emphasis on the nature of the oxidant, the toxicity, and the main advantages or disadvantages of each approach.

3.2. Improved Hummers method

In 1958, Hummers reported the most popular procedure to synthesize graphite oxide, which is based on the oxidization of graphite by using $KMnO_4$ and $NaNO_3$ in concentrated H_2SO_4 [9]. However, this method yields NO_x and is dangerous itself due to the constant explosions, which take place during the synthesis [71]. In 2010, Marcano et al. [71] reported an improved synthesis based on the Hummers method by using graphite flakes as the raw material. Graphite oxide synthesized from graphite flakes can be easily soaked and dispersed in water and could be used as the precursor for different applications due to its hydrophilic character. They detected that an improved graphite oxide with fewer defect in the basal plane can be prepared using $KMnO_4$ as oxidation agent and a 9:1 mixture of concentrated H_2SO_4 and H_3PO_4. They also reported that graphite oxide synthesized with this *Improved Hummers method* provided a greater amount of hydrophilic-oxidized graphite, likewise having a more regular structure with a greater amount of basal plane framework retained. Raman and infrared spectroscopy results indicated that graphite oxide samples obtained through both methods were almost similar, both of them showing the characteristics D and G peaks that confirmed the lattice distortion in Raman spectroscopy. FTIR-ATR spectra also confirmed the presence of functional groups. In addition, atomic force microscopy (AFM) showed that the thickness of both graphite oxides was around 1.1 nm. They confirmed with a large variety of methods, such as thermogravimetry analysis (TGA), solid-state colossal magnetoresistance (CMR), X-ray diffraction (XRD), and X-ray photoelectron spectroscopy (XPS), that graphite oxide synthesized with the *Improved Hummers method* was, if it is compared to that produced from the Hummers one, a more oxidized material, presented a more organized structure, and contained both more epoxide functionalities, and more regular carbon framework.

Brodie Method	Staudenmaier Method	Hummers Method	Modified Hummers Method	Improved Hummers Method
Oxidants: $KClO_3$, HNO_3 **Toxicity:** Yes **Disadvantages:** •Weak Toxicity. •Soft dispersability in basic solutions. •Small size, limiting thickness and providing an imperfect structure.	**Oxidants:** $KClO_3$, ($NaClO_3$), HNO_3, H_2SO_4 **Toxicity:** Yes **Disadvantages:** •Time-consuming and dangerous method. •Addition of $KClO_3$ generally takes longer than a week and CO_2 is evolved, thus making necessary to remove an inert gas. •The risk of explosions is a constant danger.	**Oxidants:** $KMnO_4$, H_2SO_4, $NaNO_3$ **Toxicity:** No (NO_x) **Advantages:** •Higher oxidation degree than that obtained in Brodie or Staudenmaier Methods. **Disadvantages:** •It is still considered that the oxidation is incomplete.	**Oxidants:** $KMnO_4$, H_2SO_4, $NaNO_3$ **Toxicity:** No (NO_x) **Advantages:** •Improved level of oxidation and, therefore, product performance. **Disadvantages:** •Separation and purification processes are tedious. •Highly time-consuming.	**Oxidants:** $KMnO_4$, H_2SO_4, H_3PO_4 **Toxicity:** No **Advantages:** •Defects in the basal plane are reduced. •Larger amount of oxidized graphite is provided. •The degree of reduction provides an equivalent level of conductivity when compared to other methods. •It is a high performance method. •Environmentally friendly, toxic gases are not generated during the preparation. •The product has a more organized structure.

Table 3. Synthesis methods of graphite oxide.

3.3. Optimization of the Improved Hummers method

The most remarkable results obtained in the optimization study of the synthesis of graphite oxide based on the method proposed by Marcano et al. [71] (*Improved Hummers method)* are summarized below. Thus, the objective was to reach the same quality product but using a lesser time-consuming experimental procedure and conducting lower production costs. With this purpose, the different stages of the *Improved Hummers method* used in the synthesis of graphite oxide were optimized. The method consists of the oxidation, in the presence of H_2SO_4 and H_3PO_4, of 3 g of graphite per 9 g of $KMnO_4$ used as the oxidizing agent in 400 ml of solution. The oxidation step is maintained at 50°C for 12 h. Later, the reaction mixture is washed twice with 200 ml of deionized water, HCl, and ethanol. Finally, the product is coagulated with 200 ml of dry diethyl ether and dried at 100°C [71].

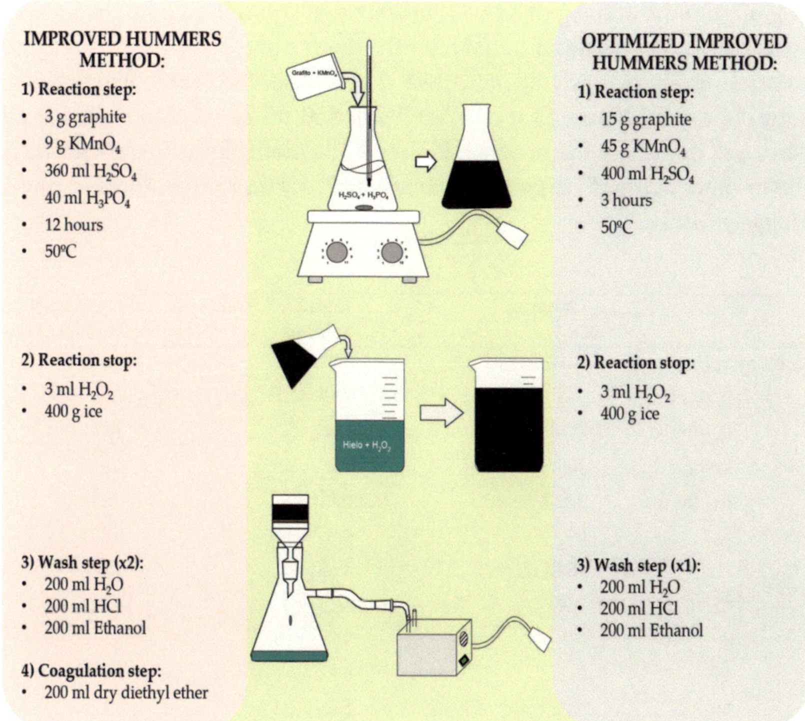

Figure 8. Differences between the *Improved Hummers* and *Optimized Improved Hummers methods*.

First, the oxidation time, which is the most time-consuming step of the whole synthesis process, was reduced. This way, the oxidation time was reduced from 12 to 3 h, whereas the other synthesis conditions were kept constant without affecting the quality of the final product. The introduction of functional groups, both at the edges and in the basal plane, was achieved in 3 h instead of the 12 h required in the original method.

Second, it was demonstrated that the coagulation step used by Marcano et al. [71] did not significantly influence over the quality and characteristic properties of the final product. Consequently, it was removed from the synthesis procedure.

On the other hand, Marcano et al. used three different products twice during the washing step: deionized water, which was used to reach the pH neutralization; HCl, which was required to remove the remaining metal from the graphite oxide, and ethanol, which was used to achieve a faster drying of the final product. We demonstrated that the quality and characteristic of the final product were not affected at all if the treatment of the cake with these three products was or not repeated. In addition, the elimination of H_3PO_4 in the synthesis procedure was considered. Similarly, this action did not alter the characteristics of the final product.

Finally, a series of tests were conducted in order to increase the amount of graphite that can be treated per batch, without altering the properties of the final product (the raw method considers 3 g of graphite and 9 g of $KMnO_4$ in 400 ml of solution). Here, the $KMnO_4$/graphite ratio (3:1) was maintained in order to not alter the degree of oxidation. This way, the amount of these materials was progressively increased. It was observed that it was possible to use up to 15 g of graphite (and hence 45 g of $KMnO_4$) in 400 ml of solution without altering the characteristics and quality of the product. **Figure 8** schematically summarizes the differences between the original method (*Improved Hummers method*) and the optimized one (*Optimized Improved Hummers method*).

		Graphite	Improved Hummers method	Optimized Improved Hummers method
Raman spectroscopy I_D/I_G		0.067	0.726	0.946
	L_a (nm)	263	26.5	20.4
DRX	L_a (nm)	41	9.6	10.1
	L_c (nm)	20	4.7	4.9
	d_{002} (nm)	0.334	0.810	0.910
	N_c	60	5.8	5.4
Elemental composition	C	98.04	48.8	51.4
	O	1.96	48.2	45.1
	S	0.0	0.7	2.8
	Cl	0	0.8	0.6
	Mn	0	1.4	0.1
SEM		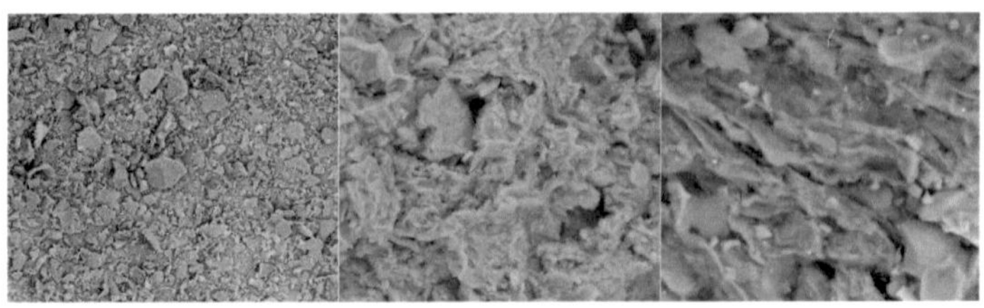		
BET	Surface area (m^2/g)	1.7	22.2	28.1
	Total pore volume (cm^3/g)	0.038	0.113	0.129

La: crystal dimension described by layer sized; **Lc:** stack height; **Nc:** number of layers in the stacking structure; **C:** carbon, **O:** oxygen, **S:** sulfur, **Cl:** chlorine; **Mn:** manganese.

Table 4. Characterization results of graphite oxide synthesized using both the *Improved Hummers* and *Optimized Improved Hummers methods.*

Table 4 lists some properties of the graphite oxide samples synthesized by the *Optimized Improved Hummers method* and those prepared from the parent one. I_D/I_G ratio, related with the structural disorder in the graphite network and inversely proportional to the sp^2 cluster average sized [72], considerably increased after graphite oxidation. Crystal dimension (La value) decreased after the incorporation of oxygenated groups, which agree with the increase in the structural disorder. In the same way, the number of graphene layers in the stacking structure (Nc) decreased after the oxidation process [73]. Elemental analysis showed an increase in the percentage of oxygen from graphite to graphite oxide due to the oxidation process. Comparing both graphite oxides, the percentage of each compound (C, O, S, Cl, and Mn) was quite similar [74]. In addition, scanning electron microscope (SEM) images showed an agglomeration of the product after the oxidation process, being several microns in size. Finally, a significant increase in the surface area was observed after graphite oxidation, because of the expansion of graphene layers [75].

The almost similar characterization values of both samples of graphite oxide demonstrated that the optimization process did not affect both quality and structure of the final product.

Acknowledgements

The present work was performed within the frame of the NANOLEAP project. This project has received funding from the European Union's Horizon 2020 research and innovation programme under grant agreement No 646397.

Author details

María del Prado Lavín López*, José Luis Valverde Palomino, María Luz Sánchez Silva and Amaya Romero Izquierdo

*Address all correspondence to: MariadelPrado.Lavin@uclm.es

Department of Chemical Engineering, University of Castilla-La Mancha, Ciudad Real, Spain

References

[1] Adler J, Pine P. Visualization techniques for modelling carbon allotropes. Computer Physics Communications. 2009;180:580–582. DOI: 10.1016/j.cpc.2008.12.014.

[2] Geim AK, Novoselov KS. The rise of graphene. Nature Materials. 2007;6:183–191. DOI: 10.1038/nmat1849.

[3] Lavin-Lopez MP, Valverde JL, Cuevas MC, Garrido A, Sanchez-Silva L, Martinez P, Romero-Izquierdo A. Synthesis and characterization of graphene: Influence of synthesis variables. Physical Chemistry Chemical Physics. 2014;16:2962–2970. DOI: 10.1039/c3cp54832e.

[4] Horing NJM. Aspects of the theory of graphene. Philosophical Transactions of the Royal Society A: Mathematical, Physical and Engineering Sciences. 2010;368:5525–5556. DOI: 10.1098/rsta.2010.0242.

[5] Novoselov KS, Geim AK, Morozov SV, Jiang D, Zhang Y, Dubonos SV, Grigorieva IV, Firsov AA. Electric field in atomically thin carbon films. Science. 2004;306:666–669. DOI: 10.1126/science.1102896.

[6] Lavin-Lopez MP, Valverde JL, Sanchez-Silva, L, and Romero A. Solvent-based exfoliation via sonication of graphitic materials for graphene manufacture. Industrial&Engineering Chemistry Research. 2016;55:845–855. DOI: 10.1021/acs.iecr.5b03502.

[7] Brodie BC. XXIII. – Researches on the atomic weight of graphite. Quarterly Journal of the Chemical Society of London. 1860;12:261–268. DOI: 10.1039/qj8601200261.

[8] Staudenmaier L. A process for the presentation of graphite acid. Reports of the German Chemical Society. 1898;31:1481–1487.

[9] Hummers Jr WS, Offeman RE. Preparation of graphitic oxide. Journal of the American Chemical Society. 1958;80:1339.

[10] Boehm HP, Clauss A, Fischer GO, Hofmann U. Thinnest carbon foils. Journal for Nature Research – Section B. 1962;17:150–153. DOI: 10.1515/znb-1962-0302.

[11] Morgan AE, Somorjai GA. Low energy electron diffraction studies of gas adsorption on the platinum (100) single crystal surface. Surface Science. 1968;12:405–425.

[12] May JW. Platinum surface LEED rings. Surface Science. 1969;17:267–270.

[13] Blakely JM, Kim JS, Potter HC. Segregation of carbon to the (100) surface of nickel. Journal of Applied Physics. 1970;41:2693–2697. DOI: 10.1063/1.1659283.

[14] Van Bommel AJ, Crombeen JE, Van Tooren A. LEED and Auger electron observations of the SiC(0001) surface. Surface Science. 1975;48:463–472. DOI: 10.1016/0039-6028(75)90419-7.

[15] Boehm HP, Setton R, Stumpp E. Nomenclature and terminology of graphite intercalation compounds. Carbon. 1986;24:241–245. DOI: 10.1016/0008-6223(86)90126-0.

[16] IUPAC - International Union of Pure and Applied Chemistry: Home [Internet]. 1997. Available from: http://goldbook.iupac.org/G02683.html. [Accessed: 2016 01-23].

[17] Lu X, Yu M, Huang H, Ruoff RS. Tailoring graphite with the goal of achieving single sheets. Proceedings of the 1998 6th Foresight Conference on Molecular Nanotechnology. 1999;10:269–272. DOI: 10.1088/0957-4484/10/3/308.

[18] Novoselov KS, Geim AK, Morozov SV, Jiang D, Katsnelson MI, Grigorieva IV, Dubonos SV, Firsov AA. Two-dimensional gas of massless Dirac fermions in graphene. Nature. 2005;438:197–200. DOI: 10.1038/nature04233.

[19] N'Diaye AT, Bleikamp S, Feibelman PJ, Michely T. Two-dimensional Ir cluster lattice on a graphene moiré on Ir(111). Physical Review Letters. 2006;97. DOI: 10.1103/PhysRevLett.97.215501.

[20] Kungliga Vetenskapsakademien – Hem [Internet]. 2010. Available from: http://www.kva.se/en/ [Accessed: 2016 01-23].

[21] Graphene Flagship [Internet]. 2013. Available from: http://graphene-flagship.eu/ [Accessed: 2016 01-24].

[22] Katsnelson MI. Graphene: carbon in two dimensions. Materials Today. 2007;10:20–27. DOI: 10.1016/s1369-7021(06)71788-6.

[23] Zhu Y, Murali S, Cai W, Li X, Suk JW, Potts JR, Ruoff RS. Graphene and graphene oxide: synthesis, properties, and applications. Advanced Materials. 2010;22:3906–3924. DOI: 10.1002/adma.201001068.

[24] Geim AK, Dubonos SV, Lok JGS, Grigorieva IV, Maan JC, Theil Hansen L, Lindelof PE. Ballistic hall micromagnetometry. Applied Physics Letters. 1997;71:2379–2381.

[25] Nair RR, Wu HA, Jayaram PN, Grigorieva IV, Geim AK. Unimpeded permeation of water through helium-leak-tight graphene-based membranes. Science. 2012;335:442–444. DOI: 10.1126/science.1211694.

[26] Boukhvalov DW, Katsnelson MI. Chemical functionalization of graphene. Journal of Physics Condensed Matter. 2009;21. DOI: 10.1088/0953-8984/21/34/344205.

[27] Whitener Jr KE, Sheehan PE. Graphene synthesis. Diamond and Related Materials. 2014;46:25–34. DOI: 10.1016/j.diamond.2014.04.006.

[28] Edwards RS, Coleman KS. Graphene synthesis: relationship to applications. Nanoscale. 2013;5:38–51. DOI: 10.1039/c2nr32629a.

[29] Vallés C, Drummond C, Saadaoui H, Furtado CA, He M, Roubeau O, Ortolani L, Monthioux M, Pénicaud A. Solutions of negatively charged graphene sheets and ribbons. Journal of the American Chemical Society. 2008;130:15802–15804. DOI: 10.1021/ja808001a.

[30] Subrahmanyam KS, Panchakarla LS, Govindaraj A, Rao CNR. Simple method of preparing graphene flakes by an arc-discharge method. Journal of Physical Chemistry C. 2009;113:4257–4259. DOI: 10.1021/jp900791y.

[31] Chen Y, Zhao H, Sheng L, Yu L, An K, Xu J, Ando Y, Zhao X. Mass-production of highly-crystalline few-layer graphene sheets by arc discharge in various H 2-inert gas mixtures. Chemical Physics Letters. 2012;538:72–76. DOI: 10.1016/j.cplett.2012.04.020.

[32] Jiao L, Zhang L, Wang X, Diankov G, Dai H. Narrow graphene nanoribbons from carbon nanotubes. Nature. 2009;458:877–880. DOI: 10.1038/nature07919.

[33] Valentini L. Formation of unzipped carbon nanotubes by CF4 plasma treatment. Diamond and Related Materials. 2011;20:445–448. DOI: 10.1016/j.diamond.2011.01.038.

[34] Yi M, Shen Z. A review on mechanical exfoliation for the scalable production of graphene. Journal of Materials Chemistry A. 2015;3:11700–11715. DOI: 10.1039/c5ta00252d.

[35] Blake P, Brimicombe PD, Nair RR, Booth TJ, Jiang D, Schedin F, Ponomarenko LA, Morozov SV, Gleeson HF, Hill EW, Geim AK, Novoselov KS. Graphene-based liquid crystal device. Nano Letters. 2008;8:1704–1708. DOI: 10.1021/nl080649i.

[36] Hernandez Y, Nicolosi V, Lotya M, Blighe FM, Sun Z, De S, McGovern IT, Holland B, Byrne M, Gun'ko YK, Boland JJ, Niraj P, Duesberg G, Krishnamurthy S, Goodhue R, Hutchison J, Scardaci V, Ferrari AC, Coleman JN. High-yield production of graphene by liquid-phase exfoliation of graphite. Nature Nanotechnology. 2008;3:563–568. DOI: 10.1038/nnano.2008.215.

[37] Ciesielski A, Samorì P. Graphene via sonication assisted liquid-phase exfoliation. Chemical Society Reviews. 2014;43:381–398. DOI: 10.1039/c3cs60217f.

[38] Bhuyan SA, Uddin N, Islam M, Bipasha FA, Hossain SS. Synthesis of graphene. International Nano Letters. 2016. DOI: 10.1007/s40089-015-0176-1.

[39] Lavin-Lopez MP, Valverde JL, Ruiz-Enrique MI, Sanchez-Silva L, Romero A. Thickness control of graphene deposited over polycrystalline nickel. New Journal of Chemistry. 2015;39:4414–4423. DOI: 10.1039/c5nj00073d.

[40] Muñoz R, Gómez-Aleixandre C. Review of CVD synthesis of graphene. Chemical Vapor Deposition. 2013;19:297–322. DOI: 10.1002/cvde.201300051.

[41] Obraztsov AN. Chemical vapour deposition: making graphene on a large scale. Nature Nanotechnology. 2009;4:212–213. DOI: 10.1038/nnano.2009.67.

[42] Emtsev KV, Bostwick A, Horn K, Jobst J, Kellogg GL, Ley L, McChesney JL, Ohta T, Reshanov SA, Röhrl J, Rotenberg E, Schmid AK, Waldmann D, Weber HB, Seyller T. Towards wafer-size graphene layers by atmospheric pressure graphitization of silicon carbide. Nature Materials. 2009;8:203–207. DOI: 10.1038/nmat2382.

[43] Bae S, Kim H, Lee Y, Xu X, Park JS, Zheng Y, Balakrishnan J, Lei T, Ri Kim H, Song YI, Kim YJ, Kim KS, Özyilmaz B, Ahn JH, Hong BH, Iijima S. Roll-to-roll production of 30-inch graphene films for transparent electrodes. Nature Nanotechnology. 2010;5:574–578. DOI: 10.1038/nnano.2010.132.

[44] Vlassiouk I, Regmi M, Fulvio P, Dai S, Datskos P, Eres G, Smirnov S. Role of hydrogen in chemical vapor deposition growth of large single-crystal graphene. ACS Nano. 2011;5:6069–6076. DOI: 10.1021/nn201978y.

[45] Kim KS, Zhao Y, Jang H, Lee SY, Kim JM, Ahn JH, Kim P, Choi JY, Hong BH. Large-scale pattern growth of graphene films for stretchable transparent electrodes. Nature. 2009;457:706–710.

[46] López GA, Mittemeijer EJ. The solubility of C in solid Cu. Scripta Materialia. 2004;51:1–5. DOI: 10.1016/j.scriptamat.2004.03.028.

[47] Ruan G, Sun Z, Peng Z, Tour JM. Growth of graphene from food, insects, and waste. ACS Nano. 2011;5:7601–7607. DOI: 10.1021/nn202625c.

[48] Sun Z, Yan Z, Yao J, Beitler E, Zhu Y, Tour JM. Growth of graphene from solid carbon sources. Nature. 2010;468:549–552. DOI: 10.1038/nature09579.

[49] Seah CM, Chai SP, Mohamed AR. Mechanisms of graphene growth by chemical vapour deposition on transition metals. Carbon. 2014;70:1–21. DOI: 10.1016/j.carbon.2013.12.073.

[50] Zhang Y, Zhang L, Zhou C. Review of chemical vapor deposition of graphene and related applications. Accounts of Chemical Research. 2013;46:2329–2339. DOI: 10.1021/ar300203n.

[51] Zhang Y, Gao T, Gao Y, Xie S, Ji Q, Yan K, Peng H, Liu Z. Defect-like structures of graphene on copper foils for strain relief investigated by high-resolution scanning tunneling microscopy. ACS Nano. 2011;5:4014–4022. DOI: 10.1021/nn200573v.

[52] Nie S, Wofford JM, Bartelt NC, Dubon OD, McCarty KF. Origin of the mosaicity in graphene grown on Cu(111). Physical Review B – Condensed Matter and Materials Physics. 2011;84. DOI: 10.1103/PhysRevB.84.155425.

[53] Rybin MG, Pozharov AS, Obraztsova ED. Control of number of graphene layers grown by chemical vapor deposition. Physica Status Solidi (C) Current Topics in Solid State Physics. 2010;7:2785–2788. DOI: 10.1002/pssc.201000241.

[54] Gao L, Ren W, Zhao J, Ma LP, Chen Z, Cheng HM. Efficient growth of high-quality graphene films on Cu foils by ambient pressure chemical vapor deposition. Applied Physics Letters. 2010;97. DOI: 10.1063/1.3512865.

[55] Ma L, Ren W, Dong Z, Liu L, Cheng H. Progress of graphene growth on copper by chemical vapor deposition: growth behavior and controlled synthesis. Chinese Science Bulletin. 2012;57:2995–2999. DOI: 10.1007/s11434-012-5335-4.

[56] Zhang Y, Gomez L, Ishikawa FN, Madaria A, Ryu K, Wang C, Badmaev A, Zhou C. Comparison of graphene growth on single-crystalline and polycrystalline Ni by chemical vapor deposition. Journal of Physical Chemistry Letters. 2010;1:3101–3107. DOI: 10.1021/jz1011466.

[57] Ferrari AC, Meyer JC, Scardaci V, Casiraghi C, Lazzeri M, Mauri F, Piscanec S, Jiang D, Novoselov KS, Roth S, Geim AK. Raman spectrum of graphene and graphene layers. Physical Review Letters. 2006;97. DOI: 10.1103/PhysRevLett.97.187401.

[58] Malard LM, Pimenta MA, Dresselhaus G, Dresselhaus MS. Raman spectroscopy in graphene. Physics Reports. 2009;473:51–87. DOI: 10.1016/j.physrep.2009.02.003.

[59] Wang YY, Ni ZH, Yu T, Shen ZX, Wang HM, Wu YH, Chen W, Wee ATS. Raman studies of monolayer graphene: the substrate effect. Journal of Physical Chemistry C. 2008;112:10637–10640. DOI: 10.1021/jp8008404.

[60] Rümmeli MH, Rocha CG, Ortmann F, Ibrahim I, Sevincli H, Börrnert F, Kunstmann J, Bachmatiuk A, Pötschke M, Shiraishi M, Meyyappan M, Büchner B, Roche S, Cuniberti G. Graphene: piecing it together. Advanced Materials. 2011;23:4471–4490. DOI: 10.1002/adma.201101855.

[61] Gong Y, Zhang X, Liu G, Wu L, Geng X, Long M, Cao X, Guo Y, Li W, Xu J, Sun M, Lu L, Liu L. Layer-controlled and wafer-scale synthesis of uniform and high-quality graphene films on a polycrystalline nickel catalyst. Advanced Functional Materials. 2012;22:3153–3159. DOI: 10.1002/adfm.201200388.

[62] Chen S, Cai W, Piner RD, Suk JW, Wu Y, Ren Y, Kang J, Ruoff RS. Synthesis and characterization of large-area graphene and graphite films on commercial Cu-Ni alloy foils. Nano Letters. 2011;11:3519–3525. DOI: 10.1021/nl201699j.

[63] Dahal A, Batzill M. Graphene-nickel interfaces: a review. Nanoscale. 2014;6:2548–2562. DOI: 10.1039/c3nr05279f.

[64] Calizo I, Teweldebrhan D, Bao W, Miao F, Lau CN, Balandin AA. Spectroscopic Raman nanometrology of graphene and graphene multilayers on arbitrary substrates. Journal of Physics: Conference Series. 2008;109:5. DOI: 10.1088/1742-6596/109/1/012008.

[65] Das A, Chakraborty B, Sood AK. Raman spectroscopy of graphene on different substrates and influence of defects. Bulletin of Materials Science. 2008;31:579–584. DOI: 10.1007/s12034-008-0090-5.

[66] Wall M. Raman spectroscopy optimizes graphene characterization. Advanced Materials and Processes. 2012;170:35–38.

[67] Suk JW, Kitt A, Magnuson CW, Hao Y, Ahmed S, An J, Swan AK, Goldberg BB, Ruoff RS. Transfer of CVD-grown monolayer graphene onto arbitrary substrates. ACS Nano. 2011;5:6916–6924. DOI: 10.1021/nn201207c.

[68] Nemanich RJ, Solin SA. First- and second-order Raman scattering from finite-size crystals of graphite. Physical Review B. 1979;20:392–401.

[69] Pei S, Cheng HM. The reduction of graphene oxide. Carbon. 2012;50:3210–3228. DOI: 10.1016/j.carbon.2011.11.010.

[70] Suk JW, Piner RD, An J, Ruoff RS. Mechanical properties of monolayer graphene oxide. ACS Nano. 2010;4:6557–6564. DOI: 10.1021/nn101781v.

[71] Marcano DC, Kosynkin DV, Berlin JM, Sinitskii A, Sun Z, Slesarev A, Alemany LB, Lu W, Tour JM. Improved synthesis of graphene oxide. ACS Nano. 2010;4:4806–4814. DOI: 10.1021/nn1006368.

[72] Akhavan O, Ghaderi E. Escherichia coli bacteria reduce graphene oxide to bactericidal graphene in a self-limiting manner. Carbon. 2012;50:1853–1860. DOI: 10.1016/j.carbon. 2011.12.035.

[73] Guerrero-Contreras J, Caballero-Briones F. Graphene oxide powders with different oxidation degree, prepared by synthesis variations of the Hummers method. Materials Chemistry and Physics. 2015;153:209–220. DOI: 10.1016/j.matchemphys.2015.01.005.

[74] Botas C, Álvarez P, Blanco C, Santamaría R, Granda M, Gutiérrez MD, Rodríguez-Reinoso F, Menéndez R. Critical temperatures in the synthesis of graphene-like materials by thermal exfoliation-reduction of graphite oxide. Carbon. 2013;52:476–485. DOI: 10.1016/j.carbon.2012.09.059.

[75] Zhao W, Kido G, Hara K, Noguchi H. Characterization of neutralized graphite oxide and its use in electric double layer capacitors. Journal of Electroanalytical Chemistry. 2014;712:185–193. DOI: 10.1016/j.jelechem.2013.11.007.

Chapter 6

Laterally Inhomogeneous Au Intercalation in Epitaxial Graphene on SiC(0 0 0 1): A Multimethod Electron Microscopy Study

Claire Mathieu, Tevfik Onur Menteş,
Emiliano Pallecchi, Andrea Locatelli,
Gilles Patriarche, Rachid Belkhou and
Abdelkarim Ouerghi

Additional information is available at the end of the chapter

http://dx.doi.org/10.5772/64076

Abstract

Epitaxial graphene is of particular interest because of its tunable electronic structure. One important approach to tune the electronic properties of graphene relays on intercalating atomic species between graphene and the topmost silicon carbide layer. Here, we investigated the morphology and electronic structure of gold-intercalated epitaxial graphene using a multitechnique approach combining spectroscopic photoemission low-energy electron microscopy (SPELEEM) for chemical and structural characterization at mesoscopic length scale and with transmission electron microscopy (STEM) at the atomic level. Deposition of gold on ex situ prepared graphene on SiC(0 0 0 1) results in the partial intercalation of Au adatoms under graphene, with the formation of a buffer layer of variable thickness. Gold has also shown to aggregate in nanometer-sized clusters lying on top of the same graphene film. X-ray photo-emission electron microscopy measurements indicate that Au induces only small changes in the doping of the graphene layer, which does not develop a quasi free-standing behavior.

Keywords: graphene, electronic properties, ARPES, LEEM/LEED, STEM

1. Introduction

Graphene is a single layer of carbon atoms arranged in a hexagonal lattice. It is one of the few structures that are stable in two dimensions as a free-standing crystal [1, 2]. Its extraordinary properties, such as unconventional two-dimensional (2D) electron gas, high carrier mobility, half-integer quantum Hall effect at room temperature, and spin transport [1], have made graphene a very promising candidate for the design of the new generation of devices, such as ultrafast electronic circuits and photodetectors. Despite significant progress in its synthesis, the development of production methods warranting fine control over film morphology and thickness, which are both crucial to determine its electronic properties, remains a considerable challenge.

Epitaxial graphene layers on silicon carbide (SiC) have been extensively studied due to their potential for large-scale production with a high crystalline quality [3, 4]. It is now well established that, in this system, the interaction between the graphene and its supporting substrate (SiC) can affect considerably the electronic properties of graphene. Indeed, it is known that the first carbon layer onto a SiC substrate is covalently bonded to the Si atoms of the substrate, and this "buffer layer" does not display graphitic electronic properties [5–8]. Moreover, the remaining unsaturated Si dangling bonds, at the buffer layer/SiC interface, induce a high intrinsic n-doping in graphene, which degrades the carrier mobility. These drawbacks pose a major obstacle to the integration of graphene/SiC in future electronic devices. A solution to this problem is provided by the passivation of the Si dangling bonds using dopants and/or decoupling the graphene layer from the SiC substrate. Recently, decoupling of graphene has been achieved by depositing molecules [9, 10] or atomic layers of Bi [11], Ge [12], F [13], and Li [14, 15]. Furthermore, Riedl et al. [16] have shown that hydrogen intercalation can induce such desired decoupling. More recently, it was demonstrated that the oxygen can partially decouple the buffer layer from the substrate and reduce the intrinsic electron doping [17]. Moreover, the intercalation of metal clusters or molecules between the graphene layers may serve to functionalize graphene.

Numerous studies have been performed to understand the intercalation of transition metals [18–20]. The study by Gierz et al. [21] claimed that the strongly interacting first carbon layer was decoupled from the SiC(0 0 0 1) substrate via gold intercalation. The shift of Dirac points due to gold intercalation was then theoretically studied by Chuang et al. [22]. A similar study performed by Premlal et al. [23], using scanning tunneling microscopy and spectroscopy (STM/STS), concluded that the intercalated gold cluster displays a new surface reconstruction and induces a possible hole-doping effect. Nonetheless, further investigations are needed to better clarify the effects on the electronic structure of graphene induced by intercalated Au.

In this chapter, we discuss the properties of Au-intercalated epitaxial graphene grown on 6H-SiC(0 0 0 1). Low-energy electron diffraction (LEED), low-energy electron microscopy (LEEM), transmission electron microscopy (TEM), X-ray photoemission electron microscopy (XPEEM), microspot angle-resolved photoemission spectroscopy (μ-ARPES), and micro-Raman spectroscopy allowed us to study the structural and electronic properties of epitaxial graphene on SiC. In particular, we focused on the effect of Au adsorption on the local morphology, structure,

http://dx.doi.org/10.5772/64076

and electronic properties of few layers graphene on SiC(0 0 0 1); as it will be shown, our work provides also information about the local distribution of Au at the interface.

2. Experimental details

The structural and electronic properties of the graphene/SiC interface were characterized using the SPELEEM III (Elmitec GmbH) microscope operating at the Nanospectroscopy beamline of the Elettra storage ring in Trieste (Italy). This instrument combines LEEM and energy-filtered XPEEM imaging with μLEED and μARPES. The μ-spot diffraction data are typically collected from areas of diameter as small as 2 μm. Such analytical methods are well-established tools for characterizing the local morphology, thickness, corrugation, and electronic structure of single layer graphene films [24, 25].

Semi-insulating on-axis SiC(0 0 0 1) substrates were used in this study. After polishing, the samples were exposed to hydrogen etching at 1600°C in order to remove surface defects. Graphene growth was carried out by annealing the substrates at 1300–1400°C under argon and silicon fluxes. This method favors the formation of large and homogeneous domains [26, 27]. During the graphitization process, the argon partial pressure was kept below $P = 2 \times 10^{-5}$ Torr. The samples were then cooled to the room temperature and transferred ex situ to the microscopes used in these studies.

Before the measurements, the graphene samples were annealed at 600°C for 30 min in ultrahigh vacuum, in order to reduce the contamination consequent to atmosphere exposure. The Au incorporation process was carried out using a post-growth deposition method. The samples were then further annealed at 800°C for 20 min in order to favor migration of gold.

In order to vary the amount of buffer layer on the surface prepared as above, we studied two samples grown under the same conditions but with different level of graphitization. As shown in the next section, sample 1 (S1) is at a later stage of graphitization in comparison with sample 2 (S2). S1 is mainly covered by 1 ML graphene, with small areas that present the characteristic of the buffer layer or bilayer graphene. On the other side, S2 presents mainly one and two graphene layers. Both samples were fully characterized by means of LEEM, μARPES, and μLEED, before and after Au deposition, respectively. Post-growth Au deposition on previously characterized graphene allows unraveling the effect of gold on the graphene layers. The Au 4f, Si 2p, and C 1s core-level images were recorded using two photon energies (hν = 200 and 360 eV) to tune the surface sensitivity. They were calibrated using the Au 4f7/2 component.

The TEM thin foil was prepared by focused ion beam (FIB). The surface was protected by an amorphous layer carbon deposited before the FIB process. The thin foils were prepared following the <1 1 –2 0> zone axis of the SiC substrate. We used a TEM/STEM microscope Jeol 2200FS working at 200 keV equipped with a spherical aberration (Cs) corrector on the STEM probe. The probe current was 50 pA with a probe size of 0.1 nm (FWHM). The convergence half-angle for the probe was 30 mrad, and the detection half-angles for the HAADF images were, respectively, 100 mrad (inner) and 170 mrad (outer).

3. Results and discussion

3.1. Growth of graphene on SiC(0 0 0 1)

We first address the structural properties of graphene, which was studied by low-energy electron diffraction. The typical LEED of the graphene sample can be seen in **Figure 1a**. The LEED pattern demonstrates that the graphene layer is well ordered and aligned with respect to the substrate, such that the basal plane unit vectors of graphene and SiC subtend an angle of 30°. The smallest hexagon is the result of a $(6\sqrt{3} \times 6\sqrt{3})R \times 30°$ reconstruction of the interfacial layer, as are the spots lying just inside the graphene pattern.

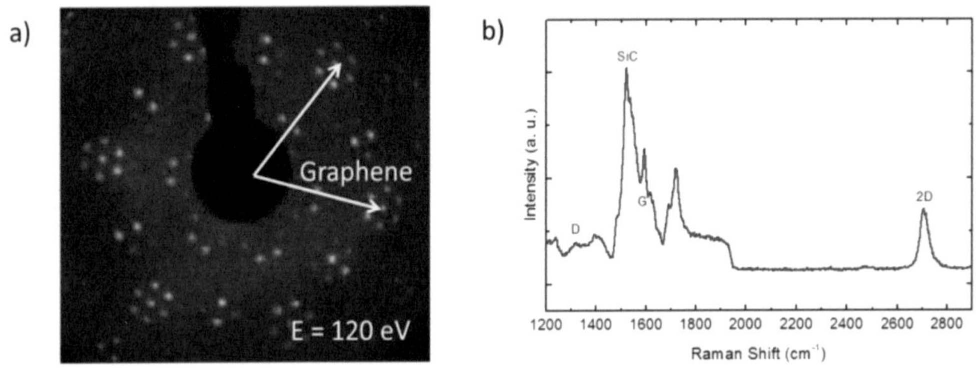

Figure 1. (a) Typical LEED image of graphene on SiC(0 0 0 1) and (b) typical micro-Raman spectra of the graphene monolayer. Contributions at the G and 2D bands are observed, together with a very low signal at the defect band D.

We further characterized our graphene by Raman spectroscopy, in **Figure 1b**, we present a typical Raman spectrum, which presents the typical features of high-quality epitaxial mono-layer graphene. Graphene contributions were identified by three main structures: (i) the D band at 1350 cm^{-1}, (ii) the G band (symmetric E2g phonon mode) at 1592 cm^{-1}, and (iii) the 2D band at 2704 cm^{-1}. For both samples, the D band, which corresponds to disorder, was weak in comparison with the G and 2D bands (double resonant electron-phonon process). The low intensity of this peak showed that there was only a small number of defect/disorder in the graphene structure. This was an indication of the high quality of the epitaxial graphene produced.

3.2. Au intercalation in epitaxial graphene on SiC(0 0 0 1)

The Au incorporation process was carried out using a post-growth deposition method. Au was deposited at a temperature of 890°C, at a rate of 10 min/layer. A total coverage of 2 ML (1 ML) was deposited on sample S1 (S2). The samples were then further annealed at 800°C for 20 min in order to favor migration of gold. We now discuss the results of LEEM, μARPES, and μLEED, characterization of both samples, before and after Au deposition.

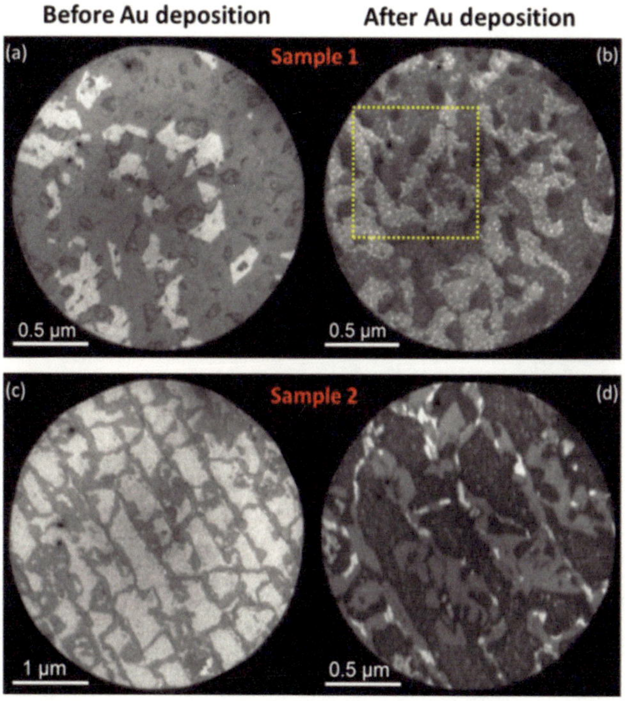

Figure 2. LEEM image of a graphene/SiC(0 0 0 1) surface, before (left) and after (right) Au deposition, with an electron beam energy of 3.90 eV. The images for the sample 1 (2) are presented at the top (bottom) of the image.

Figure 2a–d presents LEEM images obtained on the two samples before and after Au deposition. The image contrast arises from the thickness dependent reflectivity of the film. Recent quantum mechanical calculations of the IV reflectivity of graphene allow precise quantification of the graphene thickness on SiC, which vary considerably depending on the presence of buffer layers [28]. Measurement of the low energy electron reflectivity (not shown) of sample S1 before Au deposition (**Figure 1a**) has allowed us identifying the predominant presence of 1 monolayer (neutral gray regions) of graphene, accompanied with a small amount of the buffer layer (dark gray) and graphene bilayer (white contrast). After the deposition of Au (**Figure 2b**), the three regions exhibit the same contrast, suggesting that the overall thickness distribution of the sample was preserved by the Au deposition.

A notable difference between LEEM images acquired before and after Au deposition is the presence of small dots which are uniformly distributed over both single- and bilayers graphene. These structures have a maximum size of few tens of nanometers, with several of them close to (or below) the lateral resolution of our LEEM microscope. We interpret the dots as due to three-dimensional gold islands, as none of these features were observed on the sample before Au deposition. We had repeated the Au experiment on the second sample S2, with less Au (1 ML instead of 2 ML), on which the graphitization was incomplete. Indeed, the reflectivity curves confirm that the substrate was only partially covered by one and two monolayers of graphene. The rather large light gray contrast areas in **Figure 1c** are attributed to the buffer layer, thanks to the analysis of the reflectivity curves, extracted from these areas. The corresponding image recorded after Au deposition shows that the nanoclusters are once again observed over the entire field of view. The lower concentration can be explained by the total coverage of 1 ML of gold in comparison with 2 ML Au on S1. This set of images demonstrates that the clusters nucleate also over the buffer layer, as well as on the one and two graphene MLs.

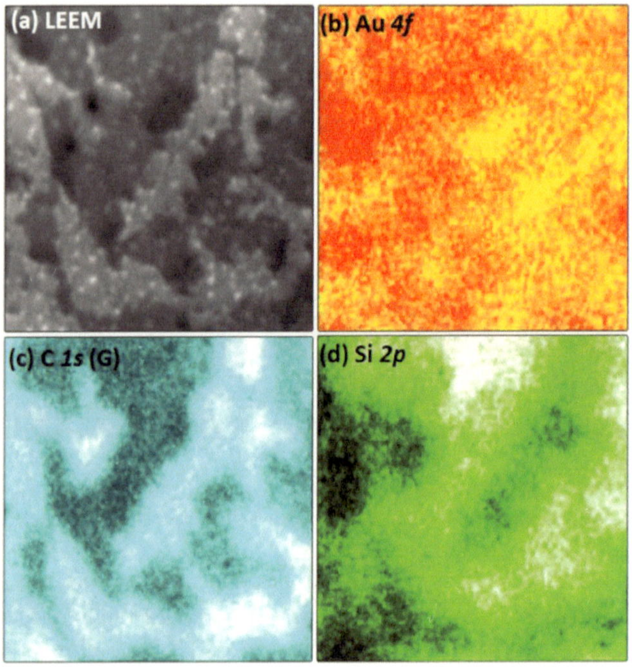

Figure 3. (a) LEEM image taken with an electron energy of 3.90 eV and (b–d) its corresponding XPEEM images recorded at the Au 4f, C 1s, and Si 2p core levels. The field of view is 0.80 μm wide. (b) Au 4f XPEEM image taken at a binding energy of 84.15 eV with hν = 200 eV. The highest (lowest) contrast is presented by the yellow (red) color. (c) C 1s XPEEM image taken at a binding energy of 284.60 eV with hν = 200 eV. The highest (lowest) contrast is presented by the white (black) color. (d) Si 2p XPEEM image, taken at a binding energy of 101.75 eV with hν = 200 eV. The highest (lowest) contrast is presented by the white (black) color.

In the following, we focus on the sample S1, corresponding to **Figure 2a** and **b**. In order to investigate the chemical properties of the sample after Au deposition, we have performed XPEEM measurements [29] at the C 1s, Au 4f, and Si 2p core levels (**Figure 3b–d**). **Figure 3a** displays a LEEM image of the sample after gold deposition, at electron energy of 3.90 eV. This image is a zoom of **Figure 2b**, marked by a dashed yellow square. **Figure 3c** is an XPEEM image at the C 1s core level, recorded at a binding energy (BE) of 284.60 eV. The BE is chosen to correspond to the maximum of the C 1s core level emission from the graphene layers. Therefore, the brightest (darkest) areas correspond to the graphene (substrate) layer. Regarding the Si 2p image, for which the maximum of intensity is attributed to the SiC contribution, recorded at a BE of 101.75 eV (hν = 200 eV), one can observe that the contrast is inverted, in comparison with the C 1s XPEEM image. Indeed, a lower Si signal is due to attenuation by the graphene layers; and the corresponding zones, on the C 1s image appear bright. We verified that these two images are in agreement with the graphene thickness, evaluated in LEEM (**Figure 3a**). The shapes of the domains observed in the LEEM image can be easily recognized in the C 1s and Si 2p maps. In addition to the C and Si distributions, **Figure 2b** displays an image of the Au 4f, recorded at 84.15 eV binding energy (hν = 200 eV). The Au 4f signal seems to be more intense at the buffer layer. The shape of the previous domains is not clearly recognizable anymore, meaning that the Au can be deposited or/and intercalated independently of the surface chemistry (substrate/graphene mono- or bilayer). The LEEM/XPEEM images do not allow a clear discrimination between whether the Au nanoclusters are intercalated between adjacent graphene layers or between the graphene and the buffer layer, or if they simply overlay on the surface. The STEM data presented below will help answering this question.

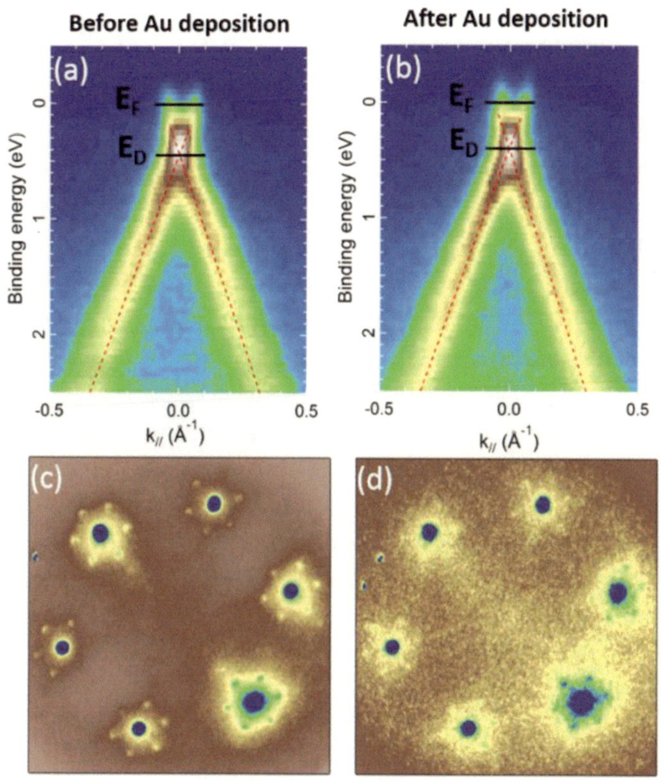

Figure 4. Top: band dispersions as a function of $k_{//}$ around the K point of the first Brillouin zone, obtained by μARPES at hν = 40 eV, performed before (a) and after (b) Au deposition. The Fermi level and the Dirac point are superimposed on the images. Bottom: 2D maps as a function of k_x and k_y, recorded for a binding energy of 0.15 eV, that is, close to the Fermi level, before (c) and after (d) gold deposition.

In order to investigate the effect of Au deposition on the electronic properties, we have performed local ARPES experiments (μARPES). **Figure 4** presents the μARPES maps, before

(**Figure 4a**) and after (**Figure 4b**) Au deposition, around the K point, perpendicular to the ΓKM direction. The 2D map of the unexposed graphene surface (**Figure 4a**) presents a shift of the Dirac point of 0.4 eV below the Fermi level. This energy shift ($\Delta E = E_D - E_F$) is nowadays well known in epitaxial graphene/SiC(0 0 0 1) [9, 17, 24] and is attributed to doping from the buffer layer [6, 7]. On the 2D μARPES map, recorded after Au deposition (**Figure 4b**), the energy difference between the Fermi level and the Dirac point is still 0.40 ± 0.02 eV. In our data, we do not observe a noticeable modulation of the Dirac point energy, contrary to what reported by Gierz et al. [21]. Nevertheless, we cannot exclude that a small shift of the Dirac point (below ~20 meV) has occurred. Moreover, Au deposition on SiC(0 0 0 1) results in p- or n-doping, depending either on the number of graphene layers, the strain at the Au/graphene or SiC/graphene interfaces [22], or the gold coverage [21]. In our case, we average the electronic information over the region that is slightly larger than that defined by the illumination, which includes both single and bilayer graphene. The local doping effect can therefore be averaged out in our data by the presence of areas with different electronic properties where Au induces p-doping and n-doping effect, respectively. However, we reckon that these antagonistic effects have to be insignificant for both cases as the Dirac cone does not get broader or split upon Au deposition.

The 2D maps of the first Brillouin zone, recorded for a BE of 0.25 eV, are presented in **Figure 4c** and **d**, close to the Fermi level, obtained before and after Au deposition, respectively. In the constant energy plots, six weak replicas of the π and π* states surrounding the primary states can be seen, as points around the upper spot (**Figure 4c**). Low-energy electron diffraction of graphene layers grown on the SiC substrate (not shown) displays a nearly commensurate superstructure with $(6\sqrt{3} \times 6\sqrt{3})R \times 30°$ unit cell with respect to the substrate because of the difference between the graphene lattice constant of 2.46 Å and that of SiC, 3.07 Å. The replicas of the π and π* states are brought about by scattering off this superstructure in a fashion similar to those in other incommensurate systems. The 2D image recorded after Au deposition (**Figure 4d**) shows that the superstructure around each K and K′ point persists. Even if the statistics are not as good as the one before Au adsorption, one can still clearly observe these satellites, which suggest that the Au deposition has not effectively decoupled the buffer layer from the substrate. The data in **Figure 4** demonstrate that the Au adsorption on this Gr/SiC(0 0 0 1) sample neither decouples the buffer layer from the substrate, nor alters its average doping.

In order to reach an atomic level characterization of the Au distribution we have performed sectional HR-TEM on the sample 2, after gold deposition. **Figure 5** shows TEM images of two distinct areas of the sample, recorded with a bright field (BF) and dark field mode (DF). These STEM images, allow accessing the crystallographic information. High-angle annular dark field (HAADF) contrast depends directly on the atomic number of the element: gold atoms appear very bright due to their high atomic number (so-called Z-contrast' imaging). The contrast of the graphene layer is very weak in the HAADF-STEM images due to its low atomic number (by comparison with the silicon). The graphene layer is easily visible in the bright field image (contrast of diffraction), as shown in **Figure 5b** and **e**.

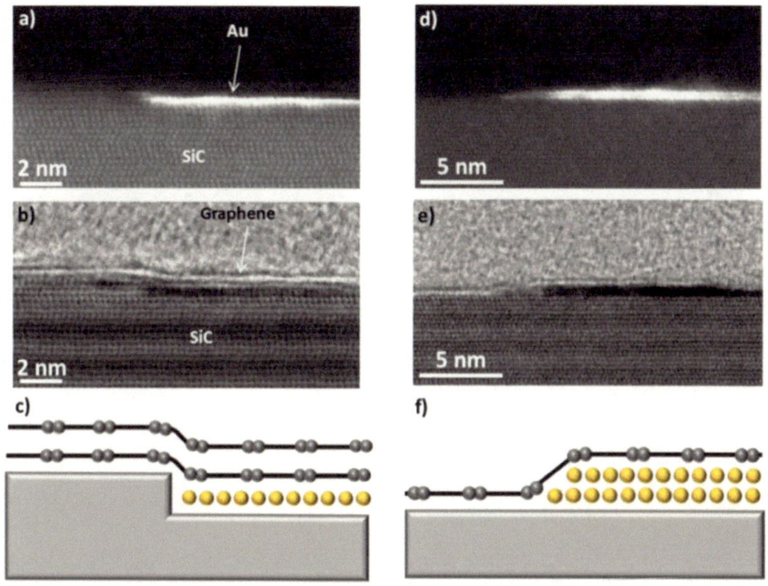

Figure 5. Cross-sectional aberration-corrected STEM images recorded in the high-angle annular dark field (HAADF) (a and d) and bright field (BF) (b and e) mode, respectively. HAADF contrast depends directly on the atomic number of the element: gold atoms appear very bright due to their high atomic number (so-called 'Z-contrast' imaging). The contrast of the graphene layer is very weak in the HAADF-STEM images due to its low atomic number (in comparison with the silicon). Layer graphene is easily visible in the bright field image (contrast of diffraction). The substrate, the Au, and graphene layers can clearly be observed. A scheme of the area is also represented. (d–f) DF and BF images with the associated scenarios, for another area of the surface. The substrate is represented by a gray square-shape. The Au and carbon atoms are represented by yellow and gray spheres.

On the DF images (**Figure 5a** and **d**), the bright contrast corresponds to the Au atoms. Moreover, the dark line above the crystalline substrate, observed on the DF image, is assigned to a graphene layer. For the first area (**Figure 5a** and **b**), combining the information extracted from the DF and BF images, respectively, we demonstrate that the Au atoms are located below the two graphene layers. These images also show that this Au insertion line stops at the step edge of the substrate, as presented in the sketch (**Figure 5c**). The BF image of the second area (**Figure 5d**) shows that two Au layers have been intercalated under a single graphene layer. Whether the graphene layer is continuous or not at the end of the Au step edge is not clear. One remaining question is how the Au penetrates into the sample. The Au atoms can penetrate the graphene layer through step edges or defects, as it has already been theoretically proposed in the case of the Si out-diffusion, on epitaxial graphene [30].

As different Au insertion mode have observed with the STEM with the HAADF imaging mode, a new set of images from another area, using once again the BF and DF modes, is presented **Figure 6**. The DF mode shows that two zones of the field of view contain Au atoms. The one on the left side shows that these atoms can pass through the graphene layer, as explained for **Figure 5**. However, the zone on the right side of the image presents a new scenario of Au adsorption. The Au atoms, as presented by the black dots of the BF, have penetrated into few layers of the substrate and a nanocluster has nucleated. The lateral size of this nanocluster can be estimated to be ~5 nm. By scanning several areas of the surface, all of these three scenarios have been observed several times, confirming that the Au deposition on the graphene layer is not a uniform process.

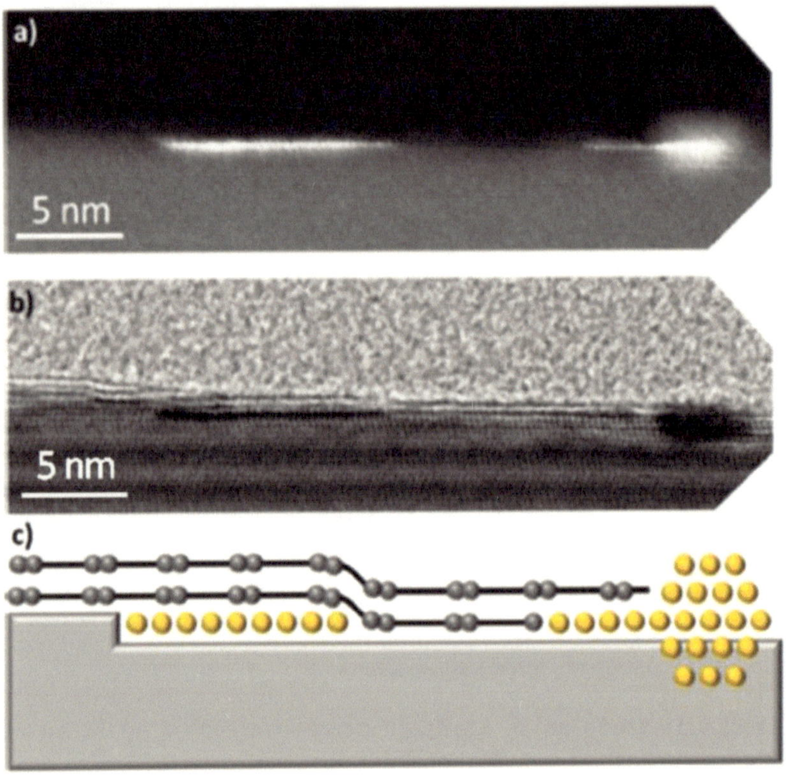

Figure 6. Cross-sectional STEM images recorded in the HAADF (top) and bright field (middle) mode, respectively. The substrate, the Au, and graphene layers can clearly be observed. A scheme of the area is also represented (bottom). The substrate is represented by a gray square shape. The Au and carbon atoms are represented by yellow and gray spheres.

The lateral distances between the gold atoms in the first layer are exactly the ones found for the silicon (or carbon) atoms of the SiC substrate underneath. Indeed, the distance measured between two adjacent Au atoms is 0.265 nm, while it is of 0.267 nm for two adjacent Si atoms in a (0 0 0 2) plane. These distances are obtained with the same value in measurements performed on numerous STEM images. The distance between two adjacent Au atoms in the second layer systematically decreases to 0.236 nm. Moreover, the distance between two adjacent gold layers is 0.255 nm, considering that the d_{0002} inter-reticular distance measured in the SiC substrate close to the surface is 0.266 nm. This value has to be compared to the experimental value of 0.252 nm [31] for the SiC bulk. Therefore, we can conclude that the strain in the SiC substrate beneath the surface is small, even when layer or nanoclusters of gold are observed.

http://dx.doi.org/10.5772/64076

The results observed in TEM and LEEM shows two main scenarios of gold migration when deposited on graphene/SiC(0 0 0 1). On the one hand, 1 or 2 ML of gold can intercalate between the substrate and the graphene layers. On the other hand, some gold atoms migrate inside the substrate to form nanoclusters. In the former case, this would lead to a change in the electronic properties, while it should not in the latter case. μ-ARPES measurements (**Figure 4**) do not present any noticeable changes. Before and after gold deposition, the Dirac point is at 0.40 eV below the Fermi level, for both cases. This equivalent doping can be explained by the fact that the gold mainly clusterizes and does not intercalate homogeneously under graphene layer. Therefore, the decoupling of the buffer layer, as observed by Gierz et al. [21], is not evidenced in our case.

4. Conclusions

In summary, we have demonstrated that the Au deposition on graphene epitaxially grown on SiC is an inhomogeneous process. The LEEM and XPEEM measurements have demonstrated that the Au nanoclusters nucleate all over the surface, independently of the surface chemistry (substrate, mono-, and bilayers of graphene). The STEM experiments have shown that the gold can diffuse under one or more graphene layers and can spread as one or two gold layers. Moreover, the gold mainly nucleates to form Au nanoclusters, which are spread all over the sample, independently of the surface chemistry. Therefore, the μARPES experiments have highlighted that the gold deposition does not induce a significant change in the electronic properties of this material and showed that the Fermi velocity of graphene remained intact in comparison with pristine graphene.

Author details

Claire Mathieu[1], Tevfik Onur Menteş[2], Emiliano Pallecchi[3], Andrea Locatelli[2], Gilles Patriarche[4], Rachid Belkhou[5] and Abdelkarim Ouerghi[4*]

*Address all correspondence to: abdelkarim.ouerghi@lpn.cnrs.fr

1 SPEC, CEA, CNRS, Paris-Saclay University, CEA Saclay, Gif-sur-Yvette Cedex, France

2 Elettra – Sincrotrone Trieste S.C.p.A., Basovizza, Trieste, Italy

3 Institut of Electronics, Microelectronics, and Nanotechnology (IEMN), Villeneuve d'Ascq Cedex, France

4 CNRS, Laboratory for Photonics and Nanostructures, Marcoussis, France

5 Synchrotron SOLEIL, Saint-Aubin, Gif-sur-Yvette Cedex, France

References

[1] Novoselov KS, Geim AK, Morozov SV, Jiang D, Zhang Y, Dubonos SV, Grigorieva IV, Firsov AA. Electric field effect in atomically thin carbon films. Science. 2004;306:666–669. doi:10.1126/science.1102896

[2] Berger C, Wu X, First PN, Conrad EH, Li X, Sprikle M, Hass J, Varchon F, Magaud L, Sadowski ML, Potemski M, Martinez G, de Heer WA. Dirac particles in epitaxial graphene films grown on SiC. Advances in Solid State Physics. 2008;47:145. doi: 10.1007/978-3-540-74325-5_12

[3] Virojanadara C, Syvajarvi M, Yakimova R, Johansson LI, Zakharov AA, Balasubramanian T. Homogeneous large-area graphene layer growth on 6H-SiC(0 0 0 1). Physical Review B. 2008;78:245403. doi:10.1103/PhysRevB.78.245403

[4] Sprinkle M, Ruan M, Hu Y, Hankinson J, Rubio-Roy M, Zhang B, Wu X, Berger C, de Heer WA. Nanoelectronics: nanoribbons on the edge. Nature Nanotechnology. 2010;5:698–699. doi:10.1038/nnano.2010.200

[5] Emtsev KV, Speck F, Seyller Th, Ley L, Riley JD. Interaction, growth and ordering of epitaxial graphene on SiC(0 0 0 1) surfaces: a comparative photoelectron spectroscopy study. Physical Review B. 2008;77:155303. doi:10.1103/PhysRevB.77.155303

[6] Ouerghi A, Marangolo M, Belkhou R, El Moussaoui S, Silly MG, Eddrief M, Largeau L, Portail M, Fain B, Sirotti F. Epitaxial graphene on 3C-SiC(1 1 1) pseudosubstrate: structural and electronic properties. Physical Review B. 2010;82:125445. doi:10.1103/PhysRevB.82.125445

[7] Varchon F, Feng R, Hass J, Li X, Ngoc Nguyen B, Naud C, Mallet P, Veuillen J-Y, Berger C, Conrad EH, Magaud L. Electronic structure of epitaxial graphene layers on SiC: effect of the substrate. Physical Review Letters. 2007;99:126805. doi:10.1103/PhysRevLett.99.126805

[8] Ouerghi A, Silly MG, Marangolo M, Mathieu C, Eddrief M, Picher M, Sirotti F, El Moussaoui S, Belkhou R. Large-area and high-quality epitaxial graphene on off-axis SiC wafers. ACS Nano. 2012;6:6075–6082. doi:10.1021/nn301152p

[9] Coletti C, Riedl C, Lee DS, Krauss B, Patthey L, von Klitzing K, Smet JH, Starke U. Charge neutrality and band-gap tuning of epitaxial graphene on SiC by molecular doping. Physical Review B. 2010;81:235401. doi:10.1103/PhysRevB.81.235401

[10] Zhou SY, Siegel DA, Fedorov AV, Lanzara A. Metal to insulator transition in epitaxial graphene induced by molecular doping. Physical Review Letters. 2008;101:086402. doi: 10.1103/PhysRevLett.101.086402

[11] Gierz I, Riedl C, Starke U, Ast CR, Kern K. Atomic hole doping of graphene. Nano Letters. 2008;8:4603–4607. doi:10.1021/nl802996s

[12] Emtsev KV, Zakharov AA, Coletti C, Forti S, Starke U. Ambipolar doping in quasifree epitaxial graphene on SiC(0 0 0 1) controlled by Ge intercaltion. Physical Review B. 2011;84:125423. doi:10.1103/PhysRevB.84.125423

[13] Walter AL, Jeon K-J, Bostwick A, Speck F, Ostler M, Seyller T, Moreschini L, Kim YS, Chang YJ, Horn K, Rotenberg E. Highly p-doped epitaxial graphene obtained by fluorine intercalation. Applied Physics Letters. 2011;98:184102. doi:10.1063/1.3586256

[14] Virojanadara C, Watcharinyanon S, Zakharov AA, Johansson LI. Epitaxial graphene on 6H-SiC and Li intercalation. Physical Review B. 2010;82:205402. doi:10.1103/PhysRevB.82.205402

[15] Virojanadara C, Zakharov AA, Watcharinyanon S, Yakimova R, Johansson LI. A low-energy electron microscopy and X-ray photo-emission electron microscopy study of Li intercalated into graphene on SiC(0 0 0 1). New Journal of Physics. 2010;12:125015. doi: 10.1088/1367-2630/12/12/125015

[16] Riedl C, Coletti C, Iwasaki T, Zakharov AA, Starke U. Quasi-free_standing epitaxial graphene on SiC obtained by hydrogen intercalation. Physical Review Letters. 2009;103:246804. doi:10.1103/PhysRevLett.103.246804

[17] Mathieu C, Lalmi B, Mentes TO, Locatelli A, Latil S, Belkhou R, Ouerghi A. Effect of oxygen adsorption on the local properties of epitaxial graphene on SiC(0 0 0 1). Physical Review B. 2012;86:035435. doi:10.1103/PhysRevB.86.035435

[18] Varykhalov A, Scholz MR, Kim TK, Rader O. Effect of noble-metal contacts on doping and band gap of graphene. Physical Review B. 2010;82:121101. doi:10.1103/PhysRevB.82.121101

[19] Nair MN, Cranney M, Vonau F, Aubel D, Le Fevre P, Tejeda A, Bertran F, Taleb-Ibrahimi A, Simon L. High van Hove singularity extension and Fermi velocity increase in epitaxial graphene functionalized by intercalated gold clusters. Physical Review B. 2012;85:245421. doi:10.1103/PhysRevB.85.245421

[20] Gao T, Gao Y, Chang C, Chen Y, Liu M, Xie S, He K, Ma X, Zhang Y, Liu Z. Atomic-scale morphology and electronic structure of manganese atomic layers underneath epitaxial graphene on SiC(0 0 0 1). ACS Nano. 2012;6(8):6562–6568. doi:10.1021/nn302303n

[21] Gierz I, Suzuki T, Weitz RT, Lee DS, Krauss B, Riedl C, Starke U, Höchst H, Smet JH, Ast CR, Kern K. Electronic decoupling of an epitaxial graphene monolayer by gold intercalation. Physical Review B. 2010;81:235408. doi:10.1103/PhysRevB.81.235408

[22] Chuang F-C, Lin W-H, Huang Z-Q, Hsu C-H, Kuo C-C, Ozolins V, Yeh V. Electronic structures of an epitaxial graphene monolayer on SiC(0 0 0 1) after gold intercalation: a first-principles study. Nanotechnology. 2011;22:275704. doi: 10.1088/0957-4484/22/27/275704

[23] Premlal B, Cranney M, Vonau F, Aubel D, Casterman D, De Souza MM, Simon L. Surface intercalation of gold underneath a graphene monolayer on SiC(0 0 0 1) studied

by scanning tunneling microscopy and spectroscopy. Applied Physics Letters. 2009;94:263115. doi:10.1063/1.3168502

[24] Locatelli A, Knox KR, Cvetko D, Mentes TO, Nino MA, Wang S, Yilmaz MB, Kim P, Osgood RM, Jr, Morgante A. Corrugation in exfoliated graphene: an electron microscopy and diffraction study. ACS Nano. 2010;4:4879. doi:10.1021/nn101116n

[25] Locatelli A, Aballe L, Mentes TO, Kiskinova M, Bauer E. Photoemission electron microscopy with chemical sensitivity: SPELEEM methods and applications. Surface and Interface Analysis. 2006;38:1554–1557. doi:10.1002/sia.2424

[26] Ouerghi A, Ridene M, Balan A, Belkhou R, Barbier A, Gogneau N, Portail M, Michon A, Latil S, Jegou P, Shukla A. Sharp interface in epitaxial graphene layers on 3C-SiC(1 0 0)/Si(1 0 0) wafers. Physical Review B. 2011;85:205429. doi:10.1103/PhysRevB.83.205429

[27] Pallecchi E, Ridene M, Kazazis D, Mathieu C, Schopfer F, Poirier W, Mailly D, Ouerghi A. Observation of the quantum Hall effect in epitaxial graphene on SiC(0 0 0 1) with oxygen adsorption. Applied Physics Letters. 2012;100:253109. doi:10.1063/1.479824

[28] Feenstra RM, Srivastava N, Gao Q, Widom M, Diaconescu B, Ohta T, Kellogg GL, Robinson JT, Vlassiouk IV. Low-energy electron reflectivity from graphene. Physical Review B. 2013;87:041406(R). doi:10.1103/PhysRevB.87.041406

[29] Knox KR, Locatelli A, Yilmaz MB, Cvetko D, Mentes TO, Nino MA, Kim P, Morgante A, Osgood RM, Jr. Making angle-resolved photoemission measurements on corrugated monolayer crystals: suspended exfoliated single-crystal graphene. Physical Review B. 2011;84:115401. doi:10.1103/PhysRevB.84.115401

[30] Sun GF, Liu Y, Rhim SH, Jia JF, Xue QK, Weinert M, Li L. Si diffusion path for pit-free graphene growth on SiC(0 0 0 1). Physical Review B. 2011;84:195455. doi:10.1103/PhysRevB.84.195455

[31] Nakagawa H, Tanaka S, Suemune I. Self_ordering of nanofacets on vicinal SiC surfaces. Physical Review Letters. 2003;91:226107. doi:10.1103/PhysRevLett.91.226107

Section 3

Application of Graphene and its Nanostructures

Chapter 7

Energy Transfer in Graphene-Based Hybrid Photosynthetic Nanostructures

Sebastian Mackowski and Izabela Kamińska

Additional information is available at the end of the chapter

http://dx.doi.org/10.5772/64300

Abstract

Energy transfer is one of the most fundamental processes at the nanoscale. Whenever a donor is placed sufficiently close to an acceptor, they could couple via electrostatic interactions and the energy is funnelled down to the acceptor. Only recently graphene, a two-dimensional sp2-hybridized carbon hexagonal lattice, has emerged as highly attractive from the point of view of potential applications in photonics and optoelectronics, as it features uniform absorption, which extends over the whole visible range down to the infrared. With the absence of fluorescence, it renders graphene as an exceptional acceptor in devices that utilize energy and/or electron transfer. In this chapter, we review recent work on the energy transfer in graphene-based assemblies involving also graphene derivatives (graphene oxide and reduced graphene oxide), as well as describe results of fluorescence studies focused on interactions between graphene and photosynthetic protein—pigment complexes. While for organic dyes the efficiency of the energy transfer is very high, in the case of the proteins, there is some shielding of chlorophylls from graphene, partially inhibiting the energy transfer. This allows for observing interesting effects associated with dependencies on the excitation energy, number of graphene layers, or the substrate that graphene is placed onto.

Keywords: graphene, reduced graphene oxide, photosynthetic complex, energy transfer, fluorescence microscopy, fluorescence imaging

1. Introduction

Non-radiative energy transfer (ET) is one of the most fundamental processes at the nanoscale. It is associated with funnelling excitation energy between molecules, quantum dots or other nanostructures [1]—in most cases—via dipole-dipole interaction. Such a scheme evolved for instance in natural photosynthesis [2] for efficient capturing and transport of the sunlight energy, and has been recently implemented in artificial light-harvesting assemblies [3]. The efficiency of ET depends on the spectral properties of a donor and an acceptor, their mutual orientation, as well as separation between them [4]. In particular, the distance dependence of the ET efficiency, which for localized dipoles scales with d^{-6}, has been exploited as a useful tool to measure lengths at the nanoscale, both statically and dynamically. In particular, the energy transfer has been considered an attractive way to control light harvesting and bio(sensing) [5–7]. As a result, various strategies had to be devised to fabricate hybrid nanostructures with well-defined morphology required for controlling the energy transfer efficiency. Among the most feasible, one can find robust nanolayers, either polymer or dielectric, deposited on a surface of nanoparticles or a substrate [8, 9], and more flexible linkers based for instance on DNA strands or biotin-streptavidin conjugation [10, 11].

Another critical parameter that influences the interaction between two dipole moments in the context of the energy transfer is the relation of their spectral properties. Namely, as shown in **Figure 1**, the absorption of one of the molecules (acceptor) has to overlap with the emission of the second molecule (donor). The larger the overlap, the higher the efficiency of the energy transfer. The final parameter that has to be considered in energy transfer geometry is the mutual orientation of the dipole moments of a donor and an acceptor. All these factors are included in the equation shown in **Figure 1**. It is important to note that the equation describes the case where both donors and acceptors are classical dipole moments.

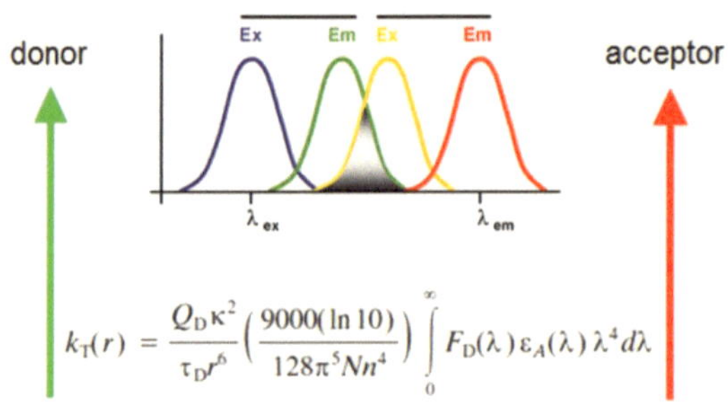

$$k_T(r) = \frac{Q_D \kappa^2}{\tau_D r^6}\left(\frac{9000(\ln 10)}{128\pi^5 N n^4}\right)\int_0^\infty F_D(\lambda)\varepsilon_A(\lambda)\lambda^4 d\lambda$$

Figure 1. Schematic representation of the energy transfer between two dipole moments of a donor (green) and acceptor (red). The energy transfer is possible only when the emission of the donor overlaps with the absorption of the acceptor.

Optical spectroscopy, and in particular fluorescence spectroscopy, provides a variety of tools to probe the energy transfer in hybrid nanostructures. It stems from the fact that the emergence energy transfer results in the decrease of the emission intensity of a donor at the expense emission intensity of an acceptor. This is in fact the most straightforward consequence energy transfer between two nanostructures. In addition to the intensity flow between donors and acceptors, another signature of the energy transfer is a shortening of the fluorescence decay time of the donor. Indeed, the energy transfer can be considered as a new channel for non-radiative recombination from the point of view of the donor, and as such it should result in shortening of the lifetime.

http://dx.doi.org/10.5772/64300

The purpose of this chapter is to review recent research carried out on hybrid nanostructures composed of graphene and graphene derivatives and naturally evolved photosynthetic complexes. Our aim is to emphasize effects that are not readily available when studying classical emitters, such as organic dyes or semiconductor nanocrystals, which have spectrally limited absorption and emission, as in contrast to these, photosynthetic complexes feature absorption that spans over the whole visible spectral range. They are also pigment-protein complexes, with chlorophyll molecules protected from the environment. Thus, the interaction between photosynthetic complexes and graphene is not immediate. However, before we describe the results obtained for photosynthetic complexes coupled with graphene, we review several key results reported for organic dyes and semiconductor nanocrystals on graphene. This is important to illustrate basic mechanisms and processes that take part in such hybrid architectures.

2. Overview of recent results

2.1. Graphene: basic properties

Graphene is nowadays one of the most intensively studied materials. Since 2004, when it was for the first time obtained by mechanical exfoliation, many research groups worldwide have focused on understanding and proving uniqueness of this one-atom thin material [12, 13]. Part of these efforts were inspired by the combination of properties rarely met in any other material, such as exceptionally high electronic and thermal conductivity, mechanical strength, unusual electronic structure and optical transmittance, impermeability to gases and many others [14].

The key property of graphene, which impacts its electronic and optical character, and is particularly important in the context of light-matter interactions, is an unusual zero bandgap structure and linear dispersion near the Brillouin zone corners. Indeed, in the case of graphene, the conduction and valence bands meet at Dirac points and in their vicinity the energy depends linearly on the wavevector [15]. Consequently, both electrons and holes mimic massless relativistic particles with effective velocity of $c^* \approx 1/300$ the speed of light.

Fully occupied valence band combined with an empty conduction band, and no energy gap between them, leads to unique electronic absorption of graphene. Remarkably, it is rather high, and for a suspended graphene monolayer is defined solely by the fine-structure constant, which translates to 2.3% absorption of incident light (**Figure 2**) [16]. In addition, the absorption of graphene shows no wavelength dependence from ultraviolet to near-infrared. Therefore, graphene sheets can be visualized using optical microscopy, as shown in **Figure 2**, and the absorption of a few-layer graphene can be roughly described as a sum of non-interacting single layers with each layer contributing 2.3% of opacity. From the point of view of hybrid assemblies where the energy transfer is exploited, graphene can be utilized as energy acceptor due to uniform absorption, but at the same time, it features no fluorescence emission. Consequently, any effects attributable to the energy transfer from any emitter to graphene will have to be probed and quantified based solely on the behaviour of energy donors.

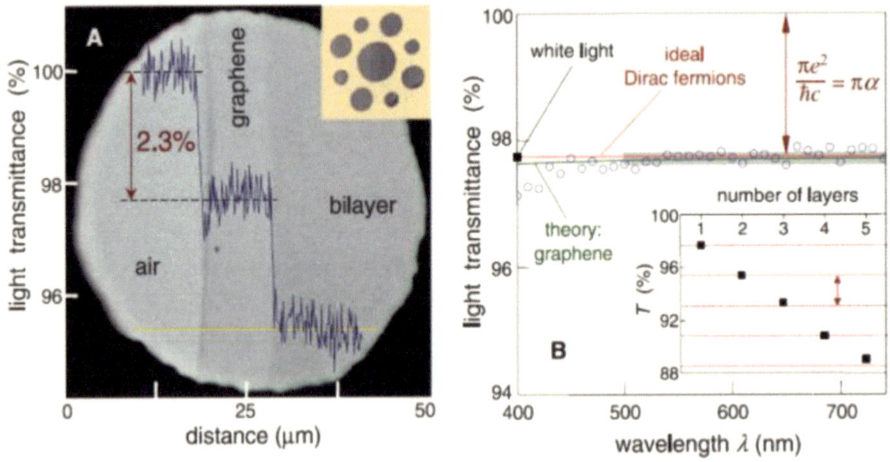

Figure 2. (A) Photograph of a 50-mm aperture partially covered by monolayer and bilayer graphene. The line-scan profile shows the intensity of transmitted white light along the yellow line. (Inset) Sample design. (B) Transmittance spectrum of single-layer graphene (open circles). The red line is the transmittance expected for two-dimensional Dirac fermions, whereas the green curve takes into account a non-linearity and triangular warping of graphene's electronic spectrum. (Inset) Transmittance of white light as a function of the number of graphene layers (squares). From Nair et al. [16]. Reprinted with permission from AAAS.

2.2. Graphene: fabrication methods

There are several ways of obtaining graphene and its derivatives, with each method holding specific advantages depending on particular application [17], but also facing limitations from the point of view of scalability, cost, reproducibility and alike.

First experiments on graphene have been carried out for the highest-quality pristine graphene obtained by mechanical exfoliation. This method, which is still arguably the leading one, allows preparing pure hexagonal carbon lattice without defects or dopants, thus exfoliated graphene has been at the centre of fundamental studies of its properties [16, 18]. It suffers, however, from small and irregular sample sizes and shapes, low throughput and high prices.

Growing demand for larger and reproducible graphene pieces has resulted in the development of other methods used for the production of two-dimensional (2D) carbon materials. Epitaxial techniques, by both sublimation and chemical vapour deposition (CVD), are the most promising due to the comparably high quality of fabricated graphene combined with high reproducibility and scalability. The first approach involves decomposition of SiC in low pressures (or ultra-high vacuum) and high temperatures. Addition of the annealing in argon atmosphere significantly improves homogeneity of graphene [19]. Upon sublimation of silicon from SiC surface, the remaining carbon atoms form graphene layers. Although the price of SiC substrate is relatively high, it could be compensated by excellent electronic parameters of graphene and performance of fabricated devices [13].

An alternative approach to produce graphene is CVD, which involves deposition of hydrocarbons onto a transition metal surface, usually copper or nickel, which works as a catalyst

http://dx.doi.org/10.5772/64300

[20, 21]. Due to the differences of solubility of carbon in Ni and Cu, the processes also differ from each other. As the solubility of carbon in Cu is rather low, the formation of graphene layers weakly depends on the actual conditions of the growth process. This allows for a better control over graphene growth, and when the monolayer is formed the deposition process stops. For this reason, large domains of single graphene layers are formed (>95% of the surface) [20]. Although graphene growth on Cu displays some disadvantages, such as wrinkles and grain boundaries, it is a relatively simple and inexpensive approach, suitable for mass production of high-quality graphene. Importantly, graphene grown on metallic surfaces can be transferred via polymer-assisted method on any arbitrary substrate, what is important for constructing graphene-based devices (photovoltaic cells, transistor, etc.) [22].

For uniform graphene layers obtained using exfoliation or CVD, hybrid assemblies for studying the energy transfer have to be constructed in a layer-by-layer geometry, where emitters are deposited onto the graphene layer (or on a polymer/oxide layer that is supposed to separate emitters from graphene).

In addition to methods that allow for growing large-area uniform graphene with well-controlled number of layers, other approaches have also been developed, based on the concept of liquid phase exfoliation. The approach of synthesizing two-dimensional carbon flakes in liquid relies on the reduction of graphene oxide (GO) [23] and key advantages thereof include solubility in aqueous and organic solvents, easy processing and surface functionalization, cheap synthesis for scalable production and relatively mild conditions of synthesis [24].

Briefly, the process usually starts with oxidizing graphite using one of the many popular oxidation methods: Hummers, Brodie or Staudenmaier [25, 26]. In the next step, due to the presence of oxygen-containing functional groups (hydroxyls, carbonyls, carboxyls or oxygen epoxides), graphite oxide can be easily exfoliated into GO via ultrasonication or mechanical stirring. Importantly, the presence of oxygen moieties distinguishes GO from graphene, and due to the predominance of sp^3- over sp^2-hybridized carbon atoms, GO is a fluorescent insulator, as opposed to graphene, which is a non-emitting conductor. However, it is an ideal precursor for the synthesis of reduced GO (rGO), also called chemical graphene. Importantly, by reducing GO, and thus restoring the sp^2-carbon network without additional components and residues [50, 51], it is possible to not only diminish fluorescence but also retrieve electrical and thermal conductivity. In contrast to graphene, rGO can be dispersed in water, and preparation of rGO flakes in solution makes it feasible to incorporate various functional groups on the surface, as required for many applications.

Among the key challenges is the establishment of reproducible methods of fabricating large-area rGO flakes and assuring the control of the number of layers, although these two factors frequently compete against each other: it is difficult to fabricate large single-layer rGO. The most commonly applied strategies are mild oxidation conditions, which promote the formation of larger flakes simply by reducing cracking of graphite flakes [27], adding a pretreatment step (an incubation in sulphuric acid with gentle stirring) before oxidation [28], or even skipping sonication to avoid breakage of flakes [29]. An important step is a separation of large flakes from aggregates and smaller sheets, and this can be performed with a high-speed centrifugation [27, 30, 31]. Also, GO might be deposited on a substrate using Langmuir-

Blodgett technique or self-assembling, before subsequent reduction, either chemically or thermally, to prevent re-aggregation of rGO flakes [29, 32].

From the point of view of the energy transfer, rGO offers important degree of freedom as compared to epitaxial graphene, namely it is possible to prepare mixtures in solution, which is often technically simpler and more suitable for—for instance—fluorescence-sensor design.

2.3. Graphene as energy acceptor

Constant absorption covering ultraviolet, visible and near infrared spectral regions (300–2500 nm), together with unusual electronic structure, renders graphene as an exceptional energy acceptor. It is possible to obtain not only energy transfer from any emitter to graphene but also the zero-energy gap of graphene implies that any interactions in graphene-based hybrid nanostructures can be investigated exclusively by studying optical properties of a donor. These properties result in a unique potential of graphene as a component of devices designed for photonics, optoelectronics, as well as photodetectors and biosensors. In particular, in recent years several studies emerged, where the energy transfer from various emitters to graphene has been investigated.

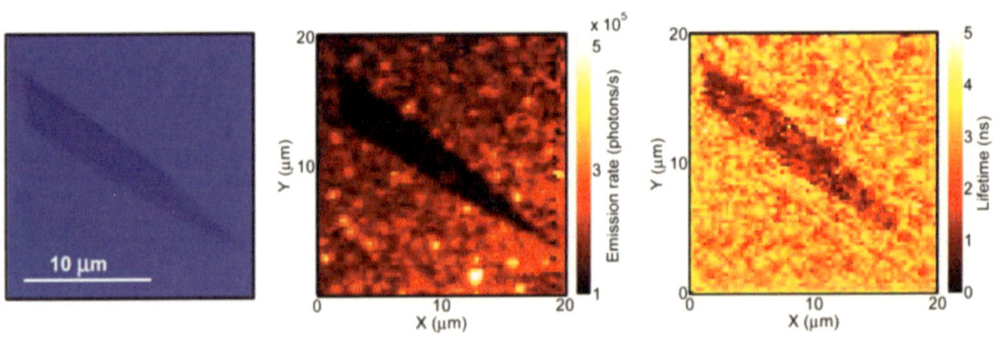

Figure 3. Images of the same area of the sample containing a monolayer graphene sheet (darker field on the left image) obtained with optical microscope, fluorescence microscope, where the emission intensity of Rhodamine dyes was measured, and fluorescence lifetime imaging microscope, where decay times of fluorescence were measured. Significant decrease of the intensity of Rhodamine on graphene is accompanied with drastic reduction of the decay time. From Gaudreau et al. [33]. Reprinted with permission from ACS.

One of the first hybrid structures where the energy transfer to graphene was investigated comprised a layer of Rhodamine molecules on a graphene flake [33], as shown in **Figure 3**. The position and shape of the graphene flake can be determined by optical microscopy and next by collecting a fluorescence image of the same area, any influence of the presence of graphene on the fluorescence properties of Rhodamine can be extracted. In this work [33], the authors included also time-resolved fluorescence lifetime-imaging microscopy. In order to study the dependence of the energy transfer on the distance between the emitters and graphene, a layer of PMMA polymer with a thickness from 5 to 20 nm was deposited on graphene. As can be seen in **Figure 3**, for molecules placed on graphene the fluorescence intensity is significantly less as compared to the reference. This decrease of emission intensity is accompanied with strong reduction of fluorescence lifetime. Both these spectral signatures are indication of the energy transfer from Rhodamine to graphene. Estimated maximum efficiency of the energy transfer was equal to 99% and it decreased with the spacer thickness. It was experimentally proved that the energy transfer to graphene scales with the distance as d^{-4}, which is different from the classic case of two interacting dipole moments. This difference is indicative of the universal value of the optical conductivity (assigned to gapless and 2D lossy media) and is in agreement with theoretical results obtained by Sebastian and Swathi [34]. There are two major consequences of these results: high efficiencies of the energy transfer indicate that in order to study energy transfer from a dye to graphene it might be critical to use a spacer, as otherwise fluorescence of the emitters can be totally quenched. There are, however, examples, where particular orientation of the molecules on a graphene surface partially inhibited the energy transfer [35]. Furthermore, weaker relation between the decay rate and distance can make it possible to investigate donor-acceptor interactions for distances unavailable for pairs of two classical dipoles.

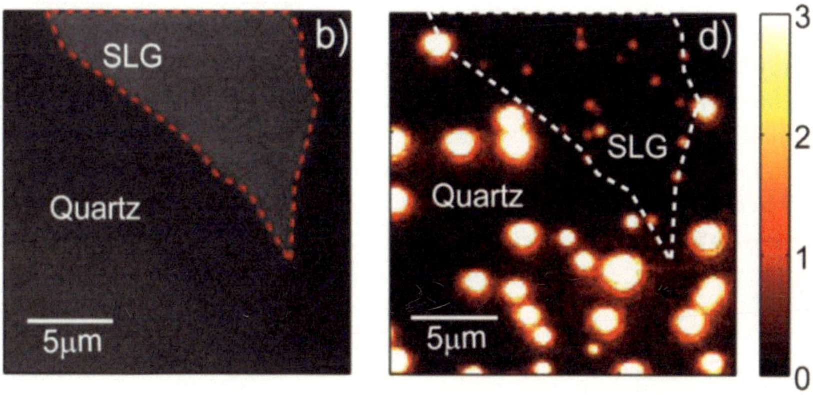

Figure 4. Optical and fluorescence images of individual nanocrystals on single-layer graphene and on the quartz substrate; (b) optical reflectivity image in the emission range of nanocrystals; (d) wide-field fluorescence image of individual CdSe-ZnS nanocrystals in the region shown in panel (b). The colour scale bar indicates the number of emitted photons (in arbitrary units) integrated over 30 s. From Chen et al. [36]. Reprinted with permission from ACS.

An important advancement into elucidating the energy transfer to graphene was experimental observation of fluorescence quenching of individual semiconductor nanocrystals [36]. For this experiment, micrometer-size sheets of graphene monolayer were used, on which highly diluted solution of CdSe/ZnS nanocrystals was deposited (**Figure 4**). By combining optical and fluorescence imaging, it was possible to correlate the differences in fluorescence intensity of individual nanocrystals with the locations where graphene was present. The result was dramatic (70-fold) quenching of fluorescence intensity for nanocrystals placed on graphene. Moreover, the quenching efficiency was found to increase with a number of graphene layers. This observation was explained using a simple model of a few-layer graphene, in which weak interactions between layers can be neglected, so the quenching factor is calculated for *n* layers of non-interacting single graphene planes. The significance of these results is in removing any ensemble averaging that is always present in experiments performed on layers, as the one described above. In addition, when layers of emitters are studied, there is always a certain thickness of this layer, across which the strength of the interaction, in this case the energy transfer, varies, sometimes considerably. As a result, an average effect is measured in fluorescence microscopy experiment, similarly as in the case of plasmonic interactions associated with metallic nanoparticles [37]. In this context, the value of single nanocrystal experiment manifests itself in the observation of suppression of fluctuations of fluorescence intensity of single nanocrystals upon deposition on graphene. This indicates the effect of graphene-induced fluorescence quenching also on the photophysics of the nanocrystals.

Very recently, both concepts were combined in a single experiment, where distance dependence of the energy transfer rate to a monolayer graphene was studied for individual emitters, both zero-dimensional CdSe/CdS nanocrystals and two-dimensional CdSe/CdS/ZnS nanoplatelets [38]. Both types of energy donors were separated from graphene with ultrasmooth dielectric film of magnesium oxide with a thickness varied from just a few Å up to several tens of nanometers. In terms of radiative energy transfer efficiency (>95%), both structures exhibited similar behaviour upon direct coupling to graphene. Important differences appear when emitters are separated from graphene with a spacer. While for zero-dimensional nanocrystals the energy transfer rate scales with a distance according to d^{-4} law, the energy transfer rate of the two-dimensional platelet decays is affected to a lesser degree. This is explained by a theoretical model, which includes the contribution of thermal distribution of free excitons in a two-dimensional quantum well at finite temperatures. The results confirm that graphene-nanocrystal hybrid structure, governed by both charge transfer and Förster-type resonant energy transfer, is a suitable system to explore the influence of exciton dimensionality and localization, as well as distance, on the energy transfer rate (**Figure 5**).

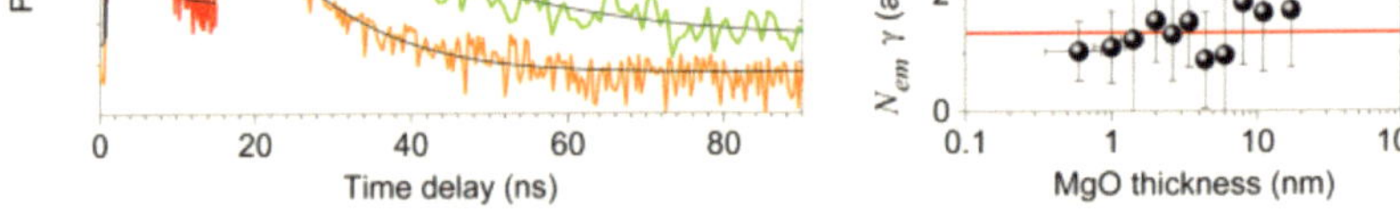

Figure 5. (a) Selected luminescence decays of individual CdSe/CdS nanocrystals separated from graphene by an MgO spacer with increasing thickness. The thin black lines are fits based on biexponential decays convoluted with the instrument response function. (b) Statistically averaged measured decay rate γ as a function of the thickness of the MgO spacer. (c) Statistically averaged product of the number of emitted photons per exciting laser pulse N_{em} and the decay rate γ. From Federspiel et al. [38]. Reprinted with permission from ACS.

Very efficient energy transfer from organic dyes to graphene-based materials has also been used to visualize the structure and determine the morphology of graphene-dye hybrids.

Fluorescence quenching microscopy is a practical tool for detecting and mapping of graphene flakes [39]. Contrary to Raman spectroscopy or optical techniques, in this case an additional component of energy donor is necessary. This is compensated by significantly increased contrast, as well as faster and more sensitive data recording of large areas compared to optical imaging (**Figure 6**) and Raman spectroscopy.

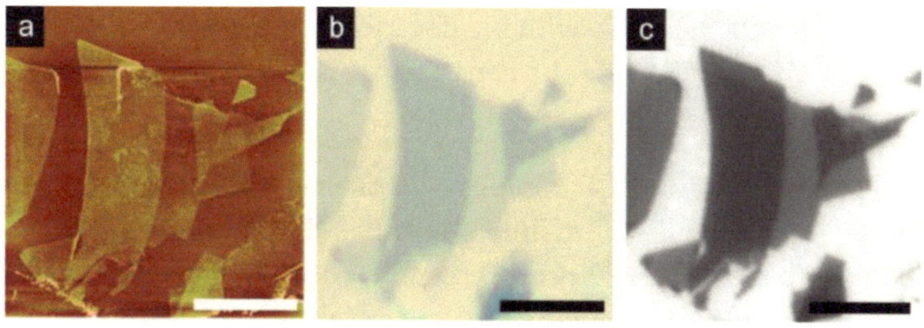

Figure 6. Images of mechanically exfoliated graphene on a SiO2/Si substrate taken by (a) AFM, (b) optical microscopy and (c) FQM using PVP/fluorescein. All scale bars = 10 μm. From Kim et al. [39]. Reprinted with permission from ACS.

Similarities between graphene and rGO are also reflected in their role as energy acceptors. However, in contrast to graphene, rGO is used most frequently as a fluorescence quencher in solutions, instead of in layer-by-layer geometries. This method, although less ordered and less controllable, may be advantageous for increasing energy/charge transfer efficiency in rGO-based assemblies. Reduced graphene oxide has been studied as an efficient fluorescence quencher of polymers [40, 41], quantum dots [42], dye-labelled aptamer [43] and also in hybrid nanostructures involving photosynthetic complexes [44]. While graphene-based hybrid structures are applied primarily for fundamental studies and to define parameters that determine energy/charge transfer, in the case of rGO composites the main focus is on potential devices and applications and optimization of their performance. Such hybrid nanostructures are considered promising for easy and relatively cheap scalable mass production of biosensors, as well as light-harvesting and optoelectronic platforms.

3. Materials and methods

In this section, we describe the structure and properties of peridinin-chlorophyll-protein (PCP), a light-harvesting pigment-protein complex, as well as graphene-based materials that are energy acceptors in our hybrid assemblies. Next, we present experimental techniques employed for investigating the energy transfer, which include—in addition to standard absorption and fluorescence spectroscopy—high-resolution confocal fluorescence microscopy coupled with time-resolved capability and spectrally resolved detection.

3.1. Peridinin-chlorophyll-protein

Peridinin-chlorophyllprotein complex from algae Dinoflagellates *Amphidinium carterae* belongs to photosynthetic complexes that are responsible for light harvesting and transferring the energy to reaction centres [2]. We focus here on aspects relevant for understanding the energy transfer between pigments comprising the PCP complex and graphene/rGO.

PCP is a water-soluble protein functioning as an antenna external to the membrane. The structure of the PCP complex, shown in **Figure 7**, has been determined with 1.3-Å resolution using X-ray crystallography [45]. In its native form, PCP consists of two chlorophyll *a* (Chl *a*) and eight peridinin (Per) molecules embedded in a protein matrix. The pigments are arranged in two almost similar clusters, with the distance between Mg atoms of the two Chl *a* in one monomer being 17.4 Å. The ratio of Per to Chl *a* of 4:1 indicates that PCP utilizes the carotenoids as its main light-harvesting pigments.

The absorption spectrum of the PCP complex, displayed in **Figure 7**, features an intense, broadband from 400 to 550 nm that is mainly due to Per absorption. The contribution of Chl *a* appears at 440 (Soret band) and 660 nm (Q_Y band). The fluorescence emission of the PCP complex originates from weakly coupled Chl molecules and it appears at 673 nm with a 30% quantum yield and a decay time constant of 4 ns, as shown by red line in **Figure 7**. Upon absorption of light, peridinins in PCP transfer their electronic excitation to Chl *a* molecules. The efficiency of this excitation energy transfer is higher than 90% as evidenced by almost ideal correspondence between absorption and fluorescence excitation spectrum. Clearly, the absorption spectrum of PCP enables the photosynthetic apparatus to harness the sunlight not only in the red spectral range but extends into the blue-green spectral region. From the point of view of the experiments described in this chapter, it is important to consider the PCP complex as a donor that can be excited at essentially any energy from 350 to 650 nm, with this excitation yielding emission at the same wavelength of 673 nm. This property distinguishes PCP, and many other photosynthetic complexes, from frequently used emitters, such as organic molecules or semiconductor quantum dots, that are much more selective in their optical characteristics.

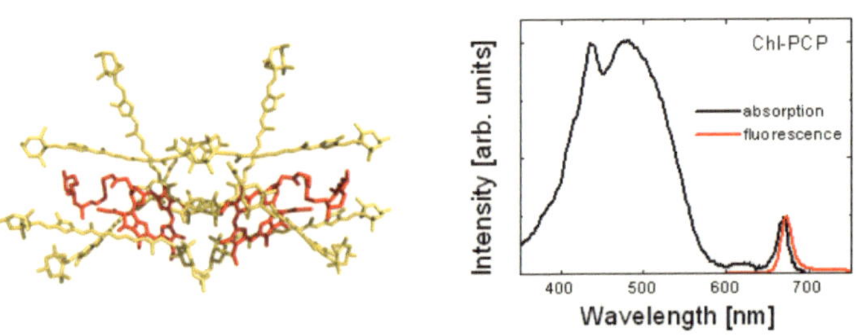

Figure 7. Pigment structure of the PCP complex together with absorption (black line) and fluorescence (red line) measured in aqueous solution at room temperature.

Previous studies of PCP complexes have been carried out on the ensemble [46, 47] and single-molecule levels [48, 49]. Transient absorption in femtosecond timescale revealed main energy transfer pathways between pigments comprising the complex, and it also was demonstrated that the interaction between the two Chl *a* molecules is relatively weak with transfer times of the order of 10 ps [47]. These findings were also corroborated with fluorescence studies of individual PCP complexes: it has been shown that it is possible to distinguish emission originating from each of the two Chl *a* molecules and using the property of sequential photobleaching of the Chl, the energy splitting between the two molecules in the monomer was determined [49].

The simplicity of the PCP complex, its water solubility that facilitates easy sample fabrication, its small size (~4 nm) and unique spectral properties have rendered this complex as a model system for fabricating hybrid nanostructures for studying interactions at the nanoscale [3]. These include in particular extensive work focused on plasmon-induced effects associated with interactions between pigments comprising the PCP complex and metallic nanostructures [37].

3.2. Graphene-based materials

Graphene oxide was synthesized from graphite powder using the modified Hummers and Offeman method described elsewhere [50, 51]. Reduced graphene oxide was prepared from graphene oxide by reduction with hydrazine. In our procedure, graphene oxide powder (2.5 mg) was dispersed in 5 ml of distilled water and placed in an ultrasound bath for 30 min. In a separate vial, 1.55 μl of 65% hydrazine monohydrate solution was added to 1 ml of distilled water. Then, 0.5 ml of the prepared hydrazine solution was added to 0.5 mg/ml-graphene oxide solution. Finally, the mixture was transferred into a round-bottomed flask, put in an oil bath, heated up and maintained at 100°C for 24 h. After this time, a clear brown solution of graphene oxide turned into black precipitate of reduced graphene oxide flakes. The final solution was washed five times with water and ethanol, and then filtered. The remaining reduced graphene oxide flakes were dried, dissolved in distilled water and left in an ultrasound bath for 1 h before further use. As estimated from XPS measurements, C/O = 7–10 and 1.7–2 ratios were measured for rGO and GO, respectively, pointing towards substantial reduction efficiency of the synthesis procedure [52]. Afterwards, rGO flakes were dispersed in distilled water, in an ultrasound bath.

Graphene substrates were purchased from Graphene Supermarket. We used 1 × 1-cm p-doped silicon wafers with a single-layer graphene deposited using chemical vapour deposition on a 285-nm thick silicon dioxide layer. The presence of a graphene monolayer on the substrates was confirmed using Raman spectroscopy.

3.3. Sample preparation

In order to study interactions between PCP and rGO, we prepared three solutions of rGO in water, one with the initial concentration of C_0 = 0.5 mg/ml, and two dilutions, 1:10 C_0 and 1:100 C_0. To prepare the samples, we mixed PCP complexes in 2% polyvinyl alcohol (PVA) with these three rGO solutions in a 1:1 ratio. The final PCP concentration in each sample was 0.2 μg/ml. In order to compare the results obtained for the rGO-containing samples, we also prepared a reference sample, where PCP and PVA concentrations were the same as above and with rGO replaced by distilled water. The layers were obtained by spin-coating solutions on pure coverslips with the rotational speed of 1200 rpm for 2 min.

For optical experiments focused on studying excitation energy dependence of the energy transfer efficiency, we used highly diluted (optical density of 0.009 at 671 nm, concentration less than 10 μM) aqueous solution of PCP complexes. Such a low concentration is very important as on the one hand it strongly reduces the inner filter effect, but this also yields a thin layer of PCP complexes on a graphene surface. As a result, we minimize the fraction of PCP that is not coupled to graphene, thus takes no part in the energy transfer.

Finally, we fabricated structures for the evaluation of the effect of polymer layer (in our case PVA) on both interaction with graphene and photostability. To this end, samples were fabricated with the concentration of PVA varying between 0.2 and 0.002%. The obtained solutions were either drop-casted or spin-coated on single-layer graphene substrates. In the case of the latter approach, the concentration of PCP had to be adjusted to be slightly higher, as spin coating strongly reduces the number of PCP complexes within the focal volume of the focused laser.

3.4. Experimental techniques

The optical properties of hybrid nanostructures comprising light-harvesting complexes and graphene-based materials were studied using absorption and fluorescence spectroscopy in the visible spectral region. Absorption spectra were obtained using Cary 50 spectrophotometer, while fluorescence in solution was measured using Fluorolog 3 spectrofluorometer. A Xenon lamp with a double grating monochromator was used for excitation and the signal was detected with a thermoelectrically cooled photomultiplier tube characterized by a dark current of less than 100 cps.

Fluorescence intensity maps were measured with an inverted fluorescence wide-field Nikon Eclipse Ti-U microscope equipped with an Andor iXon Du-888 EMCCD detector. For each sample, a series of 50 images were acquired in order to allow for reliable statistics. Every image was collected for a different sample area, which allows for minimization of any photobleaching of the PCP fluorescence. Immersion objective with a magnification of 100× (Plan Apo, Nikon) and a numerical aperture of 1.4 was used, which provides a spatial resolution of about 300 nm. As a light source, we used LED illuminators (405, 480 and 530 nm) equipped with appropriate bandpass filters. Excitation power was equal to 50 μW. Fluorescence of PCP was extracted by combining a dichroic mirror (Chroma T650lxpr) and a bandpass filter (Thorlabs FB 670-10). Fluorescence intensity maps and kinetics were collected with the electron multiplying gain of 300× and acquisition times of 0.25 or 0.5 s, depending on the experimental conditions. White-light transmission images were recorded with the same microscope, with a halogen lamp V2-A LL (12 V, 100 W) as a light source.

Spectrally and time-resolved fluorescence measurements were performed using a home-built confocal fluorescence microscope described in detail in [53]. The sample was placed on a piezoelectric translation stage. We used pulsed laser excitation at 405, 485 and 640 nm (repetition rate of 20 MHz, average power of 30 μW, power density of ~1MW/cm^{-2}). Importantly, PCP can be efficiently excited at 405 (Soret band), at 485 (Per) and at 640 nm (excited states of chlorophylls). The laser beam was focused on the sample by LMPlan 50× objective (Olympus) with a numerical aperture of 0.5. Fluorescence was first filtered by a longpass

http://dx.doi.org/10.5772/64300

filter (HQ665LP Chroma) and then the spectra were detected using Andor iDus DV 420A-BV CCD camera coupled to an Amici prism. Time-resolved measurements were performed by time-correlated single-photon-counting technique using an SPC-150 module (Becker & Hickl) with fast avalanche photodiode (idQuantique id100-20) as a detector. In order to select appropriate wavelength range, we used an additional bandpass filter (FB670/10 Thorlabs). Time resolution of the TCSPC set-up is about 100 ps.

4. Results and discussion

In this section, we describe three experiments, where interactions between the PCP complexes and both rGO and epitaxial graphene were investigated. Since the PCP complexes are soluble in water, we mixed them with rGO and investigated for the energy transfer between the light-harvesting complexes and rGO. In the case of epitaxial graphene, we focus on two aspects: the dependence of the energy transfer efficiency from PCP complexes to graphene on the excitation wavelength and the influence of sample preparation on the strength of the interactions in such hybrid nanostructures.

4.1. PCP with reduced graphene oxide

Interactions between various emitters and rGO were studied so far only in solution, where ensemble averaging can smear out subtle effects associated with the interaction between emitters and rGO. Our idea was to prepare mixtures with controlled concentration of both PCP and rGO, deposit solutions on glass coverslips and image fluorescence with high spatial resolution and high sensitivity [26]. In all experiments, the concentration of PCP was maintained constant.

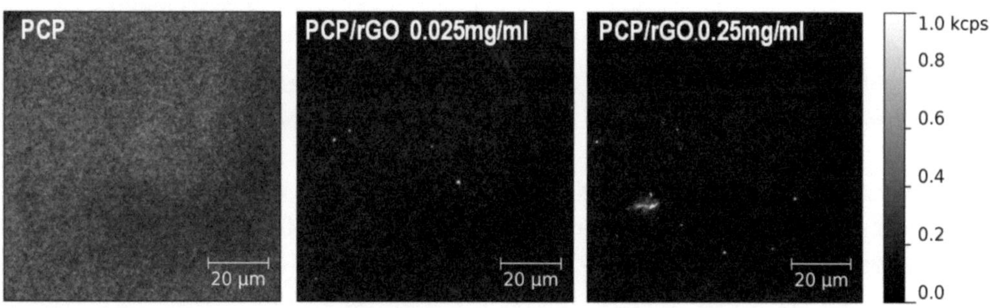

Figure 8. Fluorescence maps of (a) PCP-only reference sample and PCP/rGO mixture samples with varied rGO concentration. The excitation wavelength was 530 nm.

In **Figure 8**, we show a sequence of fluorescence maps collected for PCP complexes mixed with varied amount of rGO. The concentration of rGO was varied by orders of magnitude, so that the influence can be seen in a pronounced way. The excitation wavelength was 530 nm, but the results are qualitatively identical for the other two excitation wavelengths used in the experiment. Incorporation of rGO induces substantial changes in fluorescence images of PCP: while for the PCP-only sample the distribution of intensity is pretty much uniform, mixing the photosynthetic complexes with rGO leads to a pattern that features high-intensity spots on an otherwise uniform background. Importantly, the intensity of these isolated bright spots is approximately 10-fold enhanced as compared to the intensity measured for the reference sample (PCP-only). At the same time, the fluorescence intensity away from the bright spots in the PCP/rGO hybrid system is quenched compared to the reference. Furthermore, the number of the bright spots increases with rGO concentration. We can exclude any contribution to this emission from GO that might be present in our sample due to non-complete reduction, as the fluorescence of GO occurs in a spectral range between 350 and 550 nm, and the emission of the PCP complexes is strongly shifted to the red. This proves that for ensemble of PCP complexes, both in the reference and in the hybrid structures (PCP/rGO) studied in this work, there is no other contribution to the measured signal.

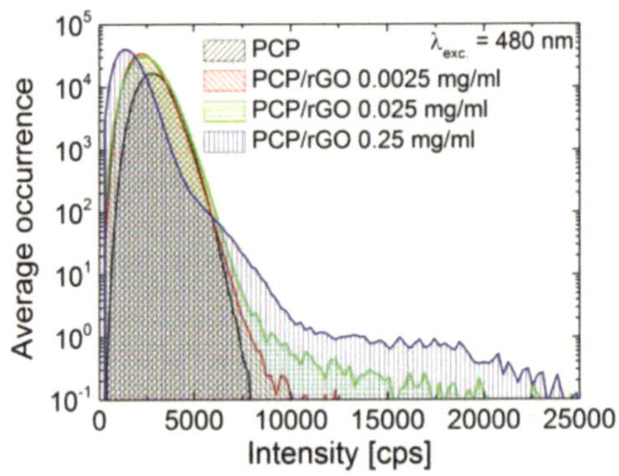

Figure 9. Histograms for PCP and PCP mixed with rGO with indicated concentrations obtained for average fluorescence intensities calculated from 50 fluorescence intensity maps for each rGO concentration. The excitation wavelength of 480 nm was used.

Statistical information about observed effects is obtained by collecting a series of more than 50 images for each sample configuration and for every excitation wavelength. In this way, we strongly reduce any possible influence of particular sample preparation or a way the experiment was carried out. Next, the fluorescence images are analysed by plotting a histogram of all the intensities measured [26]. This procedure can be applied for any single map, but also to all the maps measured for a given excitation wavelength and rGO concentration. The result of this procedure carried out of the excitation wavelength of 480 nm is displayed in **Figure 9**. The distribution of fluorescence intensity measured for the reference sample, containing only PCP complexes, features Gaussian shape, indicative of statistical distribution of concentration variation across the substrate. By contrast, the results extracted for PCP complexes mixed with rGO are more complicated. Although the majority of the intensity distribution can be approximated with Gaussian shape, similarly as in the case of the reference sample, the maximum of this distribution shifts towards lower values with increasing rGO concentration. In addition, we find considerable contribution of isolated spots spread out randomly

http://dx.doi.org/10.5772/64300

across the images, as shown in **Figure 8**, and this contribution is larger with increasing concentration of rGO in the sample. Not only the shift to lower emission intensities with incorporating rGO to the mixture as well as emergence of high-intensity spots in fluorescence images is systematic in nature, but both effects exhibit monotonic dependence on the rGO concentration in the initial solution. Indeed, for the highest rGO concentration (0.25 mg/ml), we find the largest number of bright fluorescence spots, and furthermore they exhibit the highest intensity. The results of fluorescence imaging of PCP/rGO mixtures strongly suggest that the incorporation of rGO yields both quenching of emission and formation of strongly enhanced spatially localized emitting sites. The results of fluorescence imaging show thus that the interaction between rGO and photosynthetic complexes of PCP is more complex than as discussed in a simple image of fluorescence quenching.

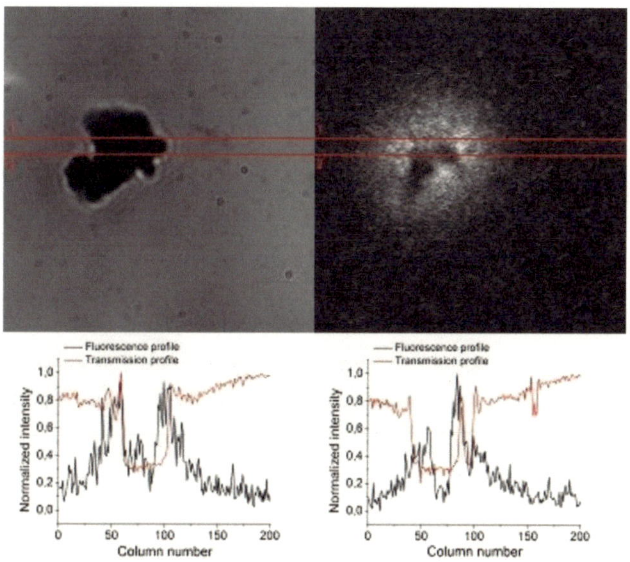

Figure 10. (Upper row) White-light transmission image of a graphene aggregate and intensity map of PCP fluorescence measured for 480-nm excitation for exactly the same location. (Lower row) Intensity profiles extracted along the cross ections marked with red lines in both images.

In order to get some insight into the possible origin of this bimodal behaviour, we also imaged a large rGO aggregate, as shown in **Figure 10**. The correspondence between transmission image and fluorescence image indicates that the same object is probed in both experiments. Even without any detailed analysis, the comparison between the two images suggests that enhanced emission occurs for PCP complexes at the edges of the flake and perhaps on its thin sections. By contrast, when the thickness of the aggregate is large, the fluorescence of PCP is efficiently quenched.

The relation between transmission and fluorescence images can be quantified using for instance Pearson correlation coefficients calculated for the two transmission-emission pair cross sections marked with red lines in **Figure 10**. For one line, we obtain a negative coefficient of −0.55, meaning that the high intensity of fluorescence correlates with dips in transmission images. A contrasting effect can be seen for the other line, where the coefficient of approximately 0.5 was obtained. Therefore, in this case, low transmission combines with quenched fluorescence. Both correlation and anti-correlation between profiles obtained for transmission and fluorescence images can be readily seen in the panels in **Figure 10**, where corresponding cross sections are plotted. These results may suggest that when a region outside the flake is considered, the correlation between emission and transmission is low. For a thin rGO flake, we find negative correlation, which could imply that PCP complexes placed at the edges of rGO experience fluorescence enhancement. Finally, for the thickest part of the flake, where both transmission and fluorescence exhibit strong decrease, the quenching is the dominant effect.

The results of fluorescence imaging of PCP/rGO hybrid nanostructures show bimodal character of the interactions. On the one hand, the fluorescence of the majority of PCP complexes is quenched; however, there are numerous localized spots characterized with considerably higher intensity. This effect depends on the rGO concentration, in particular the number of these bright spots increases with rGO concentration. While the quenching of fluorescence was observed for many graphene-based hybrid nanostructures, the enhancement is far less frequent. The exact mechanism of fluorescence enhancement is not clear, but the results show unambiguously the complexity of interactions in graphene-based photosynthetic hybrid nanostructures. Future work, which includes spectrally and temporally resolved studies, as well as experiments on large rGO flakes are in order to answer some of these questions.

4.2. Energy transfer from PCP to graphene: excitation wavelength dependence

The broad absorption of PCP complexes allows for investigating the dependence of the energy transfer efficiency on the excitation wavelength [54]. The rationale behind this experiment is that as graphene is a conductor, and contains high concentration of free carriers, it should be possible to affect the behaviour of these electrons locally using focused laser excitation. At the same time, with PCP as a donor, it is still possible to excite emission and probe the energy transfer dynamics.

In this experiment, we used epitaxial monolayer graphene transferred on 285-nm SiO_2 substrate. The optical properties of such a hybrid nanostructure were probed by time-resolved fluorescence microscopy with three excitation wavelengths of 405, 485 and 640 nm. All these wavelengths excite the emission of the PCP complexes, either via direct excitation of chlorophyll molecules or via intra-complex energy transfer from Per molecules.

In **Figure 11**, we compare fluorescence spectra measured for PCP complexes deposited on graphene with the reference. The excitation wavelength was 405 nm. This result shows substantial decrease of fluorescence intensity upon deposition of PCP on graphene, which can be tentatively attributed to the energy transfer. The decrease of fluorescence intensity is the strongest for the 405-nm excitation and the weakest for the 640-nm excitation, which again suggests that there is indeed a dependence of the energy transfer efficiency on the excitation wavelength. However, comparison of bare intensities of emission could be misleading as the

http://dx.doi.org/10.5772/64300

values might depend on many factors that can be sometimes difficult to control experimentally. However, we note that the shape of the PCP emission spectrum on graphene remains unaffected, and is identical to previously published [37], which indicates that depositing PCP on graphene leaves no measureable effect on the energy transfer pathways within the PCP complexes.

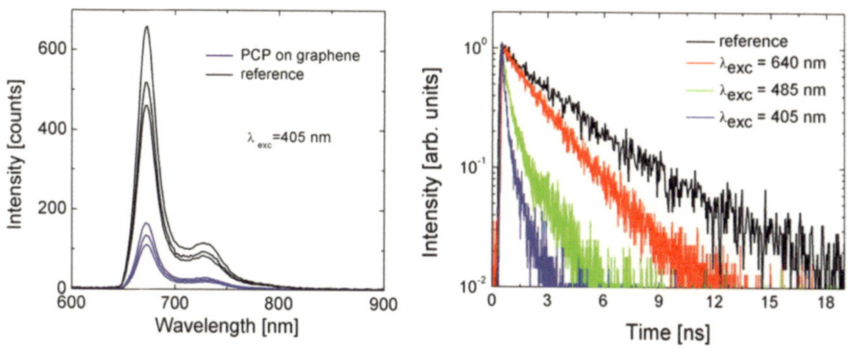

Figure 11. (Left) Fluorescence spectra of the PCP complexes deposited on graphene (blue) and on a glass substrate (black). The excitation wavelength was 405 nm. (Right) Comparison between average fluorescence transients measured for three excitation wavelengths, as indicated.

The initial assignment of the observed reduction of the emission intensity of PCP complexes deposited on graphene to the energy transfer from the chlorophylls in the photosynthetic complex to graphene is confirmed by time-resolved fluorescence spectroscopy. In **Figure 11**, we display average fluorescence transients measured for the excitation wavelengths of 405, 485 and 640 nm. The results are compared with the decay obtained for PCP complexes deposited on glass, and it has been checked that the obtained transient very weakly changes with the excitation wavelength in this case. As described in detail in [54], we find some degree of variation of fluorescence decays measured for a given excitation wavelength. It is expected, as in this experiment, that we do not control the separation between graphene and PCP complexes, and in turn introduce inhomogeneity of the interaction between the two structures across the substrate. Furthermore, it has been shown that for graphene deposited on silica, the local structure of graphene is also quite inhomogeneous with islands of high and low mobility of carriers [55]. We might therefore assume that such non-uniformity contributes to some degree to the observed spreading of fluorescence transients, although the scale of these inhomogeneities is less than 100 nm, as compared to the resolution of our microscope of about 1 μm.

Nevertheless, for both excitation wavelengths (and for 485-nm excitation as well), we observe significant shortening of fluorescence lifetime for PCP complexes deposited on graphene as compared with the reference. In addition, the 405-nm excitation yields very fast decays, while exciting with 640 nm results in considerably longer decays, regardless of the inhomogeneity of the data. It is also striking that most of them exhibit almost monoexponential behaviour, which could suggest that majority of PCP complexes within the laser spot couples to graphene with a comparable strength. In no case, however, for PCP deposited on graphene we observe long (~4-ns) decay component attributable to PCP complexes isolated from graphene.

The key conclusion from these experiments is that the energy transfer to graphene depends on the excitation wavelength. Indeed, shortening of the fluorescence decay, accompanied with a decrease of the overall fluorescence intensity, observed for all used excitation wavelengths (405, 485 and 640 nm) strongly suggests that the energy absorbed by PCP complexes is efficiently dissipated into the graphene layer. Furthermore, much shorter fluorescence decay times measured for 405-nm excitation prove that the energy transfer for this excitation wavelength is more efficient compared to 640-nm excitation. The average decay times are equal to 0.5 and 1.4 ns, respectively, what translates to the energy transfer efficiencies of 87 and 65%. This is the first experimental observation of such an effect, which distinguishes graphene as a totally unique acceptor of energy in such hybrid assemblies. The qualitative picture displaying this fact is shown in **Figure 12**.

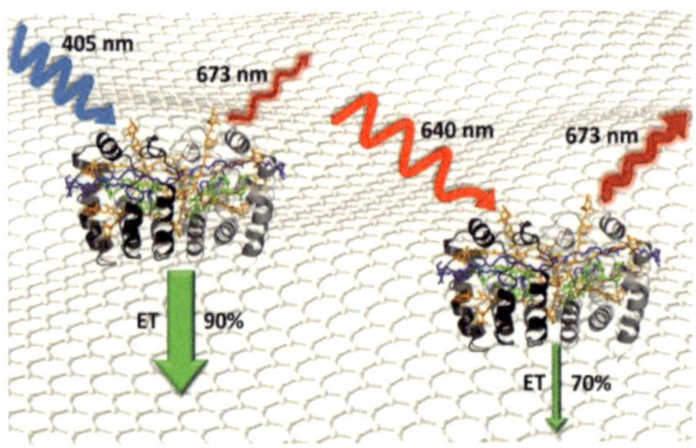

Figure 12. Schematics showing the effect of the excitation wavelength on the efficiency of the energy transfer between PCP complexes and graphene.

For molecular systems, where the energy transfer takes place between two dipole moments, the decay of a donor is independent of the excitation wavelength. This is a reminiscence of the fact that light has no effect on the surrounding of the molecules participating in the energy transfer. Apparently, for PCP complexes on graphene the situation is different. Clear influence of the excitation wavelength on the energy transfer indicates that in addition to populating PCP-excited states, laser changes also the local properties of graphene. A scenario that can explain this effect relies on the fact that focused laser excited electrons in graphene in a similar way as in metallic nanoparticle, forcing them to oscillate in a confined space defined by a monolayer of graphene on the one hand, and the size of the laser spot on the other. As a consequence, electronic excitations in chlorophylls in PCP can see graphene as a metallic nanoparticle with specific character that can influence energy dissipation.

Based on these results, we conclude that energy quenching in graphene is driven not only by dipole-dipole interaction but also by a mechanism associated with light-induced oscillations of electrons in graphene. Indeed, exciting electrons in graphene has an effect of its dissipative efficiency, which opens avenues in interfacing electronic and plasmonic character of

http://dx.doi.org/10.5772/64300

graphene in hybrid nanostructures and controls the electronic dynamics of such systems with light.

4.3. Energy transfer from PCP to graphene: influence of sample structure

An important aspect, frequently overlooked, in discussing the optical properties of hybrid nanostructures, where the interactions are distance-dependent, is the design and fabrication of layered structures. In the case of the experiment where we studied the influence of excitation energy on the efficiency of the energy transfer from PCP complexes to graphene, we used drop-casting to deposit the PCP solution on a graphene monolayer. Since it was aqueous solution, once water evaporated, PCP complexes are assumed to fall down on the graphene surface. This method, while allowing for making structures with rather well-defined geometry, results in the pigment-protein complexes being fully exposed to ambient conditions. This in turn speeds up degradation of the pigments, and the complexes as well. One of the most common ways to increase their protection against oxygen is to embed them in polymer matrix [56].

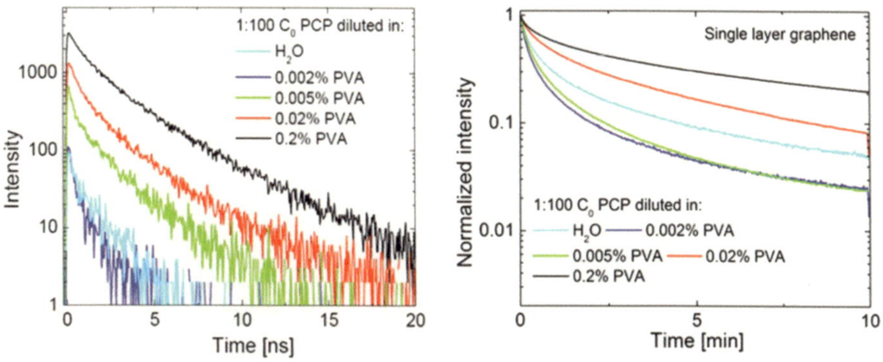

Figure 13. (Left) Fluorescence intensity decays of PCP deposited on graphene, diluted beforehand in water or in PVA aqueous solutions with varied concentrations. (Right) Corresponding kinetic curves of integrated emission intensity spectra measured in 10 min with acquisition time of 1 s. The excitation wavelength was 485 nm.

In **Figure 13**, we compare fluorescence decay curves measured for PCP complexes in water and in PVA matrix with varied concentration of the polymer, for the samples drop-casted on a single-layer graphene. The lowest concentration of PVA gives almost identical result as pure aqueous solution, in other words, the fluorescence decay is substantially shorter than the reference. This, in addition with almost monoexponential character of the decay, indicates that almost all of the PCP complexes interact with graphene, which means that the energy is transferred from chlorophylls to graphene. As the PVA concentration increases, the character of the decay changes dramatically. It is no longer monoexponential, and it features a long decay tail, reminiscent of the decay characteristic for PCP complexes that are uncoupled to graphene. This result can be understood in terms of a thicker-layer formation for a solution with higher content of PVA polymer. Such a scenario would then lead to comparatively smaller fraction of the PCP complexes that couple with graphene, most of them would be too far away from the graphene layer to experience its presence at all. This interpretation is strongly supported by the increase of total fluorescence intensity observed with increasing concentration of PVA in solution. In fact, this increase is over an order of magnitude, although the amount of the solution deposited on a substrate is in all cases identical.

There is also another important aspect that must be considered when fabricating graphene-based hybrid nanostructures. An alternative approach to prepare layers of fluorophores on graphene substrates would be through spin coating of a solution of PCP complexes in PVA matrix. The obtained layers are much more uniform than those made by simple drop-casting, and it shows no systematic dependence on the PVA concentration. Rough estimations, based on the distribution of emission intensities measured as emission spectra or decay curves, suggest that the uniformity of the layers prepared with spin coating is about a factor of two better in terms of a standard deviation, as compared to the drop-casted samples. At the same time, the concentration of PCP complexes within the focal volume of the laser would diminish considerably as compared with the drop-casting approach; thus, this parameter must be carefully adjusted.

5. Summary and conclusions

Graphene and its derivative, reduced graphene oxide, are unique energy acceptors. While not exhibiting any fluorescence, both absorb energy in the whole visible spectral region with quite uniform efficiency. As such, graphene-based materials can be considered attractive platforms for light harvesting, energy conversion and biosensing. In this chapter, we described several experimental observations obtained for hybrid nanostructures composed of natural photosynthetic complex PCP and either graphene or reduced graphene oxide. Each studied structure sheds its own light on the mechanisms and processes that are taking place in such systems. We show that by controlling the composition of the solution and sample preparation, it is possible to tune the efficiency of the energy transfer to graphene and thus determine the sensitivity of energy transfer as a probing tool for interaction with graphene. The results obtained for PCP/rGO system indicate bimodal nature of the on-going interactions: in addition to commonly observed fluorescence quenching, we find pronounced and frequent events, where the emission of the PCP complexes is substantially enhanced. Last but not least, the uniqueness of graphene as energy acceptor manifests itself in a strong dependence of the energy transfer efficiency on the excitation wavelength. This observation allows drawing a completely new picture of the excitation dynamics, and the energy transfer, in systems where the properties of either acceptors or donors can be additionally and independently controlled by light.

Acknowledgements

This research has been supported by the WELCOME project 'Hybrid Nanostructures as a Stepping Stone towards Efficient Artificial Photosynthesis' funded by the Foundation for

http://dx.doi.org/10.5772/64300

Polish Science (FNP), by project DEC-2013/10/E/ST3/00034 from the National Science Center (NCN) and by the ORGANOMET project No: PBS2/A5/40/2014 from the National Research and Development Center (NCBiR). Izabela Kamińska acknowledges support by the Mobility Plus grant 1269/MOB/IV/2015/0 from the Polish Ministry of Science and Higher Education (MNiSW) and the START stipend awarded by the Foundation for Polish Science (FNP). We also acknowledge members of the Optics of Hybrid Nanostructures Group at the Institute of Physics, Nicolaus Copernicus University in Torun, in particular Dr. Dawid Piatkowski, Magdalena Twardowska, Justyna Grzelak, Marcin Szalkowski, and Kamil Wiwatowski for their excellent work and vital contribution.

Author details

Sebastian Mackowski[1,2*] and Izabela Kamińska[1,3]

*Address all correspondence to: mackowski@fizyka.umk.pl

1 Optics of Hybrid Nanostructures Group, Faculty of Physics, Astronomy, and Informatics, Nicolaus Copernicus University, Torun, Poland

2 Baltic Institute of Technology, BalTech, Gdynia, Poland

3 NanoBioSciences, Institute of Physical and Theoretical Chemistry, TU Braunschweig, Braunschweig, Germany

References

[1] Beljonne D, Curutchet C, Scholes, GD, Silbey RJ. Beyond Förster Resonance Energy Transfer in Biological and Nanoscale Systems. Journal of Physical Chemistry B. 2009;113:6583–6599. DOI: 10.1021/jp900708f

[2] Blankenship R. Molecular Mechanisms of Photosynthesis. Wiley-Blackwell, ISBN 978-0632-04321-7, Oxford, UK. 2008; DOI: 10.1002/9780470758472

[3] Mackowski S. Hybrid Nanostructures for Efficient Light Harvesting. Journal of Physics: Condensed Matter. 2010;22:193102/1–17. DOI:10.1088/0953-8984/22/19/193102

[4] Lakowicz J. Principles of Fluorescence Spectroscopy. 3rd edition, Springer, ISBN 978-0-387-46312-4, New York City, NY, USA. 2006; DOI: 10.1007/978-0-387-46312-4

[5] Buczynska D, Bujak L, Loi MA, Brotosudarmo THP, Cogdell R, Mackowski S. Energy Transfer from Conjugated Polymer to Bacterial Light-Harvesting Complex. Applied Physics Letters. 2012;101:173703/1–4. DOI: 10.1063/1.4764082

[6] Deng X, Tang H, Jiang J. Recent Progess in Graphene-Material-Based Optical Sensors. Analytical and Bioanalytical Chemistry. 2014;406:6903–6916. DOI: 10.1007/s00216-014-7895-4

[7] Nabiev I, Rakovich A, Sukhanova A, Lukashev E, Zagidullin V, Pachenko V, Rakovich YP, Donegan JF, Rubin AB, Govorov AO. Fluorescence Quantum Dots as Artificial Antennas for Enhanced Light Harvesting and Energy Transfer to Photosynthetic Reaction Centers. Angewandte Chemie International Edition. 2010;49: 7217–7221. DOI: 10.1002/anie.201003067

[8] Zhang J, Fu Y, Chowdhury MH, Lakowicz JR. Enhanced Förster Resonance Energy Transfer on Single Metal Particle. 2. Dependence on Donor-Acceptor Separation Distance, Particle Size, and Distance from Metal Surface. Journal of Physical Chemistry C. 2007;111:11,784–11,792. DOI: 10.1021/jp067887r

[9] Lunz M, Bradley AL, Chen WY, Gerard VA, Byrne SJ, Gun'Ko YK, Lesnyak V, Gaponik N. Influence of Quantum Dot Concentration on Förster Resonant Energy Transfer in Monodispersed Nanocrystal Quantum Dot Monolayers. Physical Review B. 2010;81:205316. DOI: 10.1103/PhysRevB.81.205316

[10] Acuna GP, Moller FM, Holzmeister P, Beater S, Lalkens B, Tinnefeld P. Fluorescence Enhancement at Docking Sites of Dna-Directed Self-Assembled Nanoantennas. Science. 2002;388:506–510. DOI: 10.1126/science.1228638

[11] Lee J, Govorov AO, Kotov NA. Bioconjugates Superstructures of CdTe Nanowires and Nanoparticles: Multistep Cascade Förster Resonance Energy Transfer and Energy Channeling. Nano Letters. 2005;5:2063–2069. DOI: 10.1021/nl051042u

[12] Geim AK, Novoselov KS. The Rise of Graphene. Nature Materials. 2007;6:183–191. DOI. 10.1038/nmat1849

[13] Novoselov KS, Fal'ko VI, Colombo L, Gellert PR, Schwab MG, Kim K. A Roadmap for Graphene. Nature. 2012;490:192–200. DOI: 10.1038/nature11458

[14] Zhu BY, Murali S, Cai W, Li X, Suk JW, Potts JR, Ruoff RS. Graphene and Graphene Oxide: Synthesis, Properties, and Applications. Advanced Materials. 2010;22:3906–3924. DOI: 10.1002/adma.201001068

[15] Castro Neto AH, Guinea F, Peres NMR, Novoselov KS, Geim AK. The Electronic Properties of Graphene. Reviews of Modern Physics. 2009;81:109–162. DOI: 10.1103/RevModPhys.81.109

[16] Nair RR, Blake P, Grigorenko AN, Novoselov KS, Booth TJ, Stauber T, Peres NMR, Geim AK. Fine Structure Constant Defines Visual Transparency of Graphene. Science. 2008;320:1308. DOI: 10.1126/science.1156965

[17] Edwards RS. Coleman KS. Graphene Synthesis: Relationship to Applications, Nanoscale. 2012;5:38–51. DOI: 10.1039/C2NR32629A

[18] Novoselov KS, Geim AK, Morozov SV, Jiang D, Katsnelson MI, Grigorieva IV, Dubonos SV, Firsov AA. Two-Dimensional Gas of Massless Dirac Fermions in Graphene. Nature. 2005;438:197–200. DOI: 10.1038/nature04233

[19] Norimatsu W, Kusunoki M. Epitaxial Graphene on SiC{0001}: Advances and Perspectives. Physical Chemistry Chemical Physics. 2013;16:3501–3511. DOI: 10.1039/C3CP54523G

[20] Zhang Y, Zhang L, Zhou C. Review of Chemical Vapor Deposition of Graphene and Related Applications. Accounts of Chemical Research. 2012;46:2329–2339. DOI: 10.1021/ar300203n

[21] Voloshina E, Dedkov Y. Graphene on Metallic Surfaces: Problems and Perspectives. Physical Chemistry Chemical Physics. 2012;12:13,502–13,514. DOI: 10.1039/C2CP42171B

[22] Hallam T, Berner NC, Yim C, Duesberg GS. Strain, Bubbles, Dirt, and Folds: A Study of Graphene Polymer-Assisted Transfer. Advanced Materials Interfaces. 2014;1:1400115/1–7. DOI: 10.1002/admi.201400115

[23] Tung VC, Allen MJ, Yang Y, Kaner RB. High-Throughput Solution Processing of Large-Scale Graphene. Nature Nanotechnology. 2009;4:25–29. DOI: 10.1038/nnano.2008.329

[24] Chua CK, Pumera M. Chemical Reduction of Graphene Oxide: A Synthetic Chemistry Viewpoint. Chemical Society Reviews. 2014;43:291–312. DOI: 10.1039/C3CS60303B

[25] Dreyer DR, Park S, Bielawski CW, Ruoff RS. The Chemistry of Graphene Oxide. Chemical Society Reviews. 2010;39:228–240. DOI: 10.1039/B917103G

[26] Twardowska M, Chomicki D, Kaminska I, Niedziołka-Jönsson J, Mackowski S. Fluorescence Imaging of Hybrid Nanostructures Composed of Natural Photosynthetic Complexes and Reduced Graphene Oxide. Nanospectroscopy. 2015;1:33–39. DOI: 10.1515/nansp-2015-0002

[27] Zhao J, Pei S, Ren W, Gao L, Cheng H.-M. Efficient Preparation of Large-Area Graphene Oxide Sheet for Transparent Conductive Films. ACS Nano. 2010;4:5245–5252. DOI: 10.1021/nn1015506

[28] Jia J, Kan C.-M, Lin X, Shen X, Kim J.-K. Effects of Processing and Material Parameters on Synthesis of Monolayer Ultralarge Graphene Oxide Sheets. Carbon. 2014;77:244–254. DOI: 10.1016/j.carbon.2014.05.027

[29] Zhou X, Liu Z. A Scalable, Solution-Phase Processing Route to Graphene Oxide and Graphene Ultralarge Sheets. Chemical Communications. 2010;46:2611–2613. DOI: 10.1039/B914412A

[30] Wang X, Bai H, Shi G. Size Fractionation of Graphene Oxide Sheets by pH-Assisted Selective Sedimentation. Journal of the American Chemical Society. 2011;133:6338–6342. DOI: 10.1021/ja200218y

[31] Lin X, Shen X, Zheng Q, Yousefi N, Ye L, Mai Y.-W, Kim J.-K. Fabrication of Highly-Aligned, Conductive, and Strong Graphene Papers Using Ultralarge Graphene Oxide Sheets. ACS Nano. 2012;6:10,708–10,719. DOI: 10.1021/nn303904z

[32] Zheng Q, Shi L, Ma P.-C, Xue Q, Li J, Tang Z, Yang, W. Structure Control of Ultra-Large Graphene Oxide Sheets by the Langmuir-Blodgett Method. RSC Advances. 2013;3:4680–4691. DOI: 10.1039/C3RA22367A

[33] Gaudreau L, Tielrooij KJ, Prawiroatmodjo GEDK, Osmond J, Garcia de Abajo FJ, Koppens FHL. Universal Distance-Scaling of Nonradiative Energy Transfer to Graphene. Nano Letters. 2013;13:2030–2035. DOI: 10.1021/nl400176b

[34] Swathi RS, Sebastian KL. Long Range Resonance Energy Transfer from a Dye Molecule to Graphene has (Distance)-4 Dependence. Journal of Chemical Physics. 2009;130:086101(1–3). DOI: 10.1063/1.3077292

[35] Berner NC, Winters S, Beckes C, Yim C, Dümbgen KC, Kaminska I, Mackowski S, Cafolla AA, Hirsch A, Duesberg GS. Understanding and Optimizing the Packing Density of Perylene Bisimide Layers on CVD-Grown Graphene. Nanoscale. 2015;7:16,337–16,342. DOI: 10.1039/C5NR04772B

[36] Chen Z, Berciaud S, Nuckolls C, Heinz TF, Brus LE. Energy Transfer from Individual Semiconductor Nanocrystals to Graphene. ACS Nano. 2010;4:2964–2968. DOI: 10.1021/nn1005107

[37] Mackowski S., Metallic Nanoparticles Coupled to Photosynthetic Complexes, Smart Nanoparticles Technology , ISBN 978-953-51-0500-8, ed. A. Hashim, InTech Publishing, 2012, 3–28; DOI: 10.5772/33493

[38] Federspiel F, Froehlicher G, Nasilowski M, Pedetti S, Mahmood A, Doudin B, Park S, Lee J.-O, Halley D, Dubertret B, Gilliot P, Berciaud S. Distance Dependence of the Energy Transfer Rate from a Single Semiconductor Nanostructure to Graphene. Nano Letters. 2015;15:1252–1258. DOI: 10.1021/nl5044192

[39] Kim J, Cote LJ, Kim F, Huang J. Visualizing Graphene Based Sheets by Fluorescence Quenching Microscopy. Journal of the American Chemical Society. 2010;132:260–267. DOI: 10.1021/ja906730d

[40] Wang Y, Kurunthu D, Scott GW, Bardeen CJ. Fluorescence Quenching in Conjugated Polymers Blended with Reduced Graphene Oxide. Journal of Physical Chemistry C. 2010;114:4153–4159. DOI: 10.1021/jp9097793

[41] Ran C, Wang M, Gao W, Ding J, Shi Y, Song X, Chen H, Ren Z. Study of Photoluminescence Quenching and Photostability Enhancement of MEH-PPV by Reduced Graphene Oxide. Journal of Physical Chemistry C. 2012;116:23,053–23,060. DOI: 10.1021/jp306631y

[42] Samal M, Mohapatra P, Subbiah R, Lee C.-L, Annas B, Kim JA, Kim T, Yi DK. InP/ZnS-graphene Oxide and Reduced Graphene Oxide Nanocomposites as Fascinating

Materials for Potential Optoelectronic Applications. Nanoscale. 2013;5:9793–9805. DOI: 10.1039/C3NR02333H

[43] Zhao H, Gao S, Liu M, Chang Y, Fan X, Quan X. Fluorescent Assay for Oxytetracycline based on a Long-Chain Aptamer Assembled onto Reduced Graphene Oxide. Microchimica Acta. 2013;180:829–835. DOI: 10.1007/s00604-013-1006-7

[44] Twardowska M, Kamińska I, Wiwatowski K, Ashraf KU, Cogdell RJ, Mackowski S, Niedziółka-Jönsson J. Fluorescence Enhancement of Photosynthetic Complexes Separated from Nanoparticles by a Reduced Graphene Oxide Layer. Applied Physics Letters. 2014;104:093103/1–5. DOI: 10.1063/1.4867167

[45] Hofmann E, Wrench P, Sharples F, Hiller R, Welte W, Diederichs K. Structural Basis of Light Harvesting by Carotenoids: Peridinin-Chlorophyll-Protein from *Amphidinium carterae*. Science. 1996;272:1788–1791. DOI: 10.1126/science.272.5269.1788

[46] Akimoto S, Takaichi S, Ogata T, Nishimura Y, Yamazaki I, Mimuro M. Excitation Energy Transfer in Carotenoid Chlorophyll Protein Complexes Probed by Femtosecond Fluorescence Decays. Chemical Physics Letters. 1996;260:147–152. DOI: 10.1016/0009-2614(96)00863-9

[47] Kleima F, Hofmann E, Gobets B, van Stokkum I, van Grondelle R, Diederich K, van Amerongen H. Förster Excitation Energy Transfer in Peridinin-Chlorophyll a-Protein. Biophysical Journal. 2000;78:344–353. DOI: 10.1016/S0006-3495(00)76597-0

[48] Mackowski S, Wörmke S, Brotosudarmo T, Jung C, Hiller R, Scheer H, Bräuchle C. Energy Transfer in Reconstituted Peridinin-Chlorophyll-Protein Complexes: Ensemble and Single Molecule Spectroscopy Studies. Biophysical Journal. 2007;93:3249–3258. DOI: 10.1529/biophysj.107.112094

[49] Wörmke S, Mackowski S, Jung C, Ehrl M, Zumbusch A, Brotosudarmo T, Scheer H, Hofmann E, Hiller R, Bräuchle C. Monitoring Fluorescence of Individual Chromophores in Peridinin-Chlorophyll-Protein Complex Using Single Molecule Spectroscopy. Biochimica et Biophysica Acta – Bioenergetics. 2007;1767:956–964. DOI: 10.1016/j.bbabio.2007.05.004

[50] Pei S, Cheng H.-M. The Reduction of Graphene Oxide. Carbon. 2012;50:1972–2012. DOI: 10.1016/j.carbon.2011.11.010

[51] Mao S, Pu H, Chen J. Graphene Oxide and its Reduction: Modeling and Experimental Progress. RSC Advances. 2012;2:2643–2662. DOI:10.1039/C2RA00663D

[52] Kaminska, I, Barras A, Coffinier Y, Lisowski W, Niedziołka-Jönsson J, Woisel P, Lyskawa J, Opallo M, Siriwardena A, Boukherroub R, Szunerits S. Preparation of a Responsive Carbohydrate-Coated Biointerface based on Graphene/Azido-Terminated Tetrathiafulvalene Nanohybrid Material. ACS Applied Materials & Interfaces. 2012;4:5386–5393. DOI: 10.1021/am3013196

[53] Krajnik B, Schulte T, Piatkowski D, Czechowski N, Hofmann E, Mackowski S. SIL-based Confocal Fluorescence Microscope for Investigating Individual Nanostruc-

tures. Central European Journal of Physics. 2011;9:293–299. DOI: 10.2478/s11534-010-0098-5

[54] Mackowski S, Kaminska, I. Dependence of the Energy Transfer to Graphene on the Excitation Energy. Applied Physics Letters. 2015;107:023110/1–4. DOI: 10.1063/1.4926984

[55] Burson KM, Cullen WG, Adam S, Dean CR, Watanabe K, Taniguchi T, Kim P, Fuhrer MS. Direct Imaging of Charged Impurity Density in Common Graphene Substrates. Nano Letters. 2013;13:3576–3580. DOI: 10.1021/nl4012529

[56] Wörmke S, Mackowski S, Schaller A, Brotosudarmo T, Johanning S, Scheer H, Bräuchle C. Single Molecule Fluorescence of Native and Refolded Peridinin-Chlorophyll-Protein Complexes. Journal of Fluorescence. 2008;18:611–617. DOI: 10.1007/s10895-008-0310-9

Chapter 8

Heteroatom-Doped Graphene-Based Hybrid Materials for Hydrogen Energy Conversion

Bhaghavathi Parambath Vinayan

Additional information is available at the end of the chapter

http://dx.doi.org/10.5772/64242

Abstract

Energy is becoming a big issue in the present world due to the depletion of current fossil fuels and also due to environmental problems such as global warming and rising pollution levels. Hence, highly efficient and renewable energy materials are required to produce clean electricity. The utilization of different energy sources such as solar energy and wind energy is hampered by their fluctuation in time and non-uniform geographical distribution which are still under nascent stage. In this regard, hydrogen-based renewable energy is very promising due to higher chemical energy per mass of hydrogen (142 MJ kg^{-1}) as compared to liquid hydrocarbons (47 MJ kg^{-1}), zero emission, and its large abundancy on earth. Therefore, electrochemical conversion of hydrogen to electric energy using fuel cells would be the key technology in future. In hydrogen fuel cells, the development of highly efficient electrocatalysts mainly for sluggish cathodic oxygen reduction reaction, with inexpensive and easily available materials, is a major issue. Recent investigations suggest that chemically modified graphene support materials such as nitrogen-doped graphene can generate strong, beneficial catalyst support interactions which considerably enhance the catalyst activity and stability in fuel cells. This chapter describes the fundamental aspects of electrochemical conversion of hydrogen to electric energy using fuel cells. The chapter further explains the role of nitrogen-doped graphene nanomaterials and their hybrids with transition metal and their alloy nanoparticles in fuel cell catalysis.

Keywords: graphene, nitrogen doping, hydrogen, fuel cells, alloy nanoparticles, oxygen reduction reaction

1. Introduction

In today's world, the rapid economic progress, increase in the population, and growing human dependence on energy appliances have led to a sharp increase in the global energy production and consumption that mainly originate from the non-renewable fossil fuels. This growing energy requirement imposes the urgency of newer sustainable, renewable, and environment friendly fuels. On this basis, the research and development of renewable energy resources and its conversion are of prime importance. The efficiency of renewable energy conversion devices mainly depend upon the structure and property of the active materials used in these devices. Recent research developments show that nanomaterials, particularly two-dimensional (2D) nanomaterials, such as graphene and its derivatives, can play a significant role in efficiency enhancement and performance of energy conversion and storage devices.

1.1. Development of graphene and its analogs

Graphene is a flat single layer of carbon atoms tightly packed into a 2D honeycomb lattice. Graphene can be considered as the fundamental building block of all graphitic materials with different dimensions. Graphene or graphene analogues can be synthesized mainly by bottom-up methods and top-down methods.

1.1.1. Bottom-up synthesis methods

One of the bottom-up approaches for the synthesis of single or few layer graphene is the chemical vapor deposition (CVD) method using the catalyst metal surfaces (Cu, Ni, Ir, Ru, etc.) and methane, acetylene, CO_2, or ethylene as the carbon source in the presence of argon or hydrogen. At high temperatures (~1000°C), pyrolysis of carbon precursor gas is taking place over the catalyst surfaces and single layer or few layer graphene is deposited on metal surfaces depending upon the heating rate, heating temperature, and flow rate of gases. Single-layer graphene with high structural quality were attained by low-pressure CVD of ethylene on a hot Ir(111) surface [1]. Recently, large single-crystalline graphene domains were grown by CVD on Ru(0001) using ethylene [2]. Large-scale graphene films can also be synthesized by CVD on polycrystalline nickel substrates [3]. Epitaxial growth is the other bottom-up strategy whereby a silicon carbide (SiC) substrate is heated to high temperatures (>1100°C) to form graphene [4]. In this method, growing of a single-layer graphene and achieving large graphene domains with uniform thickness remains a challenge. In CVD and epitaxial growth, the dimensions of the formed graphene mainly depend upon the size of the substrate, and at a time, one graphene layer only can be produced. Therefore, these methods are more suitable for microscale applications such as electronic or sensors and not suitable for applications where large amounts of materials are required such as energy related.

1.1.2. Top-down synthesis methods

In the top-down synthesis, methods of graphene consist of different methods such mechanical exfoliation of graphite, graphite intercalation methods, ultrasonic cleavage of graphite, oxidation, and subsequent exfoliation/reduction of graphite.

(i) Mechanical exfoliation: This is the best top-down synthesis method to get defect-free single-layer graphene, and this method involves the peeling of the layers of graphite using a scotch tape [5]. Different approach for the mechanical exfoliation is based on the concept of anodic bonding. In this method, at a particular temperature and voltage, graphite is sealed to a borosilicate glass and the subsequent peeling procedure gives a single- or few-layer graphene sheet on the substrate [6]. Scalability toward mass production and transfer of graphene sheets from the substrates are the major challenges in the mechanical exfoliations.

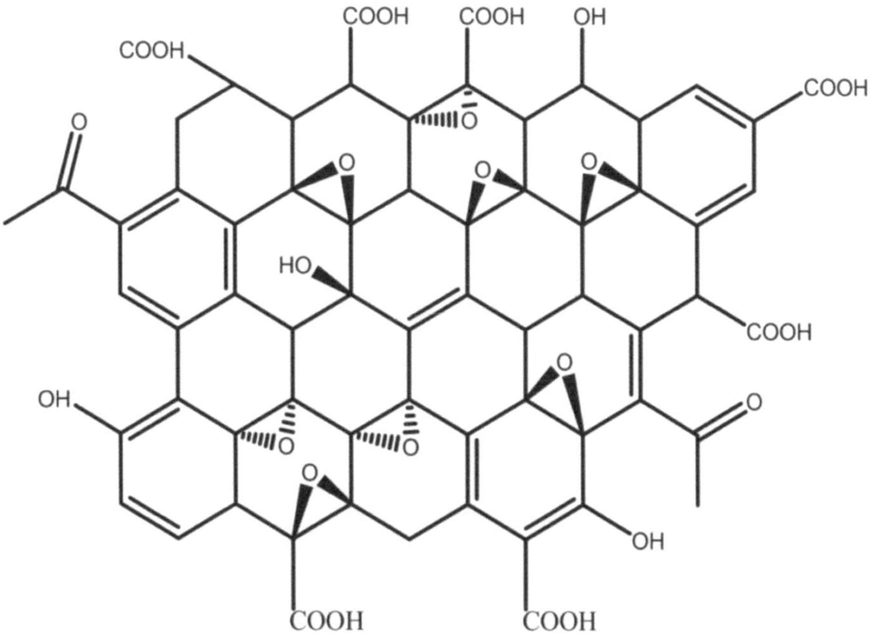

Figure 1. Schematic representation of the structure of graphite oxide.

(ii) Graphite intercalation: In this method, initially atoms or molecules such as alkali metal atoms (Li, Na, K, etc.) or acid (nitric acid, sulfuric acid, etc.) molecules are intercalated between the graphitic layers [7]. That weakens the Van der Waals interactions between the layers of graphite and facilitates the exfoliation or gives expanded graphite (EG) with partial oxidization. Giving a thermal shock at high temperatures (~1000°C) to EG provide graphite nanosheets or graphite nanoplatelets that contain multiple numbers of graphene layers (~10–50 layers).

(iii) Ultrasonic cleavage method: Here, graphite is initially suspended in particular organic solvents or surfactants (*N*-methyl-2-pyrroldinone,*N*,*N*-dimethylformamide, sodium dodecyl benzene sulfonate, etc.) and then gives an ultrasonic agitation to supply the energy to cleave the graphite [8]. The yield of getting single layer graphene at the first stages of this method is very low (~1 wt. %) and it can be increased by repeated sediment recycling. The success of

ultrasonic cleavage depends on the right selection of solvents and surfactants as well as the sonication frequency, amplitude, and time.

(iv) Graphite oxide (GO)-reduction and exfoliation: GO can be synthesized by the oxidative treatment of graphite via one of the following methods: Brodie [9], Staudenmaier [10], and Hummers [11]. Brodie's method was the addition of potassium chlorate ($KClO_3$) to the slurry of graphite in fuming nitric acid. Staudenmaier improved Brodie's method by adding the chlorate in multiple aliquots over the course of reaction in place of a single addition as Brodie had done. After that, Hummer found out an alternate oxidation method by reacting graphite with a mixture of potassium permanganate ($KMnO_4$) and concentrated sulfuric acid (H_2SO_4) [11]. The basal planes of GO are decorated with epoxide and hydroxyl groups, while edges are anchored with carboxyl and carbonyl groups as shown in **Figure 1**. Thus, GO consisting of layered structure of "graphene oxide" sheets with an interlayer spacing of ~0.84 nm. Completely oxidized GO is hydrophilic and an insulator due to the disruption of the sp^2-bonded carbon network. While partially oxidized GO is a semiconductor.

1.1.3. Chemical modifications of graphene-related nanomaterials

Surface chemical modifications of graphene are necessary for making the composites with metals, polymers, alloys, etc., in relation to different energy applications. Carbon nanomaterials can be chemically modified by different methods such as covalent functionalization, non-covalent functionalization, and substitutional doping. In covalent functionalization, functional groups are attached to the carbon surface by the formation of covalent bond formation. This type of functionalization includes treating with oxidizing agents such as H_2SO_4, HNO_3, $KMnO_4$, $K_2Cr_2O_7$, polyphosphoric acid, and H_2O_2 that bring about the attachment of different oxygen functional groups such as hydroxyl (–OH), carboxyl (–COOH), and carbonyl (–C=O) over the surface of carbon nanostructures [12, 13]. In graphene-based materials, functionalization via acid treatments leads to the agglomeration of sheets and destruction of its novel properties like high surface area and electrical conductivity [14]. At the same time, non-covalent functionalization does not make any formal chemical bond formation between the foreign molecule and carbon surface, and this includes the attaching of various species of polymers [15], polynuclear aromatic compounds [16], surfactants [17], and biomolecules [18]. The non-covalent interaction is based on Van der Waals forces or π-π stacking interactions. Substitutional doping introduces heteroatoms, such as nitrogen atoms or boron atoms, into the carbon lattice of graphene, and this type of chemical modifications changes the electronic properties of carbon nanostructures. When a nitrogen atom is doped into graphene, it usually has three common bonding configurations within the carbon lattice, including quaternary N (or graphitic N), pyridinic N, and pyrrolic N as shown in **Figure 2** [19]. Pyridinic N bonds with two C atoms gives one p electron to the π system of graphene lattice. Pyrrolic N attributes to N atoms that donate two p electrons to the π system and bond with five-membered ring, like in pyrrole. N atoms that substitute for C atoms in the hexagonal ring are quaternary N. In these nitrogen types, pyridinic N and quaternary N are sp^2 hybridized, while pyrrolic N is sp^3 hybridized. In addition to these three usual nitrogen types, N oxides of pyridinic N have also been occasionally noticed in the nitrogen-doped graphene samples.

http://dx.doi.org/10.5772/64242

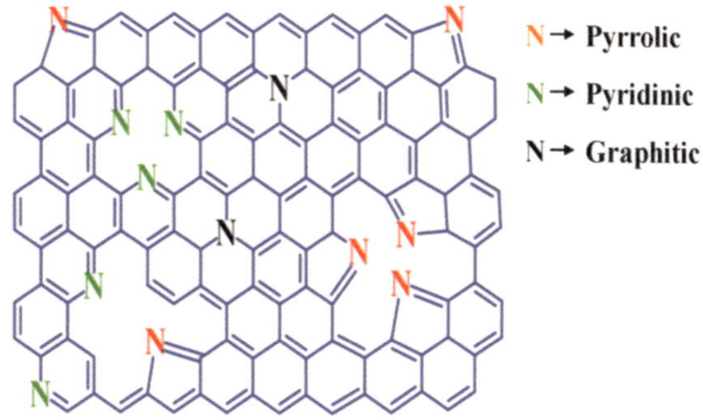

Figure 2. Schematic diagram of a nitrogen-doped graphene.

2. 2D nanomaterials for hydrogen energy conversion

2.1. Fuel cells

Fuel cells were discovered about 170 years ago by Sir William Grove. Fuel cell is a device that converts chemical energy into electrical energy. Usually electrochemical cells have two conductive electrodes called anode and cathode. The anode is defined as the electrode where oxidation occurs and the cathode is the electrode where the reduction takes place. Reactions at the anode (negative) usually take place at lower electrode potentials than that of the cathode (positive). Fuel cells are open systems, where the anode and cathode are just charge-transfer media and the active masses undergoing the redox reaction are delivered from outside the cell. To compare the power and energy densities of various electrochemical energy devices, a Ragone plot is used. The Ragone plot as shown in **Figure 3** illustrates that fuel cells can be considered as high-energy systems, whereas supercapacitors are considered to be high-power systems. Batteries have intermediate power and energy characteristics.

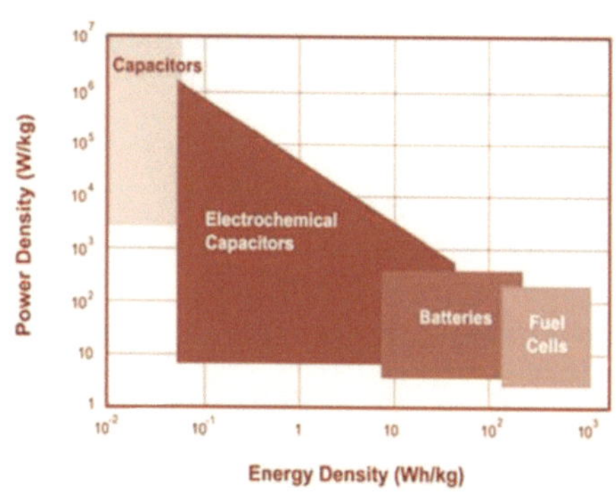

Figure 3. Ragone plot for different electrochemical energy devices [20].

2.2. Electrochemical aspects of fuel cells

The energy and power characteristics of fuel cells follow directly the thermodynamic and kinetic formulations for chemical reactions. The basic thermodynamics for a reversible electrochemical reaction is given by,

$$\Delta G = \Delta H - T\Delta S \tag{1}$$

Here ΔG is the Gibbs free energy, ΔH is the enthalpy, ΔS is the entropy, and T is the absolute temperature, with $T\Delta S$ being the heat formed with the reaction. The terms, ΔG, ΔH, and ΔS are state functions and depend only on the identity of the electrode materials and the initial and final states of the reactions. The free energy ΔG represents the net useful energy available from a given reaction, thus, the net available electrical energy from a reaction in an electrochemical cell is given by,

$$\Delta G = -nFE \tag{2}$$

where n is the number of electrons transferred per mol of reactants, F is the Faraday constant (96,400 C mol^{-1}), and E is the voltage of the cell with the specific chemical reaction. Eq. (2) represents a balance between the chemical and electric driving forces upon the ions under open circuit conditions; hence, E refers to the open circuit potential of a cell where there is no current flowing. Since ΔG = -237 kJ mol^{-1} under standard conditions for the reaction,

$$H_2 + \frac{1}{2}O_2 \rightarrow H_2O \tag{3}$$

The reversible voltage for the hydrogen-oxygen fuel cell according to Eq. (2) is E = 1.23 V.

The change of free energy (ΔG) for a given species "i" defines the chemical potential (μ_i). The chemical potential μ_i for species "i" is related to another thermodynamic quantity called chemical activity (a_i) by the defining relation:

$$\mu_i = \mu_i^o + RT \ln a_i \tag{4}$$

where μ_i^0 is a constant, which is the value of the chemical potential of species "i" in its standard conditions, R is the gas constant, and T the absolute temperature.

Assume an electrochemical cell in which the activity of species "i" is distinctive in the two electrodes: $a_i(-)$ at the negative side and $a_i(+)$ at the positive side. The difference between the chemical potentials of positive and negative electrodes can be written as

http://dx.doi.org/10.5772/64242

$$\mu_i(+) - \mu_i(-) = RT\ln\left[\frac{a_i(+)}{a_i(-)}\right] \tag{5}$$

This chemical potential difference is balanced by the electrostatic energy from Eq. (2), then

$$E = E^0 - \frac{RT}{nF}\ln\left[\frac{a_i(+)}{a_i(-)}\right] \tag{6}$$

where E^0 is the voltage under standard conditions and this relation is called the Nernst equation, which relates the measurable cell voltage to the chemical differences across an electrochemical cell. If the activity of species "*i*" in one of the electrodes is a standard reference value, the Nernst equation provides the relative electrical potential of the other electrode.

Compared to the open circuit potential at equilibrium state, the voltage drops off when current is drawn from the electrochemical cell because of irreversible losses. If more current is withdrawn from the cell, greater the losses. **Figure 4** shows a typical discharge curve of a fuel cell. There are three major types of voltage losses in the discharge curve and that give its characteristic shape. These voltage losses are (i) activation losses due to electrochemical reaction kinetics (η_{act}), (ii) ohmic losses due to ionic and electronic resistances (η_{ohmic}), and (iii) concentration losses due to limited mass transport (η_{conc}). So, by subtracting all voltage losses from the thermodynamically predicted voltage output, the real voltage output of a fuel cell can be written as

$$V = E - \eta_{act} - \eta_{ohmic} - \eta_{conc} \tag{7}$$

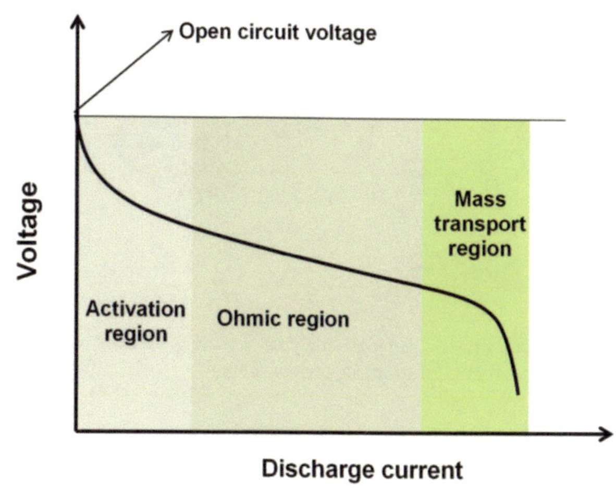

Figure 4. Discharge curve of a fuel cell.

2.3. Hydrogen fuel cells: working mechanism

Fuel cells can be classified primarily by the kind of electrolyte they employ; all of them work on the same principle. Different types of fuel cells are listed below:

- Polymer electrolyte membrane fuel cells (PEMFC)
- Alkaline fuel cells (AFC)
- Phosphoric acid fuel cells (PAFC)
- Molten carbonate fuel cells (MCFC)
- Solid oxide fuel cells (SOFC)

Among various types of fuel cell technologies, the PEMFC technology has received attention due to its high power density and efficiency (60–80%), relatively low operating temperature (less than 100°C), and the ability to respond quickly to the changing power demands. A schematic diagram of a PEMFC is given in **Figure 5(a)**. It comprises an anode, a cathode, and a polymer membrane as the proton-conducting media which separate both the electrodes. Hydrogen is supplied at the anode and oxygen is supplied at the cathode. Hydrogen oxidation is taking place at the anode and as a result, H^+ ions and electrons are produced. The H^+ ions are conducted through the proton exchange membrane and electrons through the external circuit to the cathode, where oxygen reduction reaction (ORR) is taking place and water and heat energy are produced as by-products. All the electrochemical reactions at the anode and cathode are taking place with the help of a catalyst only.

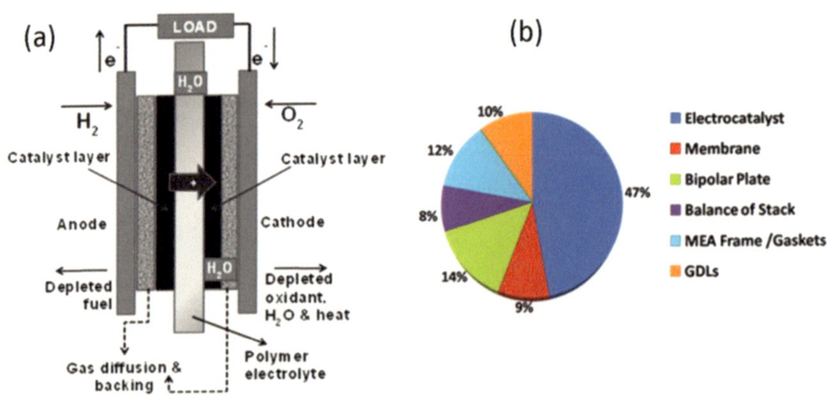

Figure 5. (a) Schematic illustration of a PEMFC. (b) Cost analysis of various fuel cell components. (Analysis data based on the US Department of Energy -Hydrogen Program Overview, 2008.)

The reactions taking place in PEMFC can be summarized as

$$\text{Anode}: 2H_2 \rightarrow 4H^+ + 4e^- \quad \left(E^+ = 0.0 \text{ V vs. SHE}\right) \tag{8}$$

$$\text{Cathode}: 4H^{+} + O_2 + 4e^{-} \rightarrow 2H_2O \left(E^{-} = 1.23 \text{ V vs. SHE}\right) \tag{9}$$

(SHE stands for standard hydrogen electrode)

The net reaction can be written as

$$H_2 + \tfrac{1}{2} O_2 \rightarrow H_2O \quad \left(E_{cell} = 1.23 \text{ V}\right) \tag{10}$$

2.4. Role of electrocatalysts in hydrogen fuel cells

For a wide-scale commercialization, the PEMFCs should overcome several challenges including two critical barriers, namely cost and durability. At present, platinum (Pt) is the best electrocatalyst in PEMFCs in light of the fact that it is adequately reactive at low temperatures for bonding with hydrogen and oxygen intermediates and catalyzes the electrode reactions to form the final products. Nevertheless, the high cost and insufficiency of Pt limit the practical implementations of PEMFC for static and automobile applications. **Figure 5(b)** shows the cost (in percentage) of different components within PEMFC. To date, 78% of the total cost of a fuel cell is due to the membrane electrode assembly of which 47% is contributed by the catalysts alone. At present, with the help of modern technology and nano materials, a PEMFC performance with maximum power densities of ≈1 W cm^{-2} at the cell voltage of ≈ 0.65 V has been achieved using the Pt loading (anode and cathode) of 0.4–0.5 mg_{Pt} cm^{-2} [21, 22]. This PEMFC performance equals to fivefold reduction in the amount of Pt was used in the early of 2000. At present, the research world is looking for to reduce the Pt-specific power density lesser than that of 0.2 g_{Pt} kW^{-1} at the cell voltages of ≥0.65 V. According to the US department of energy (DOE), the key target is to reduce the fuel cell cost to $30/kW with 5000 h minimum durability. For reducing the PEMFC cost, the following approaches are implemented such as (i) alloying of Pt catalyst with 3d-transition metals (TM) to enhance the catalytic activity, (ii) maximum utilization of Pt by reducing the particle size and usage of proper catalyst support materials, and (iii) considerable reduction of mass-transport-induced voltage losses at high current densities with proper engineering of electrodes, diffusion media, and flow fields [21].

In PEMFC, sluggish kinetics of ORR causes a large over-potential at cathode side in low temperatures. Therefore, development of highly active cathode electrocatalysts is essential. Alloying Pt with 3d transition metals (TM), especially Co, Fe, Ni, and Cr, has been demonstrated to increase the ORR activity [23]. In particular, Pt-Co alloys have been found to exhibit enhanced catalytic activity toward the ORR [23–27]. The modified electronic structure of the Pt–TM alloy composite affects the Pt-Pt bond distance resulting in strong adsorption of the oxygen molecules and weak adsorption of ORR blocking OH^{-} radicals over these alloy catalyst nanoparticles [23, 28]. In addition, it is essential to have a uniform dispersion and control over the size of catalyst particles for high activity [21, 29]. Metal catalyst nanoparticles have a very high tendency to agglomerate due to large metal-metal cohesion energy and which in turn leads to a decrease in the electrochemical surface area (ECSA). To sort out this issue, it is important to search for a right catalyst supporting material that has a high surface area, excellent electronic conductivity, and strong affinity toward catalyst particles, on which a uniform dispersion of Pt NPs can be accomplished. Catalyst supports also take a role in altering the geometry and electronic properties of the catalysts particles during their growth. A highly conductive support can enable effective electron collection and their transfer to the collector electrode. It is required to have a good porosity in support materials for better mass transportation of the fuel and oxidant to the densely scattered triple phase boundaries. Carbon supports in presently using commercial platinum electrocatalysts (Pt/C) have a tendency of electrochemical oxidation during long-time operations in acidic environment, which leads to Pt nanoparticles agglomeration or detachment from the support material leading to the degradation of fuel cell performance [30].

2.5. Graphene-based fuel cell electrocatalysts

Since graphene is a two-dimensional planar sheet with open structure, there is the possibility of both sides of graphene being utilized for supporting the catalyst nanoparticles and which is expected to be a promising catalyst support material in the future. Graphite oxide is commonly used as the starting material for the bulk preparation of graphene-based fuel cell electrocatalysts. The various oxy functional groups over the surface of graphite oxide help for its dissolution in various solvents and serve as the anchors or nucleating centers for the controlled growth of catalyst nanoparticles on it. Different techniques have been reported for the decoration of platinum nanoparticles (Pt NPs) on graphene for fuel cell applications. Jafri et al. have synthesized graphene nanoplatelet-supported Pt NPs by reducing platinum chloride with $NaBH_4$ on thermally exfoliated graphene nanoplatelets [31]. Pt-graphene was also synthesized by in situ reduction of graphite oxide and platinum chloride using different polyol methods and $NaBH_4$ methods [32, 33]. In various synthesis methods of Pt-graphene, the dispersed nanoparticles exhibit a wide size range, a non-uniform spatial distribution, and the synthesis method does not provide significant control over the metal loading. This shows that the decoration of Pt NPs on graphene is a challenge in terms of controlling the particle size, distribution, and keeping the inherent properties of graphene such as high surface area and electronic conductivity intact. Therefore, the objectives of devising a synthesis method for Pt-graphene are related to fuel cell applications, which (a) yields high performance, (b) incurs low cost, (c) requires minimum resources, and (d) has optimum-sized catalyst particles (~3–5 nm) dispersed uniformly while preserving the inherent properties of graphene is of prime concern. Different graphene-based hybrid electrocatalyst materials such as graphene/Pt metal, graphene/Pt alloys, graphene/metal oxides, and heteroatom-doped graphene, etc. are effective in maximizing the ORR performance of PEMFC and have been illustrated in **Figure 6**.

The early reports of the PEMFC performance using Pt-graphene as ORR catalysts were not competitive with the commercial Pt/Vulcan carbon electrocatalysts. In the early studies, Seger et al. reported a maximum power density value of ~160 mW cm^{-2} for Pt-graphene electrocatalyst as synthesized by $NaBH_4$ reduction method [32]. A fuel cell performance of ~350 mA cm $^{-2}$ at 0.6 V was shown by Si et al. for Pt-graphene cathode electrocatalyst [34]. The main reason for this comparatively lower performance is the decrease in the surface area due to restacking of graphene sheets and poor electronic conductivity because of the inappropriate synthesis procedures.

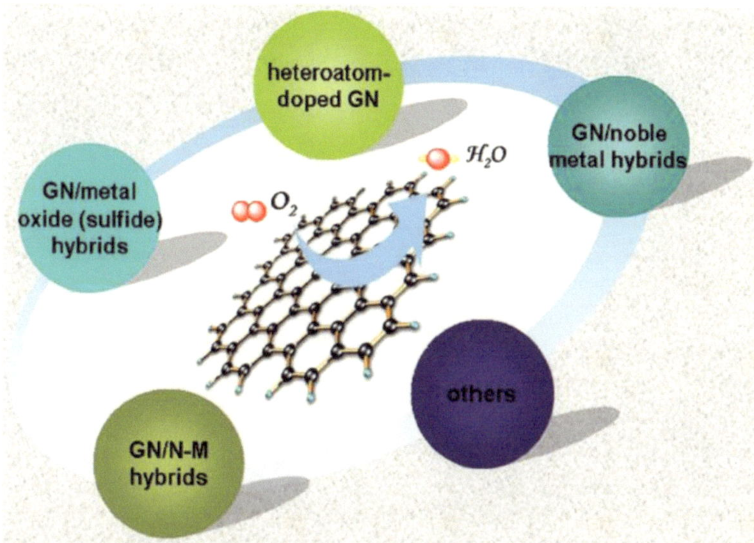

Figure 6. Various graphene-based nanomaterials as catalysts for ORR. (Reprinted from ref. [35] with permission of The Royal Society of Chemistry).

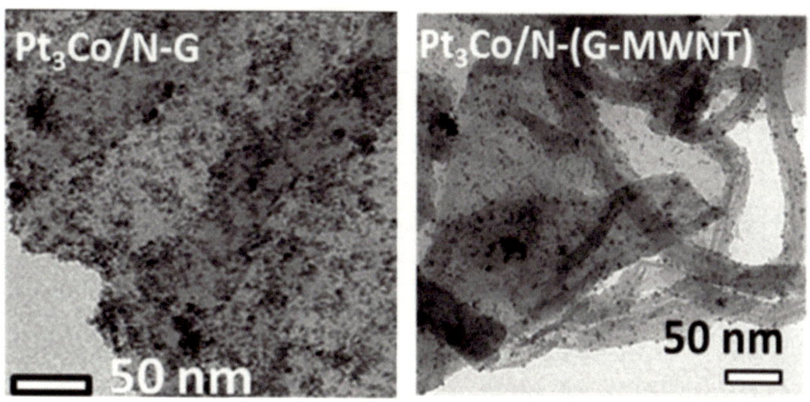

Figure 7. TEM image of Pt_3Co/nitrogen-doped graphene (Pt_3Co/NG), Pt_3Co/nitrogen-doped (graphene-MWNT) hybrid structure (Pt_3Co/N-(G-MWNT)) [25, 26].

To overcome the problem of restacking, incorporation of metal nanoparticles and nanotubes was used during the synthesis of graphene itself and these nanoparticles/nanotubes act as the spacers in between the graphene layers, which prevent further agglomeration of the sheets [36, 37]. Further to enhance the electrical conductivity and electrochemical activity of graphene, heteroatom doping (N, B, S, or P) is a very promising method [38–42].

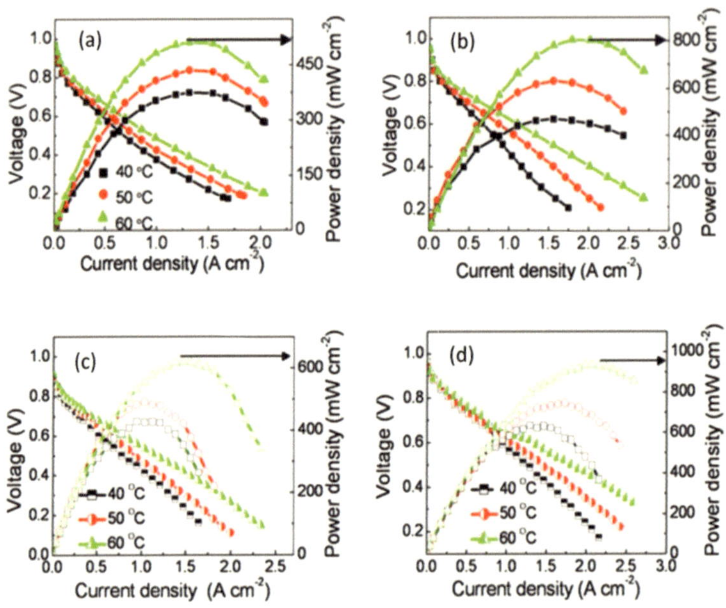

Figure 8. Polarization curves of cathode electrocatalysts (a) Pt/N-G, and (b) P_3Co/N-G, (c) Pt/N-(G-MWNT), and (d) Pt_3Co/N-(G-MWNT) with Pt/C as anode electrocatalyst at three different temperatures (40, 50, 60°C) without any back pressure [25, 26].

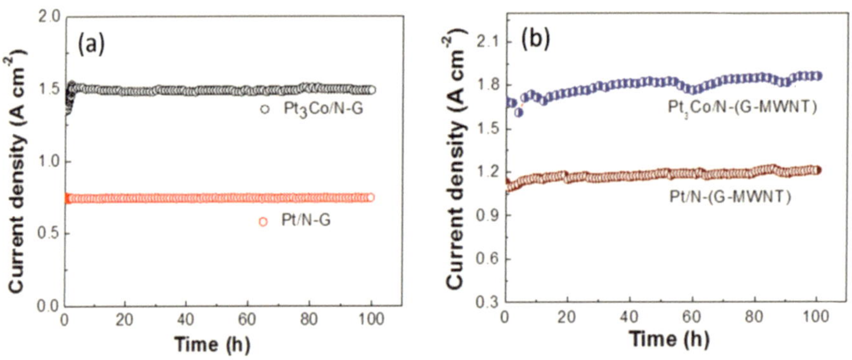

Figure 9. (a–b) Stability of PEMFC with (a) Pt/N-G, and (b) P_3Co/N-G, (c) Pt/N-(G-MWNT), and (d) Pt_3Co/N-(G-MWNT) as cathode electrocatalysts and Pt/C anode electrocatalyst at 0.5V in 60°C temperature with no back pressure [25, 26].

Figure 7 illustrates the uniform distribution of Pt_3Co alloy nanoparticles over the surface of nitrogen-doped graphene and graphene-multiwalled carbon nanotubes (MWNT) hybrid structures. The PEMFC performance polarization curves for Pt or Pt_3Co alloy catalysts nanoparticle dispersed nitrogen-doped graphene (NG) and nitrogen-doped (graphene-MWNT) hybrid catalysts have been shown in **Figure 8**. The corresponding performance values of the electrocatalysts are summarized in **Table 1**. The commercial Pt/C was used as the anode catalysts for all cathode catalysts. A catalyst loading of 0.25 and 0.4 mg_{pt} cm^{-2} were sustained at the anode and the cathode, respectively. **Figure 9(a)** and **(b)** compares the stability of the above electrocatalysts for 100 h at 0.5 V in 60°C temperature with no back pressure.

Cathode electrocatalyst	ECSA (m^2 g^{-1})	Temperature = 60°C	
		Current density at 0.6 V (mA cm^{-2})	Maximum power density (mW cm^{-2})
Pt/N-G	57.9	665	512
Pt_3Co/N-G	48.5	1110	805
Pt/N-(G-MWNT)	79.2	804	783
Pt_3Co/N-(G-MWNT)	98.5	1344	935

Table 1. PEMFC performance of Pt and Pt_3Co alloy catalysts using nitrogen doped graphene.

The superior PEMFC performance exhibited by Pt_3Co/nitrogen-doped nano-carbon cathode electrocatalysts (Pt_3Co/N-G and Pt_3Co/N-(G-MWNT)) can be viewed from two perspectives as shown in **Figure 10**, that is (i) the catalytic activity of Pt_3Co alloy nanoparticles and (ii) the role played by nitrogen doping in carbon support nanomaterials.

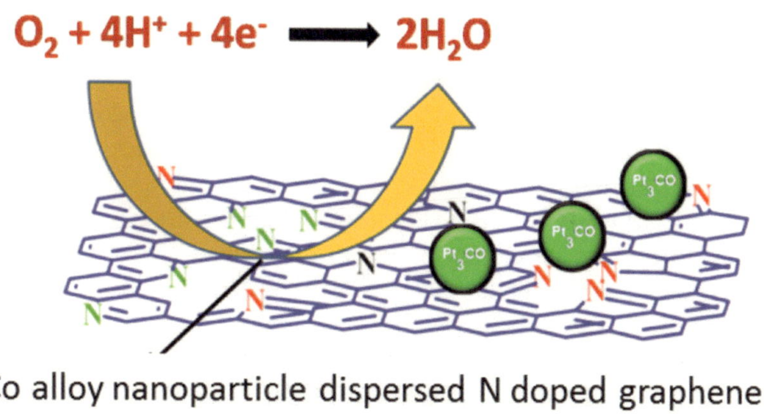

Figure 10. ORR in Pt_3Co alloy nanoparticle dispersed nitrogen-doped graphene.

In ORR electrocatalysts, two opposing effects should counter balance each other such as comparatively strong adsorption energy of oxygen and ORR reaction intermediates (O_2^-, O_2^{2-}) and at the same time, low adsorption of ORR blocking species such as OH^- anions and oxides [21]. For Pt metal surfaces the d-band center is closer to the Fermi level and they firmly adsorb both oxygen and ORR-blocking OH^- species and that restricts the availability of free Pt sites for ORR [43]. There is a shift in Pt d-band center from Fermi level and a reduction of Pt-Pt bond distance can be created as a result of the alloying of Pt with transition metals [44]. Despite the fact that more shifts in d-band center from Fermi level makes a less adsorption of OH-species and at the same time the ORR rate is less because of the feeble adsorption of oxygen.

Consequently, the movement in d-band center ought to be ideal in a manner that to balance the solid adsorption of oxygen and weak adsorption by OH^- blocking species. The shift in *d*-band center is optimum for Pt_3TM (TM = Fe, Co, Ni) electrocatalysts in such a way that to give high ORR activity as compared to other Pt-TM alloys or pure Pt [45]. Also the uniform dispersion of catalyst particles with appropriate particle size (~3–5 nm) again enhances the ORR activity of the electrocatalysts.

The formation of an atomic charge density and asymmetry in spin density on nitrogen-doped graphene system encourages the charge exchange from the carbon support to the adsorbing oxygen molecule and results in the creation of an superoxide ion ($O_2^{\bullet -}$) [25, 46]. This debilitates the O–O bond and helps for its easy dissociation. In other words nitrogen doping helps for the $4e^-$ transfer reaction rather than $2e^-$ transfer reaction in ORR reaction [38]. Moreover, the doping of nitrogen atoms into the graphene lattice reinforces the bonding between the catalysts nanoparticles and the support that not just aides in a uniform dispersion of catalyst particles on the carbon support additionally prevents their self-agglomeration during long PEMFC operation [47]. In nitrogen-doped (graphene-MWNT) hybrid structures, the nitrogen doping also increases the electrical conductivity of graphene support materials along with highly conductive carbon nanotubes [25].

3. Future outlook and challenges

The gathering of many unique properties of graphene-related 2D nanomaterials such as large surface area, excellent conductivity, good mechanical, and chemical stability along with low cost and accessibility for mass production has opened up a new research area in the field of materials science. Chemical modification of graphene-related materials especially by nitrogen doping and incorporation of Pt and Pt alloys can significantly increase the catalytic activity and durability toward oxygen electro-reduction. Even though the following things need to be investigated in depth for future studies such as (a) the exact nature of electrocatalytically active sites facilitating the ORR in N-doped graphene and origin of their catalytic reactivity, (b) design of facile and large-scale synthesis approaches, for N-doped graphene and their hybrids with metals and alloy nanoparticles. With respect to N-doped graphene synthesis, the technique should be capable of the large incorporation of electrochemically active N functional groups to the support. Regarding N-doped graphene hybrids, the method should ensure the precise tuning of the size, morphology and compositions of electrocatalyst particles for getting high ORR catalytic activity and strong bonding between N-doped support and catalyst nanoparticles for long-term durability, and (c) investigation of alloying effect of Pt with 3d transition metals (Co, Fe, Ni, Cu, etc.) supported on N-doped graphene towards ORR by DFT modeling and electrochemical experiments.

Acknowledgements

B.P. Vinayan acknowledges the Alexander von Humboldt Foundation for research funding.

Author details

Bhaghavathi Parambath Vinayan

Address all correspondence to: vinayan.parambath@kit.edu

Helmholtz Institute Ulm for Electrochemical Energy Storage (HIU), Ulm, Germany

References

[1] Coraux J, N'Diaye AT, Busse C, Michely T. Structural coherency of graphene on Ir(111). Nano Lett. 2008;8:565–70. DOI: 10.1021/nl0728874

[2] Sutter PW, Flege JI, Sutter EA. Epitaxial graphene on ruthenium. Nat Mater. 2008;7:406–11. DOI:10.1038/nmat2166

[3] Kim KS. Large-scale pattern growth of graphene films for stretchable transparent electrodes. Nature. 2009;457:706–10. DOI:10.1038/nature07719

[4] Berger C, Song Z, Li T, Li X, Ogbazghi AY, Feng R, et al. Ultrathin epitaxial graphite: 2D electron gas properties and a route toward graphene-based nanoelectronics. J Phys Chem B. 2004;108:19912–6. DOI: 10.1021/jp040650f

[5] Geim A, Novoselov K. The rise of graphene. Nat Mater. 2007;6:183–91. DOI:10.1038/nmat1849

[6] Shukla A, Kumar R, Mazher J, Balan A. Graphene made easy: High quality, large-area samples. Solid State Commun. 2009;149:718–21. DOI:http://dx.doi.org/10.1016/j.ssc.2009.02.007

[7] Park S, Ruoff RS. Chemical methods for the production of graphenes. Nat Nano. 2009;4:217–24. DOI:10.1038/nnano.2009.58

[8] Cai M, Thorpe D, Adamson DH, Schniepp HC. Methods of graphite exfoliation. J Mater Chem. 2012;22:24992–5002. DOI: 10.1039/c2jm34517j

[9] Brodie BC. On the atomic weight of graphite. Philos Trans R Soc Lond. 1859;149:249–59. DOI: 10.2307/108699

[10] Staudenmaier L. Verfahren zur Darstellung der Graphitsäure. Ber. Dtsch. Chem. Ges. 1898;31: 1481–1487. DOI: 10.1002/cber.18980310237

[11] Hummers W, Ofeman R. Preparation of graphitic oxide. J Am Chem Soc. 1958;80:1339. DOI: 10.1021/ja01539a017

[12] Vaisman L, Wagner HD, Marom G. The role of surfactants in dispersion of carbon nanotubes. Adv Colloid Interface Sci. 2006;128–130:37–46. DOI: http://dx.doi.org/10.1016/j.cis.2006.11.007

[13] Rajalakshmi N, Ryu H, Shaijumon MM, Ramaprabhu S. Performance of polymer electrolyte membrane fuel cells with carbon nanotubes as oxygen reduction catalyst support material. J Power Sources. 2005;140:250–7. DOI: 10.1016/j.jpowsour.2004.08.042

[14] Georgakilas V, Otyepka M, Bourlinos AB, Chandra V, Kim N, Kemp KC, et al. Functionalization of graphene: covalent and non-covalent approaches, derivatives and applications. Chem Rev (Washington, DC, USA). 2012;112:6156–214. DOI: 10.1021/cr3000412

[15] Andrews R, Jacques D, Qian D, Rantell T. Multiwall carbon nanotubes: synthesis and application. Acc Chem Res. 2002;35:1008–17. DOI: 10.1021/ar010151m

[16] Wu G, Li L, Li J-H, Xu B-Q. Methanol electrooxidation on Pt particles dispersed into PANI/SWNT composite films. J Power Sources. 2006;155:118–27. DOI: http://dx.doi.org/10.1016/j.jpowsour.2005.04.035

[17] Matarredona O, Rhoads H, Li Z, Harwell JH, Balzano L, Resasco DE. Dispersion of single-walled carbon nanotubes in aqueous solutions of the anionic surfactant NaDDBS. J Phys Chem B. 2003;107:13357–67. DOI: 10.1021/jp0365099

[18] Katz E, Willner I. Biomolecule-functionalized carbon nanotubes: applications in nanobioelectronics. ChemPhysChem. 2004;5:1084–104. DOI:10.1002/cphc.200400193

[19] Wang H, Maiyalagan T, Wang X. Review on recent progress in nitrogen-doped graphene: synthesis, characterization, and its potential applications. ACS Catal. 2012;2:781–94. DOI: 10.1021/cs200652y

[20] Winter M, Brodd RJ. What are batteries, fuel cells, and supercapacitors? Chem Rev (Washington, DC, USA). 2004;104:4245–70. DOI: 10.1021/cr020730k

[21] Gasteiger HA, Kocha SS, Sompalli B, Wagner FT. Activity benchmarks and requirements for Pt, Pt-alloy, and non-Pt oxygen reduction catalysts for PEMFCs. Appl Catal B: Environ. 2005;56:9–35. DOI: 10.1016/j.apcatb.2004.06.021

[22] Gröger O, Gasteiger HA, Suchsland J-P. Review—electromobility: batteries or fuel cells? J Electrochem Soc. 2015;162:A2605–A22. DOI: 10.1149/2.0211514jes

[23] Stamenkovic VR, Mun BS, Arenz M, Mayrhofer KJJ, Lucas CA, Wang G, et al. Trends in electrocatalysis on extended and nanoscale Pt-bimetallic alloy surfaces. Nat Mater. 2007;6:241–7. DOI:10.1038/nmat1840

[24] Toda T, Igarashi H, Uchida H, Watanabe M. Enhancement of the electroreduction of oxygen on Pt alloys with Fe, Ni, and Co. J Electrochem Soc. 1999;146:3750–6. DOI: 10.1149/1.1392544

[25] Vinayan BP, Ramaprabhu S. Platinum-TM (TM = Fe, Co) alloy nanoparticles dispersed nitrogen doped (reduced graphene oxide-multiwalled carbon nanotube) hybrid

structure cathode electrocatalysts for high performance PEMFC applications. Nanoscale. 2013;5:5109–18. DOI: 10.1039/c3nr00585b

[26] Vinayan BP, Nagar R, Rajalakshmi N, Ramaprabhu S. Novel platinum–cobalt alloy nanoparticles dispersed on nitrogen-doped graphene as a cathode electrocatalyst for PEMFC applications. Adv Funct Mater. 2012;22:3519–26. DOI: 10.1002/adfm.201102544

[27] Vinayan BP, Jafri RI, Nagar R, Rajalakshmi N, Sethupathi K, Ramaprabhu S. Catalytic activity of platinum–cobalt alloy nanoparticles decorated functionalized multiwalled carbon nanotubes for oxygen reduction reaction in PEMFC. Int J Hydrogen Energy. 2012;37:412–21. DOI: http://dx.doi.org/10.1016/j.ijhydene.2011.09.069

[28] Wang C, Markovic NM, Stamenkovic VR. Advanced platinum alloy electrocatalysts for the oxygen reduction reaction. ACS Catal. 2012;2:891–8. DOI: 10.1021/cs3000792

[29] Stephens IEL, Bondarenko AS, Gronbjerg U, Rossmeisl J, Chorkendorff I. Understanding the electrocatalysis of oxygen reduction on platinum and its alloys. Energy Environ Sci. 2012;5:6744–62. DOI: 10.1039/C2EE03590A

[30] Wu J, Yuan XZ, Martin JJ, Wang H, Zhang J, Shen J, et al. A review of PEM fuel cell durability: degradation mechanisms and mitigation strategies. J Power Sources. 2008;184:104–19. DOI:10.1016/j.jpowsour.2008.06.006

[31] Jafri RI, Arockiados T, Rajalakshmi N, Ramaprabhu S. Nanostructured Pt dispersed on graphene-multiwalled carbon nanotube hybrid nanomaterials as electrocatalyst for PEMFC. J Electrochem Soc. 2010;157:B874–B9. DOI: 10.1149/2.097311jes

[32] Seger B, Kamat PV. Electrocatalytically Active graphene-platinum nanocomposites. Role of 2-D carbon support in PEM fuel cells. J Phys Chem C. 2009;113:7990–5. DOI: 10.1021/jp900360k

[33] Li Y, Tang L, Li J. Preparation and electrochemical performance for methanol oxidation of pt/graphene nanocomposites. Electrochem Commun. 2009;11:846–9. DOI: 10.1016/j.elecom.2009.02.009

[34] Si Y, Samulski ET. Exfoliated graphene separated by platinum nanoparticles. Chem Mater. 2008;20:6792–7. DOI: 10.1021/cm801356a

[35] Zhu C, Dong S. Recent progress in graphene-based nanomaterials as advanced electrocatalysts towards oxygen reduction reaction. Nanoscale. 2013;5:1753–67. DOI: 10.1039/c2nr33839d

[36] Vinayan BP, Nagar R, Ramaprabhu S. Synthesis and investigation of mechanism of platinum–graphene electrocatalysts by novel co-reduction techniques for proton exchange membrane fuel cell applications. J Mater Chem. 2012;22:25325–34. DOI: 10.1039/C2JM33894G

[37] Vinayan BP, Nagar R, Raman V, Rajalakshmi N, Dhathathreyan KS, Ramaprabhu S. Synthesis of graphene-multiwalled carbon nanotubes hybrid nanostructure by

strengthened electrostatic interaction and its lithium ion battery application. J Mater Chem. 2012;22:9949–56. DOI: 10.1039/C2JM16294F

[38] Vinayan BP, Diemant T, Behm RJ, Ramaprabhu S. Iron encapsulated nitrogen and sulfur co-doped few layer graphene as a non-precious ORR catalyst for PEMFC application. RSC Adv. 2015;5:66494–501. DOI: 10.1039/C5RA09030J

[39] Vinayan BP, Schwarzburger NI, Fichtner M. Synthesis of a nitrogen rich (2D-1D) hybrid carbon nanomaterial using a MnO_2 nanorod template for high performance Li-ion battery applications. J Mater Chem A. 2015;3:6810–8. DOI: 10.1039/C4TA05642F

[40] Vinayan BP, Zhao-Karger Z, Diemant T, Chakravadhanula VSK, Schwarzburger NI, Cambaz MA, et al. Performance study of magnesium-sulfur battery using a graphene based sulfur composite cathode electrode and a non-nucleophilic Mg electrolyte. Nanoscale. 2016;8:3296–306. DOI: 10.1039/C5NR04383B

[41] Vinayan BP, Sethupathi K, Ramaprabhu S. Facile synthesis of triangular shaped palladium nanoparticles decorated nitrogen doped graphene and their catalytic study for renewable energy applications. Int J Hydrogen Energy. 2013;38:2240–50. DOI: http://dx.doi.org/10.1016/j.ijhydene.2012.11.091

[42] Sahoo M, Sreena KP, Vinayan BP, Ramaprabhu S. Green synthesis of boron doped graphene and its application as high performance anode material in Li ion battery. Mater Res Bull. 2015;61:383–90. DOI: http://dx.doi.org/10.1016/j.materresbull.2014.10.049

[43] Shin J, Choi J-H, Cha P-R, Kim SK, Kim I, Lee S-C, et al. Catalytic activity for oxygen reduction reaction on platinum-based core-shell nanoparticles: all-electron density functional theory. Nanoscale. 2015;7:15830–9. DOI: 10.1039/C5NR04706D

[44] Stamenkovic VR, Fowler B, Mun BS, Wang G, Ross PN, Lucas CA, et al. Improved oxygen reduction activity on Pt3Ni(111) via increased surface site availability.Science. 2007;315:493–7. DOI: 10.1126/science.1135941

[45] van der Vliet DF, Wang C, Tripkovic D, Strmcnik D, Zhang XF, Debe MK, et al. Mesostructured thin films as electrocatalysts with tunable composition and surface morphology. Nat Mater. 2012;11:1051–8. DOI:10.1038/nmat3457

[46] Geng D, Chen Y, Chen Y, Li Y, Li R, Sun X, et al. High oxygen-reduction activity and durability of nitrogen-doped graphene. Energy Environ Sci. 2011;4:760–4. DOI: 10.1039/C0EE00326C

[47] Zhou Y, Neyerlin K, Olson TS, Pylypenko S, Bult J, Dinh HN, et al. Enhancement of Pt and Pt-alloy fuel cell catalyst activity and durability via nitrogen-modified carbon supports. Energy Environ Sci. 2010;3:1437–46. DOI: 10.1039/C003710A

Chapter 9

Porous Graphene Materials for Energy Storage and Conversion Applications

Kimal Chandula Wasalathilake, Godwin Ayoko and Cheng Yan

Additional information is available at the end of the chapter

http://dx.doi.org/10.5772/63554

Abstract

Porous graphene materials possess a unique structure with interconnected networks, high surface area, and high pore volume. Because of the combination of its remarkable architecture and intrinsic properties, such as high mechanical strength, excellent electrical conductivity, and good thermal stability, porous graphene has attracted tremendous attention in many fields, such as nanocomposites, lithium batteries, supercapacitors, and dye-sensitized solar cells. This chapter reviews synthesis methods, properties, and several key applications of porous graphene materials.

Keywords: porous graphene, synthesis, surface area, Li batteries, supercapacitors

1. Introduction

Porous materials are generally referred to materials containing pores or voids with different shapes and sizes. These porous structures have demonstrated unique properties and emerged as attractive candidates for a wide range of applications in medicine, catalysis, sensors, adsorbents, and energy storage and conversion [1–10]. Particularly, porous carbon is an exceptional material with a low density and high specific strength. It is also capable of bonding with other atoms through its sp, sp^2, and sp^3 hybrid orbitals. Among various carbon materials, graphene has received enormous attention because of its high surface area (2630 m^2/g), exceptional thermal conductivity (5000 W/m.K), high Young's modulus (1.0 TPa), and chemical stability. Studies have shown that it has a high intrinsic carrier mobility of 2×10^5 $cm^2/V.s$ and an excellent electrical conductivity of 10^6 S/cm at room temperature [11–13]. Graphene is a two-dimensional hexagonal lattice of sp^2 hybridized carbon atoms and since its discovery in 2004, significant efforts have been put in exploring its potential applications. Various synthesis methods have been developed to produce graphene including epitaxial growth of graphene on metal or SiC substrates [14, 15], chemical vapor deposition (CVD) [16–18], chemical reduction [19, 20], thermal reduction [21, 22], electrochemical synthesis [23, 24], and liquid phase exfoliation [25, 26]. However, because of the strong π-π stacking and van der Waals interactions between graphene sheets, the experimentally obtainable surface area is far below the theoretical value. To overcome this problem, increasing effort has been put to transforming graphene into porous structures to achieve higher surface area [27–29]. Along with the inherent properties of graphene, porous graphene has a clear edge over other porous carbon materials. For example, the excellent electrical conductivity can be used as a perfect current collector for the rapid diffusion of electrons/ions while its high mechanical strength provides mechanical stability to the porous framework. These unique properties make porous graphene a highly promising material for energy storage and conversion applications like lithium-ion batteries (LIBs), lithium-sulfur (Li-S) batteries, supercapacitors, the dye-sensitized solar cells (DSSCs), and fuel cells.

2. Synthesis of porous graphene

According to the standard specified by the International Union of Pure and Applied Chemistry (IUPAC), microporous materials have pore diameters of less than 2 nm, mesoporous materials have pore diameters between 2 and 50 nm, and macroporous materials have pore diameters of greater than 50 nm. There are basically two main methods, which can be used to fabricate porous graphene materials. These are the template and template-free methods, which will be described in greater details below.

2.1. Template approach

Template synthesis is an effective method for the transformation of graphene into porous graphene. It uses various inorganic and organic structures as templates for the transformation. Depending on the required size and morphology of pores, the appropriate template could be selected. This method can be divided into two categories: (1) soft-template method and (2) hard-template method.

2.1.1. Soft-template methods

Different kinds of amphiphilic molecules, such as surfactants and copolymers are used as structure directing agents under mild operating conditions in the soft-template methods. A bottom-up approach has been used for the preparation of mesoporous materials with two-dimensional (2D) sandwich structure consisting of graphene layers and mesoporous silica with the use of cationic surfactant, cetyltrimethyl ammonium bromide (CTAB) [30]. The presence of oxygen functionalized groups makes graphene oxide (GO) sheets negatively charged. CTAB has the capability of electrostatically adsorbing and self-assembling onto the surface of negatively charged GO in alkaline solution. The GO-based silica hybrid is formed after the hydrolysis of silicon precursor, tetraethylorthosilicate (TEOS), and removal of CTAB. Thermal annealing at high temperature gives graphene-based silica sheets. The adsorption data have indicated a high specific area of 980 m²/g. In another method, mesostructured graphene-based SnO_2 composite is prepared by hydrothermally treating a suspension of GO, CTAB, and $SnCl_4$ [31].

Two-dimensional ordered mesoporous carbon nanosheets have been prepared by low molecular weight phenolic resols on graphene sheets using a triblock copolymer called Pluronic F-127 as the structure-directing agent [32]. After mixing an aqueous GO dispersion with the above prepolymer, hydrothermal treatment and further thermal annealing were carried out to prepare mesoporous carbon/graphene composite. It was reported that, Brunauer-Emmett-Teller (BET) surface area decreases with increase in the GO ratio in the composite. In another case, the same hydrothermally driven low-concentration micelle assembly approach was used with the help of anodic aluminum oxide (AAO) membranes to provide a large surface area [33]. After the hydrothermal treatment, AAO membrane was carbonized at 400–500°C for 2 h in argon atmosphere, followed by further carbonization at 700°C for 2 h in the same environment. Finally, mesoporous graphene sheets were obtained by dissolving the AAO substrate (**Figure 1**). The TEM images suggest that these nanosheets displayed ordered mesostructures, having an average pore size of 9 nm and wall thickness of 4 nm. Wen et al. [34] used a dual template method with Pluronic F-127 as the soft template and SiO_2 as the hard template to fabricate three-dimensional graphene-based hierarchically porous carbon (3DGHPC). Carbonization was carried out to convert the layer of coated polymers on SiO_2 spheres to carbon phase and simultaneously reduce GO. Finally, the 3DGHPC was obtained by treating as-prepared composite with 10% HCl to remove the SiO_2 template followed by plenty of washing with Deionized (DI) water and drying at 50°C for 24 h. The as-prepared 3DGHPC displayed a specific area of 384.4 m^2/g with a pore volume of 0.73 cm^3/g.

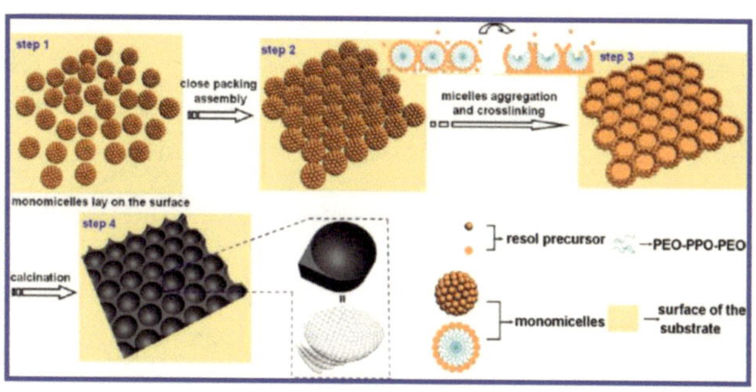

Figure 1. Schematic representation of the formation of ordered mesoporous graphene nanosheets [33]. Reprinted with the permission of the American Chemical Society.

2.1.2. *Hard-template methods*

When preparing porous graphene by hard template method, the template should initially be prepared. This includes the preparation of hard template itself and functionalization of its surface to get the required properties. Then, depending on the requirement, the template should be coated with graphene or GO. The final step is the selective removal of the template without destroying its structure.

Huang et al. [35] used methyl group grafted silica spheres as a hard template to prepare nanoporous graphene foams. These graphene foams had pore sizes of 30–120 nm and ultrahigh pore volumes of 4.3 cm^3/g. The surface area was reported to be 851 m^2/g. Hydrophobic surface of methyl group grafted silica spheres interacts with the hydrophobic basal planes of GO to induce self-assembled lamellar like structures. Choi et al. [36] were able to use polystyrene (PS) colloidal particles as sacrificial templates to synthesize macroporous embossed chemically modified graphene (CMG) sheets with an average pore size of 2 μm. Initially, free-standing PS/CMG film was made by vacuum filtration of a mixed suspension of CMG and PS. PS particles were then removed to generate 3D macropores. Three-dimensional macroscopic graphene foams (GFs) were made by the chemical vapor deposition (CVD) method using nickel (Ni) foam as the 3D scaffold template followed by the removal of the template by hot HCl [37–43]. In 2011, Cheng et al. [44] reported a flexible 3D GF using template directed CVD. The as-prepared GF had a specific surface area, up to 850 m^2/g, corresponding to an average number of layers of ~3. Poly methyl methacrylate (PMMA) can be used as a hard template to prepare macroporous graphene materials. Chen et al. [45] fabricated macroporous bubble graphene film by PMMA directed ordered assembly method. GO was mixed with the PMMA suspension and vacuum filtration was conducted to make a sandwich type assembly of the PMMA spheres and GO. Composite film was then peeled off from the filter, air dried and calcinated at 800°C to remove the template and reduce GO. As-prepared macroporous graphene film has a specific surface area of 128.2 m^2/g with an average pore diameter of 107.3 nm.

2.2. Template-free approach

In the template-free approach, defects are introduced in the graphene basal planes by different methods. Chemical etching or chemical activation is one such method which had been used extensively to prepare porous carbon materials. It is an effective and relatively easy method to fabricate porous graphene sheets without using any template.

Zhu et al. [46] produced porous carbon by a simple activation with KOH of microwave exfoliated GO (MEGO) and thermally exfoliated GO (TEGO). A mixture of the MEGO and KOH was thermally treated for 1 h at 800°C in a tube furnace in argon atmosphere at a pressure of 400 torr. Pores ranging from ~1 to ~10 nm were generated in the carbon matrix by the activation with KOH. The activation of carbon with KOH proceeds as, $6KOH + C \leftrightarrow 2K + 3H_2 + 2K_2CO_3$, followed by the decomposition of K_2CO_3 and reaction of $K/K_2CO_3/CO_2$ with carbon [46, 47].

Porous graphene hybrids can also be produced by thermally treating a mixture of graphene and porous components [48–55]. Rui et al. [48] produced a V_2O_5/rGO composite by thermal pyrolysis of a hybrid of vanadium oxide (VO) and rGO at the temperature of 350°C for 30 min under a heating rate of 10°C/min in air. In the thermal pyrolysis process, reduced VO (rVO) is converted into polycrystalline V_2O_5 porous spheres ranging from 200 to 800 nm.

Apart from using organic and inorganic species to carry out the template-free approaches to produce porous graphene, the amphiphilic nature of GO itself can also be used to fabricate foam-like structures of macroscopic graphene. The pore sizes of these 3D macroscopic structures are in the range of submicrometer to several micrometers. Because of macroscopic nature, they possess high mechanical strength, compressibility, excellent conductivity, and adsorption characteristics [56–59]. Xu et al. [60] prepared a self-assembled graphene hydrogel (SGH) by heating the GO dispersion sealed in a Teflon-lined autoclave at 180°C for 12 h. The hydrothermally reduced GO had a well-defined 3D interconnected porous network (**Figure 2**). The framework of SGH was assembled on partial overlapping of flexible graphene sheets because of π-π stacking interactions. The as-prepared SGH showed excellent mechanical strength and a good electrical conductivity of 5×10^3 S/cm. Later, the same research group reported a highly conductive graphene hydrogel which was reduced by hydrazine hydrate or hydrogen iodide to improve the conductivity by further removing its residual oxygenated groups [61].

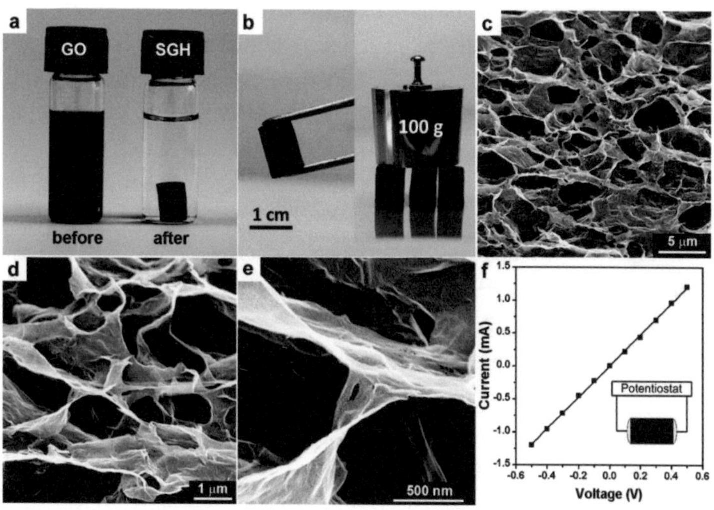

Figure 2. (a) Photographs of a 2 mg/ml homogeneous GO aqueous dispersion before and after hydrothermal reduction at 180°C for 12 h; (b) photographs of a strong SGH allowing easy handling and supporting weight; (c-e) SEM images with different magnifications of the SGH interior microstructures; (f) room temperature *I-V* curve of the SGH exhibiting Ohmic characteristic, inset shows the two-probe method for the conductivity measurements [60]. Reprinted with the permission of the American Chemical Society.

3. Applications of porous graphene materials

Unique porous structure of graphene along with its superior properties makes graphene a potential candidate for energy storage and conversion applications. The following sections review several key applications of porous graphene in LIBs, Li-S batteries, supercapacitors, and the dye-sensitized solar cells.

3.1. Lithium-ion batteries

Lithium-ion battery has a widespread increasing demand because of its high energy density, flexibility, low maintenance, and longer lifespan compared with other battery technologies [62]. To further increase the energy density, charging efficiency, and cycle life of lithium-ion batteries, it is essential to look at new electrode materials that have good lithium storage capability. Porous graphene with exceptional properties holds a great potential as an electrode material for the lithium-ion battery. The high surface area of graphene can significantly increase the diffusion of lithium ions and electrons. Furthermore, the superior electrical conductivity provides a good conductive network within the electrodes. Graphene can construct a 3D framework with a strong tolerance to the volume change of electrochemically active materials during charge-discharge cycles [63].

An anode material for Li-ion battery was made by hierarchical mesoporous and macroporous carbon using the spinodal decomposition of a mesophase pitch (MP) carbon precursor and polystyrene as a soft template [64]. Scanning Electron Microscope (SEM) images of this structure revealed a 3D bicontinuous network of macropores and according to Hg porosimetry the average macropore size was recorded as 100 μm. The first reversible capacity of 470 mAh/g was recorded at a discharge-charge rate of C/5. When discharge-charge rates were increased to 1 and 5 C, reversible capacities of 320 and 200 mAh/g were obtained. Yang et al. [30] managed to synthesize a graphene-based mesoporous carbon anode which performed better than previous graphitic anode. Two-dimensional sandwich like graphene structure increases the surface area while each nanosheet acts as a mini-current collector. They facilitate the rapid transportation of electrons during charge-discharge cycles. At the rate of C/5, its reversible capacity stabilized at 770 mAh/g. When the discharge-charge rates were increased to 1 and 5 C, the reversible capacities recorded 540 and 370 mAh/g, respectively.

Graphene materials loaded with macroporous structures have shown positive results as anode materials for the Li-ion batteries. Mn_3O_4-graphene [65], Co_3O_4-graphene [66], and Fe_3O_4-graphene [67–69] have been studied extensively as potential anode materials for Li-ion materials. Chen et al. [69] reported a 3D graphene-Fe_3O_4 hybrid prepared by chemical reduction of the GO in the presence of Fe_3O_4 nanoparticles. The as-prepared hybrid was tested as an anode material for LiBs and exhibited capacities of 990 and 730 mAh/g at current densities of 800 and 1600 mA/g, respectively.

3.2. Li-sulfur batteries

For more than 20 years, the Li-ion battery has dominated the rechargeable battery market for portable devices and it is still the best choice for electric vehicles. But, when it comes to the electrical performance, a significant improvement is less likely as the performance of the Li-ion battery has almost reached its theoretical limits [70, 71]. Therefore, Li-S battery is considered as one of the potential candidates to replace the Li-ion battery as the next generation rechargeable battery. Sulfur is considered as the 10th most abundant element in the Earth. When employed as a cathode, it has a high specific capacity of 1675 mAh/g and it can deliver a specific energy of 2600 Wh/kg. However, several key issues have prevented the practical applications of Li-S batteries so far. The issues which need to be addressed are (i) poor electrical conductivity of sulfur and its final discharge products (Li_2S/Li_2S_2); (ii) large volume change of sulfur electrode during electrochemical cycling; and (iii) dissolution of polysulfides, intermediate reactant products in the organic electrolyte leading to deposition of Li_2S_2/Li_2S at the electrode interface. To overcome these drawbacks, extensive researches have been carried out to use graphene materials as scaffolds for cathodes in Li-S batteries [72–76].

For the first time, Wang et al. [77] synthesized a sulfur-graphene (S-GNS) composite by heating a mixture of graphene nanosheets and elemental sulfur. The electrochemical performance of the battery was unsatisfactory as S-GNS electrode contained only 17.6 wt% sulfur. Wang et al. [78] improved the performance of this cathode by increasing the sulfur content up to 44.5 wt % using the same synthesis method. The reversible capacities of the electrode were recorded as 662 mAh/g at 1 C and 391 mAh/g at 2 C after 100 cycles.

Kim et al. [79] produced mesoporous graphene-silica composite (m-GS) as a cathode structure to host sulfur for Li-S batteries. With the help of the ternary cooperative assembly of triblock copolymer (P123), silica precursor and graphene, porous silica structure was made parallel to graphene sheets. Sulfur was infiltrated into the mesoporus structure by melt diffusion at 155°C for 12 h. S intercalated graphite oxide cathode was made by in situ sulfur reduction and intercalation of graphite oxide [80]. By heating a mixture of S_8 and graphite oxide at 600°C under vacuum, would break large molecules of S_8 into S_2 and in the meantime reduce graphite oxide to graphene. Interplanar distance of the carbon matrix allows S_2 to intercalate into GO. To minimize the capacity decay, surface S_8 could be removed by CS_2. This specified cathode was able to maintain a reversible capacity of 880 mAh/g after 200 cycles.

To obtain better electrochemical performance, Zhang et al. [81] created dense nanopores on the surface of graphene nanosheets by chemically activating hydrothermally reduced graphene oxide (rGO). Sulfur was infiltrated into the KOH-activated graphene hydrogels by the melt diffusion method (**Figure 3**). The rGO hydrogel served as a trap for soluble polysulfides.

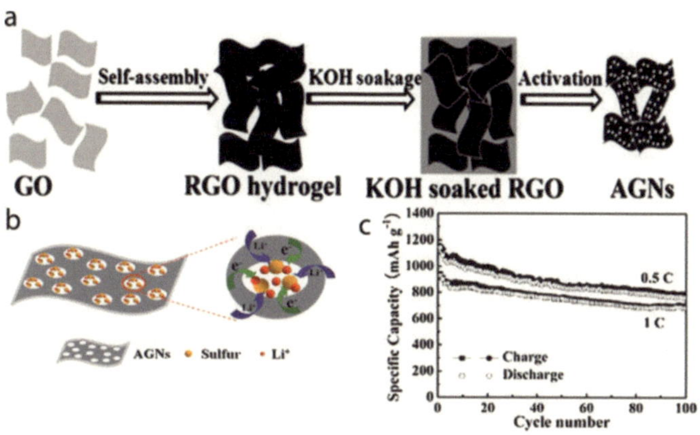

Figure 3. (a) Schematic representation of the preparation route of activated graphene nanosheets through self-assembly of GO, ion diffusion and chemical activation strategy. (b) Proposed scheme for the constrained electrochemical reaction process of the graphene/sulfur composite. (c) Cycling performances of graphene/sulfur composite electrode at 0.5 C and 1 C [81]. Reprinted with the permission of the Royal Society of Chemistry.

According to results from nitrogen sorption measurements, the surface area of the mesoporous system was 2313 m^2/g and the mean value of nanopores was 3.8 nm. At 0.5 C and 1 C, the graphene/sulfur composite electrode delivered high reversible capacities of 1143 and 927 mAh/g, respectively.

Evers and Nazar [82] prepared a graphene-sulfur cathode material with a sulfur loading of 87 wt% by a simple one pot method. A mixture of GO and soluble polysulfide was oxidized *in situ* as a one pot reaction. Because of the formation of insulating Li_2S layer, the initial discharge capacity of 705 mAh/g at 0.2 C decreased drastically after 50 cycles. S/rGO composite material for Li-S battery cathode was made by concurrently oxidizing sulfide and reducing GO [83]. In this method, Na_2S and Na_2SO_3 were mixed with the GO solution. The composite material is obtained by the reduction of GO by Na_2S. The composite with a sulfur loading of 63.6 wt% delivered a reversible capacity of 804 mAh/g after 80 cycles at 0.186 C and 440 mAh/g after 500 cycles at 0.75 C. Gao et al. [84] prepared a sulfur cathode composed of sulfur nanoparticles wrapped in graphene by using $Na_2S_2O_3$ as a precursor of sulfur. In acidic medium, $Na_2S_2O_3$ can also serve as a reducing agent of GO. Polyvinylpyrrolidone (PVP) was used to prevent the S particles from aggregation and to keep the sulfur particles at submicrometer range. By using $(NH_4)_2S_2O_3$ as a sulfur precursor, Xu et al. made a graphene-encapsulated sulfur composite. In this synthesis method, a mild reducing agent, urea, was used to reduce GO.

3.3. Supercapacitors

Supercapacitor is another major alternative solution for the energy storage applications. Supercapacitors have higher power densities than batteries and also higher energy densities than dielectric capacitors [46, 85–88]. The first attempt to use graphene as a supercapacitor was done by Rouff et al. [89] in 2008. In that method, GO was reduced by using hydrazine hydrate and the surface area as measured by BET method was 705 m^2/g. Specific capacitances of 135 and 99 F/g were obtained in aqueous and organic electrolytes, respectively. However, strong π-π stacking and van der Waals attractions among inter layers cause irreversible agglomeration to form graphite, resulting a decrease in surface area which may hinder the diffusion of the electrolyte. Therefore, making graphene in to a highly open porous structure is an effective way to increase the accessible surface area and the specific capacitance.

Zhu et al. [46] were able to make a carbon based supercapacitor by chemically activating the microwave exfoliated GO (MEGO) and thermally exfoliated GO (TEGO) using the KOH to obtain surface area values up to 3100 m^2/g and a high electrical conductivity of 500 S/m with a C/O atomic ratio of 35 . The specific capacitance values calculated from the charge-discharge curves were 165, 166, and 166 F/g at current densities of 1.4, 2.8, and 5.7 A/g, respectively.

Chen et al. [90] discovered a route to convert noncovalent functionalized graphene to a graphene-activated carbon composite by chemically activating with the KOH, which consisted of a specific surface area of 798 m^2/g. Stable graphene colloids absorbed by oligomers of p-phenylene diamne (PPD) were converted to a graphene-activated carbon composite by the KOH activation annealing method. The KOH activation created micro/mesopores in the activated carbon covered on graphene whereas pores in activated carbon also contributed the high surface area of the composite. The as-prepared graphene composite exhibited a specific capacitance of 122 F/g and energy density of 6.1 Wh/kg in aqueous electrolyte. Maximum energy density values of 52.2 and 99.2 Wh/kg were obtained in 1-ethyl-3-methylimidazolium tetrafluoroborate ($EMIMBF_4$) electrolyte at room temperature and 80°C, respectively.

Zhang et al. [91] introduced a method to produce porous 3D graphene-based bulk materials with ultrahigh specific area of 3523 m^2/g and excellent bulk conductivity (up to 303 S/m) by *in-situ* hydrothermal polymerization/carbonization of a mixture of industry carbon sources and the GO followed by KOH activation (**Figure 4**). The carbon sources used in this method were biomass, phenol-formaldehyde (PF), polyvinyl alcohol (PVA), sucrose, cellulose, and lignin. Graphene-PF composite material gave the highest specific capacitance values of 202 F/g in 1 M $TEABF_4$/AN and 231 F/g in neat $EMIMBF_4$ electrolyte systems, respectively.

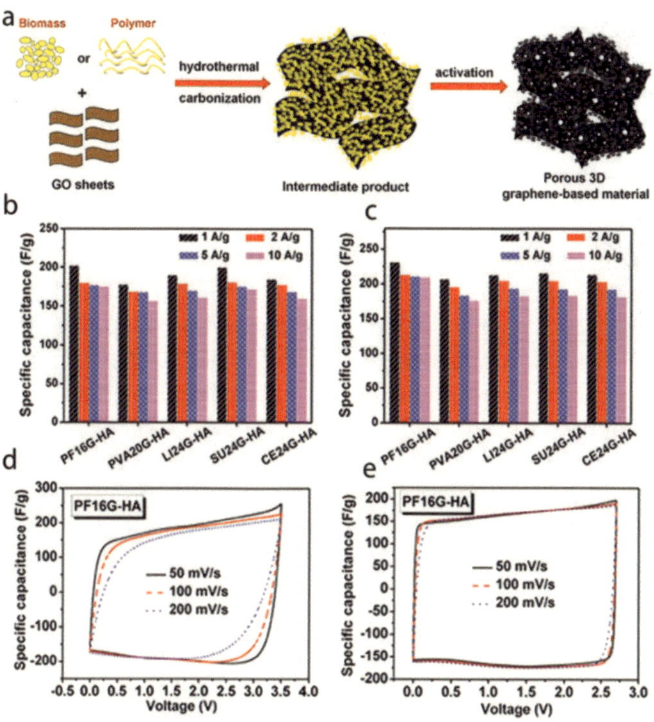

Figure 4. (a) Schematic representation of the synthesis procedure of the porous three-dimensional graphene-based materials. Galvanostatic charge/discharge test results of supercapacitors based on the optimized porous 3D graphene-based materials in (b) 1 M $TEABF_4$/AN and (c) neat $EMIMBF_4$ electrolytes under different current densities. CV curves of PF16G-HA based supercapacitor under different scan rates in (d) 1 $MTEABF_4$/AN and (e) neat $EMIMBF_4$ electrolyte. Reproduced with permission [91], Copyright 2013 NPG.

The electrochemical performance of carbon based materials can be enhanced by doping carbon network with nitrogen and boron [92–95]. Nitrogen and boron co-doped 3D graphene aerogel

(BN-GA) was fabricated by using the GO and ammonia boron trifluoride (NH_3BF_3) [96]. The interconnected framework of graphene nanosheets had a surface area of 249 m^2/g with a macroporous structure. BN-GAs were directly processed into thin electrodes without destroying the 3D continuous frameworks and used in all-solid-state supercapacitors (ASSSs). Because of the unique structure and strong synergetic effects of nitrogen and boron co-doping, a specific capacitance of 62 F/g and energy density of 8.65 Wh/kg were obtained. 3D graphene aerogels with mesoporous silica frameworks (GA-SiO_2) were fabricated by the hydrolysis of TEOS with graphene aerogel and CTAB as the soft template [97]. Graphene aerogel-mesoporous carbon (GA-MC) with a surface area of 295 m^2/g was generated by infiltrating a sucrose solution into the GA-SiO_2 followed by carbonization at 700°C for 3 h in argon. The as-prepared GA-MC exhibited a specific capacitance of 226 F/g when it was used as a supercapacitor.

3.4. Dye-sensitized solar cells

The dye-sensitized solar cells are among third generation photovoltaic devices that are cost-effective and highly efficient. It consists of a mesoporous TiO_2 photoanode with a dye to increase light absorption, a counter electrode (CE), and electrolyte. The CE should reduce redox species, which are used to regenerate the sensitizer after electron ejection. To increase the efficiency of DSSC, it is essential to select a CE material with low sheet resistance, high catalytic activity for the reduction of redox species, excellent chemical stability, and low cost. Recently, graphene-based CEs have been studied extensively as a potential cost-effective replacement for platinum based CEs.

Compared with other graphene-based materials, functionalized or doped graphene exhibits exceptional electocatalytic activity. In 2012, Xu et al. prepared Hemin, an iron-containing porphyrin functionalized rGO by microwave irradiation [98]. The Hemin-rGO hybrid exhibited a power conversion efficiency (PCE) value of 2.45 %. Yen et al. [99] reported a nitrogen-doped graphene prepared using a hydrothermal method. The nitrogen-doped domains on the graphene surface act as electroactive sites, which have selectivity for redox species in the reduction reaction. The as-prepared nitrogen-doped graphene CE exhibited the PCE value of 4.75%. Xue et al. [100] managed to prepare 3D nitrogen-doped graphene foams with a nitrogen content of 7.6% freeze drying the GO foams followed by annealing at 800°C in ammonia/argon mixture for 1 h. Because of the high content of nitrogen, the PCE value of 7.07% was obtained.

4. Conclusion

In conclusion, owing to its high surface area, unique pore structure, and remarkable electrochemical performances, porous graphene has attracted great attention in the fields of energy storage and conversion. However, there are several key issues, which need to be addressed. The precise control of pore size, pore morphology, and wall thickness is necessary for the assembly of hierarchically structured porous graphene materials. Introduction of different sizes of pores into graphene matrix is essential to produce porous graphene materials to obtain synergic effects of different pores. With increasing research efforts in the field, we believe that there would be significant advances in the synthesis and application of porous graphene in the near future, benefiting development of high performance energy conversion and storage devices. This research was partially supported under the Australian Research Council Discovery Project (DP150101717). Kimal Chandula Wasalathilake acknowledges the QUTPRA scholarship from the Queensland University of Technology.

Author details

Kimal Chandula Wasalathilake, Godwin Ayoko and Cheng Yan*

*Address all correspondence to: c2.yan@qut.edu.au

School of Chemistry, Physics and Mechanical Engineering, Science and Engineering Faculty, Queensland University of Technology (QUT), Brisbane, Australia

References

[1] Davis ME. Ordered porous materials for emerging applications. Nature. 2002;417(6891):813–821.

[2] Cooper AI. Porous materials and supercritical fluids. Advanced Materials. 2003;15(13): 1049–1059. DOI: 10.1002/adma.200300380.

[3] Zhang H,Cooper AI. Synthesis and applications of emulsion-templated porous materials. Soft Matter. 2005;1(2):107–113. DOI: 10.1039/B502551F.

[4] Lee J, Kim J,Hyeon T. Recent progress in the synthesis of porous carbon materials. Advanced Materials. 2006;18(16):2073–2094. DOI: 10.1002/adma.200501576.

[5] Lu AH, Schüth F. Nanocasting: a versatile strategy for creating nanostructured porous materials. Advanced Materials. 2006;18(14):1793–1805. DOI: 10.1002/adma.200600148.

[6] White RJ, Budarin V, Luque R, Clark JH, Macquarrie DJ. Tuneable porous carbonaceous materials from renewable resources. Chemical Society Reviews. 2009;38(12):3401–3418. DOI: 10.1039/B822668G.

[7] White RJ, Luque R, Budarin VL, Clark JH, Macquarrie DJ. Supported metal nanoparticles on porous materials. Methods and applications. Chemical Society Reviews. 2009;38(2):481–494. DOI: 10.1039/B802654H.

[8] Thomas A. Functional materials: from hard to soft porous frameworks. Angewandte Chemie International Edition. 2010;49(45):8328–8344. DOI: 10.1002/anie.201000167.

[9] Bae Y-S, Snurr RQ. Development and evaluation of porous materials for carbon dioxide separation and capture. Angewandte Chemie International Edition. 2011;50(49):11586–11596. DOI: 10.1002/anie.201101891.

[10] Boissiere C, Grosso D, Chaumonnot A, Nicole L, Sanchez C. Aerosol route to functional nanostructured inorganic and hybrid porous materials. Advanced Materials. 2011;23(5):599–623. DOI: 10.1002/adma.201001410.

[11] Geim AK, Novoselov KS. The rise of graphene. Nature Material 2007;6(3):183–191.

[12] Geim AK. Graphene: status and prospects. Science. 2009;324(5934):1530–1534. DOI: 10.1126/science.1158877.

[13] Novoselov KS, Falko VI, Colombo L, Gellert PR, Schwab MG,Kim K. A roadmap for graphene. Nature. 2012;490(7419):192–200.

[14] Berger C, Song Z, Li T, Li X, Ogbazghi AY, Feng R, Dai Z, Marchenkov AN, Conrad EH, First PN,de Heer WA. Ultrathin epitaxial graphite: 2D electron gas properties and a route toward graphene-based nanoelectronics. The Journal of Physical Chemistry B. 2004;108(52):19912–19916. DOI: 10.1021/jp040650f.

[15] Edwards RS,Coleman KS. Graphene film growth on polycrystalline metals. Accounts of Chemical Research. 2013;46(1):23–30. DOI: 10.1021/ar3001266.

[16] Wu B, Geng D, Guo Y, Huang L, Xue Y, Zheng J, Chen J, Yu G, Liu Y, Jiang L, Hu W. Equiangular hexagon-shape-controlled synthesis of graphene on copper surface. Advanced Materials. 2011;23(31):3522–3525. DOI: 10.1002/adma.201101746.

[17] Xue Y, Wu B, Jiang L, Guo Y, Huang L, Chen J, Tan J, Geng D, Luo B, Hu W, Yu G, Liu Y. Low temperature growth of highly nitrogen-doped single crystal graphene arrays by chemical vapor deposition. Journal of the American Chemical Society. 2012;134(27): 11060–11063. DOI: 10.1021/ja302483t.

[18] Li X, Cai W, An J, Kim S, Nah J, Yang D, Piner R, Velamakanni A, Jung I, Tutuc E, Banerjee SK, Colombo L,Ruoff RS. Large-area synthesis of high-quality and uniform graphene films on copper foils. Science. 2009;324(5932):1312–1314. DOI: 10.1126/science.1171245.

[19] Pei S,Cheng H-M. The reduction of graphene oxide. Carbon. 2012;50(9):3210–3228. DOI: http://dx.doi.org/10.1016/j.carbon.2011.11.010.

[20] Park S,Ruoff RS. Chemical methods for the production of graphenes. Nature Nano. 2009;4(4):217–224.

[21] Zhang C, Lv W, Xie X, Tang D, Liu C, Yang Q-H. Towards low temperature thermal exfoliation of graphite oxide for graphene production. Carbon. 2013;62:11–24. DOI: http://dx.doi.org/10.1016/j.carbon.2013.05.033.

[22] Liu Y-Z, Chen C-M, Li Y-F, Li X-M, Kong Q-Q, Wang M-Z. Crumpled reduced graphene oxide by flame-induced reduction of graphite oxide for supercapacitive energy storage. Journal of Materials Chemistry A. 2014;2(16):5730–5737. DOI: 10.1039/C3TA15082H.

[23] Liu J, Yang H, Zhen SG, Poh CK, Chaurasia A, Luo J, Wu X, Yeow EKL, Sahoo NG, Lin J,Shen Z. A green approach to the synthesis of high-quality graphene oxide flakes via electrochemical exfoliation of pencil core. RSC Advances. 2013;3(29):11745–11750. DOI: 10.1039/C3RA41366G.

[24] Zhao M-Q, Zhang Q, Huang J-Q,Wei F. Hierarchical nanocomposites derived from nanocarbons and layered double hydroxides – properties, synthesis, and applications. Advanced Functional Materials. 2012;22(4):675–694. DOI: 10.1002/adfm.201102222.

[25] Cui X, Zhang C, Hao R, Hou Y. Liquid-phase exfoliation, functionalization and applications of graphene. Nanoscale. 2011;3(5):2118–2126. DOI: 10.1039/C1NR10127G.

[26] Nicolosi V, Chhowalla M, Kanatzidis MG, Strano MS, Coleman JN. Liquid exfoliation of layered materials. Science. 2013;340(6139). DOI: 10.1126/science.1226419–18.

[27] Han S, Wu D, Li S, Zhang F,Feng X. Porous graphene materials for advanced electrochemical energy storage and conversion devices. Advanced Materials. 2014;26(6):849–864. DOI: 10.1002/adma.201303115.

[28] Jiang L,Fan Z. Design of advanced porous graphene materials: from graphene nanomesh to 3D architectures. Nanoscale. 2014;6(4):1922–1945. DOI: 10.1039/C3NR04555B.

[29] Yan Z, Yao W, Hu L, Liu D, Wang C, Lee C-S. Progress in the preparation and application of three-dimensional graphene-based porous nanocomposites. Nanoscale. 2015;7(13):5563–5577. DOI: 10.1039/C5NR00030K.

[30] Yang S, Feng X, Wang L, Tang K, Maier J, Müllen K. Graphene-based nanosheets with a sandwich structure. Angewandte Chemie. 2010;122(28):4905–4909. DOI: 10.1002/ange.201001634.

[31] Yang S, Yue W, Zhu J, Ren Y, Yang X. Graphene-based mesoporous SnO_2 with enhanced electrochemical performance for lithium-ion batteries. Advanced Functional Materials. 2013;23(28):3570–3576. DOI: 10.1002/adfm.201203286.

[32] Wang L, Sun L, Tian C, Tan T, Mu G, Zhang H, Fu H. A novel soft template strategy to fabricate mesoporous carbon/graphene composites as high-performance supercapacitor electrodes. RSC Advances. 2012;2(22):8359–8367. DOI: 10.1039/C2RA20845H.

[33] Fang Y, Lv Y, Che R, Wu H, Zhang X, Gu D, Zheng G,Zhao D. Two-dimensional mesoporous carbon nanosheets and their derived graphene nanosheets: synthesis and efficient lithium ion storage. Journal of the American Chemical Society. 2013;135(4): 1524–1530. DOI: 10.1021/ja310849c.

[34] Wen X, Zhang D, Yan T, Zhang J,Shi L. Three-dimensional graphene-based hierarchically porous carbon composites prepared by a dual-template strategy for capacitive

deionization. Journal of Materials Chemistry A. 2013;1(39):12334–12344. DOI: 10.1039/C3TA12683H.

[35] Huang X, Qian K, Yang J, Zhang J, Li L, Yu C, Zhao D. Functional nanoporous graphene foams with controlled pore sizes. Advanced Materials. 2012;24(32):4419–4423. DOI: 10.1002/adma.201201680.

[36] Choi BG, Yang M, Hong WH, Choi JW,Huh YS. 3D macroporous graphene frameworks for supercapacitors with high energy and power densities. ACS Nano. 2012;6(5):4020–4028. DOI: 10.1021/nn3003345.

[37] Cao X, Shi Y, Shi W, Lu G, Huang X, Yan Q, Zhang Q,Zhang H. Preparation of novel 3D graphene networks for supercapacitor applications. Small. 2011;7(22):3163–3168. DOI: 10.1002/smll.201100990.

[38] Dong X-C, Xu H, Wang X-W, Huang Y-X, Chan-Park MB, Zhang H, Wang L-H, Huang W,Chen P. 3D Graphene–cobalt oxide electrode for high-performance supercapacitor and enzymeless glucose detection. ACS Nano. 2012;6(4):3206–3213. DOI: 10.1021/nn300097q.

[39] Maiyalagan T, Dong X, Chen P,Wang X. Electrodeposited Pt on three-dimensional interconnected graphene as a free-standing electrode for fuel cell application. Journal of Materials Chemistry. 2012;22(12):5286–5290. DOI: 10.1039/C2JM16541D.

[40] Yong Y-C, Dong X-C, Chan-Park MB, Song H, Chen P. Macroporous and monolithic anode based on polyaniline hybridized three-dimensional graphene for high-performance microbial fuel cells. ACS Nano. 2012;6(3):2394–2400. DOI: 10.1021/nn204656d.

[41] Qiu H, Dong X, Sana B, Peng T, Paramelle D, Chen P,Lim S. Ferritin-templated synthesis and self-assembly of Pt nanoparticles on a monolithic porous graphene network for electrocatalysis in fuel cells. ACS Applied Materials & Interfaces. 2013;5(3):782–787. DOI: 10.1021/am3022366.

[42] Wang M, Fu L, Gan L, Zhang C, Rümmeli M, Bachmatiuk A, Huang K, Fang Y, Liu Z. CVD Growth of large area smooth-edged graphene nanomesh by nanosphere lithography. Scientific Reports. 2013;3:1238. DOI: 10.1038/srep01238. http://www.nature.com/articles/srep01238#supplementary-information

[43] Chen Z, Xu C, Ma C, Ren W, Cheng H-M. Lightweight and flexible graphene foam composites for high-performance electromagnetic interference shielding. Advanced Materials. 2013;25(9):1296–1300. DOI: 10.1002/adma.201204196.

[44] Chen Z, Ren W, Gao L, Liu B, Pei S,Cheng H-M. Three-dimensional flexible and conductive interconnected graphene networks grown by chemical vapour deposition. Nature Material. 2011;10(6):424–428. DOI: http://www.nature.com/nmat/journal/v10/n6/abs/nmat3001.html#supplementary-information

[45] Chen C-M, Zhang Q, Huang C-H, Zhao X-C, Zhang B-S, Kong Q-Q, Wang M-Z, Yang Y-G, Cai R, Sheng Su D. Macroporous 'bubble' graphene film via template-directed

ordered-assembly for high rate supercapacitors. Chemical Communications. 2012;48(57):7149–7151. DOI: 10.1039/C2CC32189K.

[46] Zhu Y, Murali S, Stoller MD, Ganesh KJ, Cai W, Ferreira PJ, Pirkle A, Wallace RM, Cychosz KA, Thommes M, Su D, Stach EA, Ruoff RS. Carbon-based supercapacitors produced by activation of graphene. Science. 2011;332(6037):1537–1541. DOI: 10.1126/science.1200770.

[47] Lillo-Ródenas MA, Cazorla-Amorós D, Linares-Solano A. Understanding chemical reactions between carbons and NaOH and KOH: an insight into the chemical activation mechanism. Carbon. 2003;41(2):267–275. DOI: http://dx.doi.org/10.1016/S0008-6223(02)00279-8.

[48] Rui X, Zhu J, Sim D, Xu C, Zeng Y, Hng HH, Lim TM,Yan Q. Reduced graphene oxide supported highly porous V_2O_5 spheres as a high-power cathode material for lithium ion batteries. Nanoscale. 2011;3(11):4752–4758. DOI: 10.1039/C1NR10879D.

[49] Chen Y, Wang Q, Zhu C, Gao P, Ouyang Q, Wang T, Ma Y,Sun C. Graphene/porous cobalt nanocomposite and its noticeable electrochemical hydrogen storage ability at room temperature. Journal of Materials Chemistry. 2012;22(13):5924–5927. DOI: 10.1039/C2JM16825A.

[50] Lee JM, Kim IY, Han SY, Kim TW,Hwang S-J. Graphene nanosheets as a platform for the 2D ordering of metal oxide nanoparticles: mesoporous 2D aggregate of anatase TiO_2 nanoparticles with improved electrode performance. Chemistry – A European Journal. 2012;18(43):13800–13809. DOI: 10.1002/chem.201200551.

[51] Yan J, Sun W, Wei T, Zhang Q, Fan Z, Wei F. Fabrication and electrochemical performances of hierarchical porous $Ni(OH)_2$ nanoflakes anchored on graphene sheets. Journal of Materials Chemistry. 2012;22(23):11494–11502. DOI: 10.1039/C2JM30221G.

[52] Yang S, Sun Y, Chen L, Hernandez Y, Feng X, Müllen K. Porous iron oxide ribbons grown on graphene for high-performance lithium storage. Scientific Reports. 2012;2:427. DOI: 10.1038/srep00427. http://www.nature.com/articles/srep00427#supplementary-information.

[53] Zheng M, Qiu D, Zhao B, Ma L, Wang X, Lin Z, Pan L, Zheng Y,Shi Y. Mesoporous iron oxide directly anchored on a graphene matrix for lithium-ion battery anodes with enhanced strain accommodation. RSC Advances. 2013;3(3):699–703. DOI: 10.1039/C2RA22702A.

[54] Liu J, Cai H, Yu X, Zhang K, Li X, Li J, Pan N, Shi Q, Luo Y,Wang X. Fabrication of graphene nanomesh and improved chemical enhancement for Raman Spectroscopy. The Journal of Physical Chemistry C. 2012;116(29):15741–15746. DOI: 10.1021/jp303265d.

[55] Yan Y, Yin Y-X, Xin S, Guo Y-G, Wan L-J. Ionothermal synthesis of sulfur-doped porous carbons hybridized with graphene as superior anode materials for lithium-ion batteries. Chemical Communications. 2012;48(86):10663–10665. DOI: 10.1039/C2CC36234A.

[56] Nardecchia S, Carriazo D, Ferrer ML, Gutierrez MC, del Monte F. Three dimensional macroporous architectures and aerogels built of carbon nanotubes and/or graphene: synthesis and applications. Chemical Society Reviews. 2013;42(2):794–830. DOI: 10.1039/C2CS35353A.

[57] Li C,Shi G. Functional gels based on chemically modified graphenes. Advanced Materials. 2014;26(24):3992–4012. DOI: 10.1002/adma.201306104.

[58] Chabot V, Higgins D, Yu A, Xiao X, Chen Z, Zhang J. A review of graphene and graphene oxide sponge: material synthesis and applications to energy and the environment. Energy & Environmental Science. 2014;7(5):1564–1596. DOI: 10.1039/C3EE43385D.

[59] Wu D, Zhang F, Liang H, Feng X. Nanocomposites and macroscopic materials: assembly of chemically modified graphene sheets. Chemical Society Reviews. 2012;41(18):6160–6177. DOI: 10.1039/C2CS35179J.

[60] Xu Y, Sheng K, Li C, Shi G. Self-assembled graphene hydrogel via a one-step hydrothermal process. ACS Nano. 2010;4(7):4324–4330. DOI: 10.1021/nn101187z.

[61] Zhang L,Shi G. Preparation of highly conductive graphene hydrogels for fabricating supercapacitors with high rate capability. The Journal of Physical Chemistry C. 2011;115(34):17206–17212. DOI: 10.1021/jp204036a.

[62] Tarascon JM, Armand M. Issues and challenges facing rechargeable lithium batteries. Nature. 2001;414(6861):359–367.

[63] Aleksandrzak M,Mijowska E, Graphene and its derivatives for energy storage. In: Graphene Materials. 2015, A. Tiwari and M. Syväjärvi. Hoboken, NJ, USA John Wiley & Sons, Inc. p. 191–224.

[64] Adelhelm P, Hu YS, Chuenchom L, Antonietti M, Smarsly BM,Maier J. Generation of hierarchical meso- and macroporous carbon from mesophase pitch by spinodal decomposition using polymer templates. Advanced Materials. 2007;19(22):4012–4017. DOI: 10.1002/adma.200700699.

[65] Wang H, Cui L-F, Yang Y, Sanchez Casalongue H, Robinson JT, Liang Y, Cui Y,Dai H. Mn_3O_4–graphene hybrid as a high-capacity anode material for lithium ion batteries. Journal of the American Chemical Society. 2010;132(40):13978–13980. DOI: 10.1021/ja105296a.

[66] Li B, Cao H, Shao J, Li G, Qu M,Yin G. Co3O4@graphene composites as anode materials for high-performance lithium ion batteries. Inorganic Chemistry. 2011;50(5):1628–1632. DOI: 10.1021/ic1023086.

[67] Zhou G, Wang D-W, Li F, Zhang L, Li N, Wu Z-S, Wen L, Lu GQ,Cheng H-M. Graphene-wrapped Fe_3O_4 anode material with improved reversible capacity and cyclic stability for lithium ion batteries. Chemistry of Materials. 2010;22(18):5306–5313. DOI: 10.1021/cm101532x.

[68] Zhuo L, Wu Y, Wang L, Ming J, Yu Y, Zhang X, Zhao F. CO_2-expanded ethanol chemical synthesis of a Fe_3O_4@graphene composite and its good electrochemical properties as anode material for Li-ion batteries. Journal of Materials Chemistry A. 2013;1(12):3954–3960. DOI: 10.1039/C3TA01388J.

[69] Chen W, Li S, Chen C, Yan L. Self-assembly and embedding of nanoparticles by in situ reduced graphene for preparation of a 3D graphene/nanoparticle aerogel. Advanced Materials. 2011;23(47):5679–5683. DOI: 10.1002/adma.201102838.

[70] Goodenough JB,Kim Y. Challenges for rechargeable Li batteries. Chemistry of Materials. 2010;22(3):587–603. DOI: 10.1021/cm901452z.

[71] Manthiram A. Materials challenges and opportunities of lithium ion batteries. The Journal of Physical Chemistry Letters. 2011;2(3):176–184. DOI: 10.1021/jz1015422.

[72] Kim H, Lim H-D, Kim J, Kang K. Graphene for advanced Li/S and Li/air batteries. Journal of Materials Chemistry A. 2014;2(1):33–47. DOI: 10.1039/C3TA12522J.

[73] Yu M, Li R, Wu M, Shi G. Graphene materials for lithium–sulfur batteries. Energy Storage Materials. 2015;1:51-73. DOI: http://dx.doi.org/10.1016/j.ensm.2015.08.004.

[74] Manthiram A, Fu Y, Chung S-H, Zu C, Su Y-S. Rechargeable lithium–sulfur batteries. Chemical Reviews. 2014;114(23):11751–11787. DOI: 10.1021/cr500062v.

[75] Manthiram A, Chung S-H, Zu C. Lithium–sulfur batteries: progress and prospects. Advanced Materials. 2015;27(12):1980–2006. DOI: 10.1002/adma.201405115.

[76] Lin Z,Liang C. Lithium-sulfur batteries: from liquid to solid cells. Journal of Materials Chemistry A. 2015;3(3):936–958. DOI: 10.1039/C4TA04727C.

[77] Wang J-Z, Lu L, Choucair M, Stride JA, Xu X, Liu H-K. Sulfur-graphene composite for rechargeable lithium batteries. Journal of Power Sources. 2011;196(16):7030–7034. DOI: http://dx.doi.org/10.1016/j.jpowsour.2010.09.106.

[78] Wang B, Li K, Su D, Ahn H,Wang G. Superior electrochemical performance of sulfur/graphene nanocomposite material for high-capacity lithium–sulfur batteries. Chemistry – An Asian Journal. 2012;7(7):1637–1643. DOI: 10.1002/asia.201200004.

[79] Kim KH, Jun Y-S, Gerbec JA, See KA, Stucky GD, Jung H-T. Sulfur infiltrated mesoporous graphene–silica composite as a polysulfide retaining cathode material for lithium–sulfur batteries. Carbon. 2014;69:543–551. DOI: http://dx.doi.org/10.1016/j.carbon.2013.12.065.

[80] Zheng S, Wen Y, Zhu Y, Han Z, Wang J, Yang J, Wang C. In situ sulfur reduction and intercalation of graphite oxides for Li-S battery cathodes. Advanced Energy Materials. 2014;4(16):n/a-n/a. DOI: 10.1002/aenm.1400482–1 to 1400482–9.

[81] Ding B, Yuan C, Shen L, Xu G, Nie P, Lai Q, Zhang X. Chemically tailoring the nanostructure of graphene nanosheets to confine sulfur for high-performance lithium-sulfur batteries. Journal of Materials Chemistry A. 2013;1(4):1096–1101. DOI: 10.1039/C2TA00396A.

[82] Evers S, Nazar LF. Graphene-enveloped sulfur in a one pot reaction: a cathode with good coulombic efficiency and high practical sulfur content. Chemical Communications. 2012;48(9):1233–1235. DOI: 10.1039/C2CC16726C.

[83] Sun H, Xu G-L, Xu Y-F, Sun S-G, Zhang X, Qiu Y, Yang S. A composite material of uniformly dispersed sulfur on reduced graphene oxide: aqueous one-pot synthesis, characterization and excellent performance as the cathode in rechargeable lithium-sulfur batteries. Nano Research. 2012;5(10):726–738. DOI: 10.1007/s12274-012-0257-7.

[84] Gao X, Li J, Guan D,Yuan C. A scalable graphene sulfur composite synthesis for rechargeable lithium batteries with good capacity and excellent columbic efficiency. ACS Applied Materials & Interfaces. 2014;6(6):4154–4159. DOI: 10.1021/am4057979.

[85] Simon P, Gogotsi Y. Materials for electrochemical capacitors. Nature Material 2008;7(11):845–854.

[86] Liu C, Li F, Ma L-P,Cheng H-M. Advanced materials for energy storage. Advanced Materials. 2010;22(8):E28–E62. DOI: 10.1002/adma.200903328.

[87] Zhai Y, Dou Y, Zhao D, Fulvio PF, Mayes RT, Dai S. Carbon materials for chemical capacitive energy storage. Advanced Materials. 2011;23(42):4828–4850. DOI: 10.1002/adma.201100984.

[88] Zhang X, Zhang H, Li C, Wang K, Sun X, Ma Y. Recent advances in porous graphene materials for supercapacitor applications. RSC Advances. 2014;4(86):45862–45884. DOI: 10.1039/C4RA07869A.

[89] Stoller MD, Park S, Zhu Y, An J, Ruoff RS. Graphene-based ultracapacitors. Nano Letters. 2008;8(10):3498–3502. DOI: 10.1021/nl802558y.

[90] Chen Y, Zhang X, Zhang H, Sun X, Zhang D, Ma Y. High-performance supercapacitors based on a graphene-activated carbon composite prepared by chemical activation. RSC Advances. 2012;2(20):7747–7753. DOI: 10.1039/C2RA20667F.

[91] Zhang L, Zhang F, Yang X, Long G, Wu Y, Zhang T, Leng K, Huang Y, Ma Y, Yu A,Chen Y. Porous 3D graphene-based bulk materials with exceptional high surface area and excellent conductivity for supercapacitors. Scientific Reports. 2013;3:1408. DOI: 10.1038/srep01408. http://www.nature.com/articles/srep01408#supplementary-information.

[92] Jeong HM, Lee JW, Shin WH, Choi YJ, Shin HJ, Kang JK,Choi JW. Nitrogen-doped graphene for high-performance ultracapacitors and the importance of nitrogen-doped sites at basal planes. Nano Letters. 2011;11(6):2472–2477. DOI: 10.1021/nl2009058.

[93] Wu Z-S, Ren W, Xu L, Li F,Cheng H-M. Doped graphene sheets as anode materials with superhigh rate and large capacity for lithium ion batteries. ACS Nano. 2011;5(7):5463–5471. DOI: 10.1021/nn2006249.

[94] Wang S, Iyyamperumal E, Roy A, Xue Y, Yu D,Dai L. Vertically aligned BCN nanotubes as efficient metal-free electrocatalysts for the oxygen reduction reaction: a synergetic effect by co-doping with boron and nitrogen. Angewandte Chemie International Edition. 2011;50(49):11756–11760. DOI: 10.1002/anie.201105204.

[95] Qiu Y, Zhang X,Yang S. High performance supercapacitors based on highly conductive nitrogen-doped graphene sheets. Physical Chemistry Chemical Physics. 2011;13(27): 12554–12558. DOI: 10.1039/C1CP21148J.

[96] Wu Z-S, Winter A, Chen L, Sun Y, Turchanin A, Feng X,Müllen K. Three-dimensional nitrogen and boron co-doped graphene for high-performance all-solid-state supercapacitors. Advanced Materials. 2012;24(37):5130–5135. DOI: 10.1002/adma.201201948.

[97] Wu Z-S, Sun Y, Tan Y-Z, Yang S, Feng X,Müllen K. Three-dimensional graphene-based macro- and mesoporous frameworks for high-performance electrochemical capacitive energy storage. Journal of the American Chemical Society. 2012;134(48):19532–19535. DOI: 10.1021/ja308676h.

[98] Xu C, Li J, Wang X, Wang J, Wan L, Li Y, Zhang M, Shang X,Yang Y. Synthesis of hemin functionalized graphene and its application as a counter electrode in dye-sensitized solar cells. Materials Chemistry and Physics. 2012;132(2–3):858–864. DOI: http://dx.doi.org/10.1016/j.matchemphys.2011.12.025

[99] Yen M-Y, Hsieh C-K, Teng C-C, Hsiao M-C, Liu P-I, Ma C-CM, Tsai M-C, Tsai C-H, Lin Y-R, Chou T-Y. Metal-free, nitrogen-doped graphene used as a novel catalyst for dye-sensitized solar cell counter electrodes. RSC Advances. 2012;2(7):2725–2728. DOI: 10.1039/C2RA00970F.

[100] Xue Y, Liu J, Chen H, Wang R, Li D, Qu J,Dai L. Nitrogen-doped graphene foams as metal-free counter electrodes in high-performance dye-sensitized solar cells. Angewandte Chemie International Edition. 2012;51(48):12124–12127. DOI: 10.1002/anie.201207277.

Chapter 10

Defects in Graphene and its Derivatives

Soumyajyoti Haldar and Biplab Sanyal

Additional information is available at the end of the chapter

http://dx.doi.org/10.5772/64297

Abstract

The experimental realization of graphene along with its unique properties in 2004 triggered huge scientific researches in the field of graphene and other two-dimensional (2D) materials. The experimental preparation processes of these materials are prone to defect formation. These defects affect the properties of the pristine system, which can be beneficial or detrimental from the application point of view. In this book chapter, we discuss a few cases of defects in 2D materials such as graphene and its derivatives and their roles in applications.

Keywords: defects in graphene and its derivatives, graphene defects, hybrid materials, gas sensing, *ab initio* theory, magnetism

1. Introduction

The (re)discovery [1, 2] of graphene—a single layer of carbon atoms arranged in a honeycomb lattice—in 2004 by Novoselov et al. has triggered a new aspect of research in two-dimensional (2D) materials [3, 4]. Although the existence of materials with their properties governed by their 2D units was well known for quite some time [5, 6], it is the experimental realization of a single layer graphene has showed that it is possible to exfoliate stable 2D materials from the 3D solids exhibiting various fascinating properties.

A huge number of crystalline solid-state materials having different mechanical, electronic, and transport properties exist from which stable 2D materials can be created due to the presence of weak interaction between the layers [7]. 2D allotropes (e.g., silicene, graphyne, germanene), compounds (e.g., graphane, hexagonal boron nitride, transition metal di-chalcogenides) are the few examples of 2D materials. These 2D materials have the potential for a wide range of applications due to the interesting electronic and structural properties [2, 8–12].

To exploit these various properties, the samples have to be made in a scalable way. Chemical vapor deposition (CVD) has become a very common method for large-scale fabrication. Nonetheless, the CVD samples inevitably contain defects, for example, edges, hetero structures, grain boundaries, vacancies, and interstitial impurities [13–15]. These defects can be seen very easily in transmission electron microscopy (TEM) experiments [16] or scanning tunnelling microscopy (STM) experiments [17]. **Figure 1a, b** shows experimental STM and TEM images of an isolated single vacancy in graphene. In the STM image, the single vacancy can be seen as a blob because of increased local density of states. These states appear due to the presence of dangling bonds around the single vacancy.

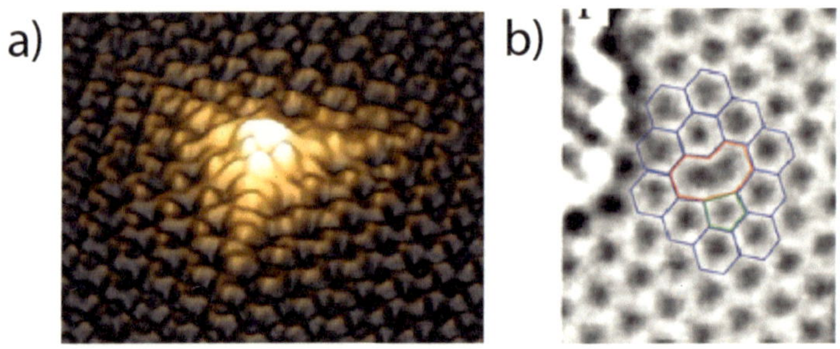

Figure 1. (a) Experimental STM image of single isolated vacancy in graphene. Reprinted with permission from Ugeda et al. [17], copyright (2010) by the American Physical Society. (b) Experimental TEM image of reconstructed single vacancy with atomic configurations. Reprinted (adapted) with permission from Meyer et al. [16], copyright (2008) by the American Chemical Society.

In general, these defects manipulate the properties of the materials and hence their avoidance or deliberate engineering requires a thorough understanding. In one hand, defects can be detrimental to device properties [13], but on the other hand, especially at the nanoscale, defects can bring new functionalities which could be utilized for applications [18, 19].

In this book chapter, we address a few cases of defects in 2D materials such as graphene and its derivatives. We show how one can tune the various properties of the pristine materials with the control insertion of defects in these systems and use them in various applications.

2. Theoretical methods

We have mainly used *ab initio* density functional theory-based methods to calculate various properties of defected 2D materials such as graphene and its derivatives in general. In this section, we will provide a brief introduction to the theoretical methods used.

2.1. Density functional theory

Various different properties of these many-body systems are described by the wave functions associated with it. These wave functions are governed by the time-dependent Schrödinger equation

$$\hat{H}\psi = E\psi \tag{1}$$

where $\hat{H}$ is the Hamiltonian of the many-body systems and represents the energy operator, and E is the total energy of the system. However, one needs various approximations to solve the Schrödinger equation for all kinds of systems.

In density functional theory (DFT), the electron density $n(\vec{r})$ is used to obtain the solution of the Schrödinger equation. The core concept of the DFT is given by two theorems of Hohenberg and Kohn [20], where they showed that the properties of interacting systems can be obtained exactly by the ground state electron density, $n_0(\vec{r})$. Following the two theorems, the total energy of the system can be written as follows:

$$E\left[n(\bar{r})\right] = F\left[n(\bar{r})\right] + \int V_{ext}(\bar{r})n(\bar{r})\, d\bar{r} \tag{2}$$

Where functional F represents kinetic energy and all electron-electron interactions. Functional F does not depend on the external potential, and hence, it is same for all the systems. However, Hohenberg-Kohn theorem does not provide any solution toward the exact form of the functional F.

Kohn and Sham [21] gave a way around to obtain the functional F by replacing the interacting many-body system with a non-interacting system consisting of a set of one electron functions (orbitals) while keeping the same ground state. According to the Kohn-Sham formalism, the total energy functional can be written as follows:

$$E\left[n(\bar{r})\right] = T_S\left[n(\bar{r})\right] + \int V_{ext}(\bar{r})n(\bar{r})\, d\bar{r} + \frac{1}{2}\iint \frac{n(\bar{r}_1)n(\bar{r}_2)}{\left|\bar{r}_1 - \bar{r}_2\right|} d\bar{r}_1 d\bar{r}_2 + E_{XC}\left[n(\bar{r})\right] \tag{3}$$

Where T_S is the kinetic energy term of the non-interacting electrons, and V_{ext} is the external potential. The third term in the above equation is the Hartree term representing the classical Coulomb interactions between electrons, and the last term is known as exchange-correlation energy (E_{XC}), which contains all the many-body effects. The formalism of Kohn-Sham is an exact theory. If the form of the E_{XC} is exactly known, then using this formalism, one can calculate the exact ground state of the interacting many-body system.

In reality, the exact form of the exchange-correlation is not trivial, and hence, it is necessary to model the form of the exchange-correlation. Different forms of exchange-correlation can be constructed depending upon various level of approximation, for example, local density approximation (LDA) [20, 22, 23], generalized gradient approximation (GGA) [24–26], hybrid functionals (a mixture of Hartree-Fock and DFT functionals) etc. It is also important to remember that the implementation of single-particle Kohn-Sham equation is not trivial due to the complex behavior of wave functions in different spatial region, for example, in the core and in the valence region. To describe this complex wave function, a complete basis function is needed which can be of different form, for example, plane waves, localized atomic-like orbitals, Gaussian functions etc.

3. Manipulation of properties of graphene with defects

As mentioned in the introduction, an immense amount of scientific activities is going on in the field of graphene research because of its special properties [27, 28]. However, the lack of band gap limits the usage of graphene in electronic device applications. Therefore, the modification and tuning of graphene properties to open up an energy gap have become a cutting edge research interest among the scientific community. In this section, we show a few examples of manipulating the properties of graphene and hybrid systems with graphene.

3.1. Magnetic impurities in graphene/graphane interface

Graphane—another 2D material—is hydrogenated graphene, where each carbon atom is attached with a hydrogen atom. Unlike graphene, this material is an insulating system with sp^3 hybridization resulting in a large band gap. It is one of the materials, which was first predicted by *ab initio* theory [29] and then latter synthesized in the experiments [30]. Depending upon the concentration of hydrogenation in graphene, semimetal to metal to insulator transition is observed [31]. It has been shown that patterning graphene with partial hydrogenation leads to modification of graphene properties, for example, conducting channels, band gap opening, quantum dots, and magnetically coupled interfaces [31–36]. As a potential material for spintronic applications, graphene/graphane interfaces are of particular interest as these interfaces can mimic the edge properties that can be seen in zigzag or armchair graphene nanoribbons [37–41]. Hence, it will be interesting to study the effect of Fe adatom, as a representative of magnetic impurities, in these hybrid 2D superlattice structures [42].

Figure 2 shows the two different graphene-graphane superlattice structures considered in our calculation. The hydrogen atoms are removed along the diagonal (edge) of the graphane to create armchair (zigzag) graphene-graphane superlattices. We have considered three, five, and seven rows of channel widths for both configurations. In order to find out the stable adsorption site in the graphene channel, we have placed Fe adatom in different places. The analysis of formation energy indicates that the preferred adsorption site for Fe in armchair channel is at the hollow site of graphene channel equidistant from the interface. However, for the zigzag channel, the Fe adatom prefers to bind at a hollow site closer to the interface. Further analysis of energetics as a function of channel width shows that with increasing channel width, the binding energy remains almost constant in the zigzag channel and it decreases in the armchair channel. The calculated value of total magnetic moments for all the systems are ~2.0 μ_B, which is similar to the value of total magnetic moment of Fe adatom substitutionaly placed in graphene [43]. However, the value of onsite local moments is different in both channels and is ~0.5 μ_B higher in the zigzag channel. Our result shows that the binding energy of Fe adatom in the zigzag channels is higher than the binding energy of Fe adatom on pristine graphene by ~0.2 eV. Hence, we can conclude that the mixed sp^2–sp^3 character of graphene-graphane superlattice helps a strong binding of Fe adatom.

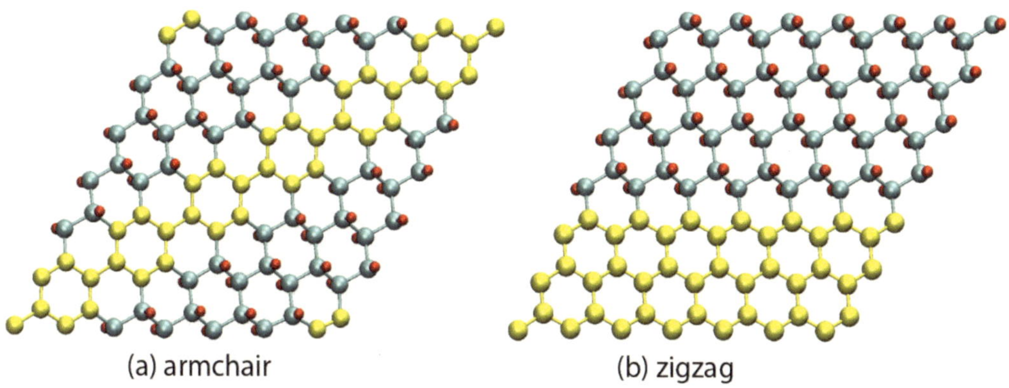

Figure 2. Representative decorations of (a) armchair and (b) zigzag channel in graphane. Reprinted with permission from Haldar et al. [42], copyright (2012) American Physical Society.

Figure 3 shows the total density of states for a single Fe adatom placed on a three-row armchair channel. The analysis of site projected DOS shows that the Fe *d* spin-down electrons induce states below the Fermi energy and reduces the gap quite significantly. Similar features can be observed in the higher row channels although the value of the gap depends on the width.

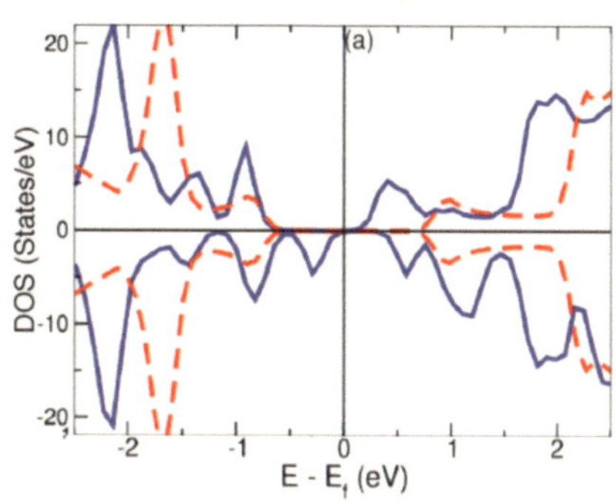

Figure 3. Total DOS of Fe adatom in three-row armchair channel (blue solid line). Total DOS for pristine channel is shown in red dashed line. Reprinted with permission from Haldar et al. [42], copyright (2012) American Physical Society.

The spin density plot of Fe adatom adsorbed in three-row armchair and zigzag channels is shown in **Figure 4**. For armchair channel, it is quite evident from the figure that most of the spin-up density is localized on Fe. The interaction of Fe *d* orbitals with the p_z orbitals of the surrounding C atoms induces polarization in the surrounding C ring, and it induces negative moment on the C atoms. However, in the zigzag channel, the spin density is delocalized and the effect of Fe adatoms can be seen up to fourth nearest neighbor along the interface. The absorption of Fe adatom in this case reduces the onsite magnetic moments of the edge C atoms. The maximum reduction of this onsite moments can be seen on the nearest site and it can be upto ~15%. In the zigzag channel, the Fe adatom surrounding C atoms ring is not antiferromagnetically ordered and only three C atoms from the same sublattice show significant spin-down densities.

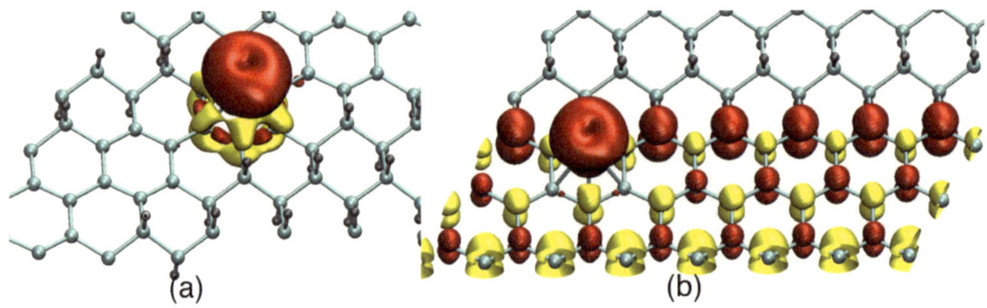

Figure 4. The spin density plot for Fe adatom adsorbed on the three row-armchair and zigzag channel. Reprinted with permission from Haldar et al. [42], copyright (2012) American Physical Society.

We have also calculated the magnetic interactions of two Fe adatoms in these channels. Our result indicates that two Fe adatoms in the armchair channel interact very weakly and hence the exchange energy is also very small favoring an antiferromagnetic interaction. In contrast to the armchair channel, the two Fe adatoms in the zigzag channel interact strongly. A very strong ferromagnetic coupling can be observed in this case between the Fe adatoms, and consequently, they have significantly higher exchange energy compared to the armchair channel.

3.2. Improvement in gas sensing activities

Graphene has also potential application toward the gas sensing properties. This is mainly due to two of the following facts

1. The two-dimensional nature of graphene that consists of only surface and no volume. This feature of graphene enhances the effects of surface dopants.

2. Graphene has a very high conductivity and electrical noise, which enables to detect very small signal changes due to gas molecule absorptions.

Experiments have demonstrated the application of graphene as a solid-state gas sensing device and especially in the detection of single gas molecule, for example, NO_2 [44]. Gaseous molecules act as electron donors or acceptors and modify the carrier density of graphene. Hence, it changes electrical resistance of graphene. Therefore, by measuring the electrical resistance changes, graphene can be used as a gas sensing device [44, 45]. On a pristine graphene lattice, NO_2 molecules are physisorbed. However, chemisorption affects the conduction electron much more than the physisorption. Pristine graphene surface does not have dangling bonds that can chemisorb the gas molecules. However, the presence of defects

http://dx.doi.org/10.5772/64297

can make chemisorption stronger. Hence, in order to increase the gas sensing properties of graphene, one needs to understand the reaction of gas molecules with defected graphene. In this work, we have created defects in graphene using ion beams and studied the gas-sensing properties using current-time measurements, Raman spectroscopy, and gated conductivity characterization [46].

In this study, the graphene flakes were created using the mechanical exfoliation technique on heavily doped Si substrates containing 300 nm SiO_2 top layer. Electron beam lithography was used to fabricate the electrical contacts on device. The defects were created in the pristine graphene by irradiation with 30 keV Ga^+ ions in a vacuum chamber under ~10^{-6} mbar pressure. We have irradiated 20 × 20 μm^2 area and one single irradiation consists of an ion dose of ~10^{12} ions cm^{-2}. We have used a mixture of N_2 and 100-ppm NO_2 gasses as target gas and N_2 gas as a purging gas.

We have used Raman spectroscopy (514 nm wavelength) and atomic force microscopy experiments to determine the thickness of the graphene flakes. From the shape of 2D peak, one can determine the number of layers. **Figure 5** shows the evolution of the Raman spectra with respect to the ion irradiation of graphene. For comparison, the Raman spectrum of the pristine graphene is also shown. Our analysis shows that the graphene flake in this study has a bilayer structure. The D-peak appears at 1352 cm^{-1}, which indicates the formation of defects in graphene. The breathing modes of *sp*^{2} rings cause the appearance of D-peak and only the presence of defects activates it. The intensity of D-peak increases further after the second irradiation and also D′-peak appears at 1626 cm^{-1} which suggest an increase of defects in graphene.

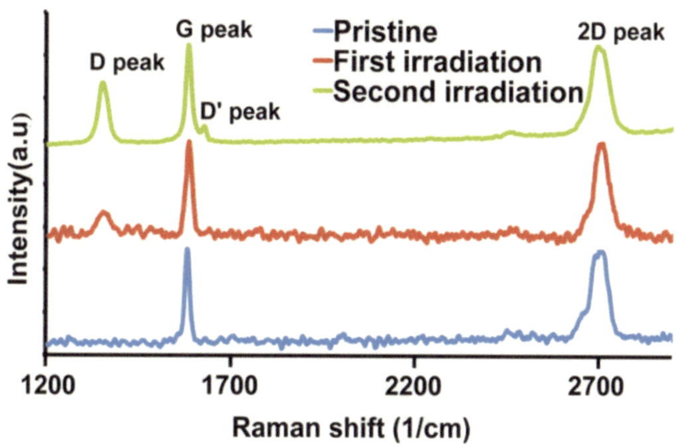

Figure 5. Evolution of Raman spectra with respect to ion irradiation of graphene. Reprinted with permission from Hajati et al. [46], copyright (2012) IOP Publishing.

We have performed gated conductivity experiments to measure the gas sensing properties. **Figure 6** shows the normalized conductance (G/G_0) responses during the exposure of 100 ppm NO_2 in N_2 at room temperature. The conductance of graphene before the exposure is denoted by G_0. The exposure of NO_2 increases the conductance. In pristine graphene, the electrons are transferred from graphene to NO_2 molecules thus increasing the hole density in graphene. A faster response in changing conductance can be observed when the defected graphene (after first irradiation) is exposed to NO_2 gas. These show higher sensitivity to NO_2 gas when compared to the pristine graphene. However, the gas sensing properties decrease after the second irradiation due to the increase of defects, which increases the number of scattering states. Hence, it reduces the conductance.

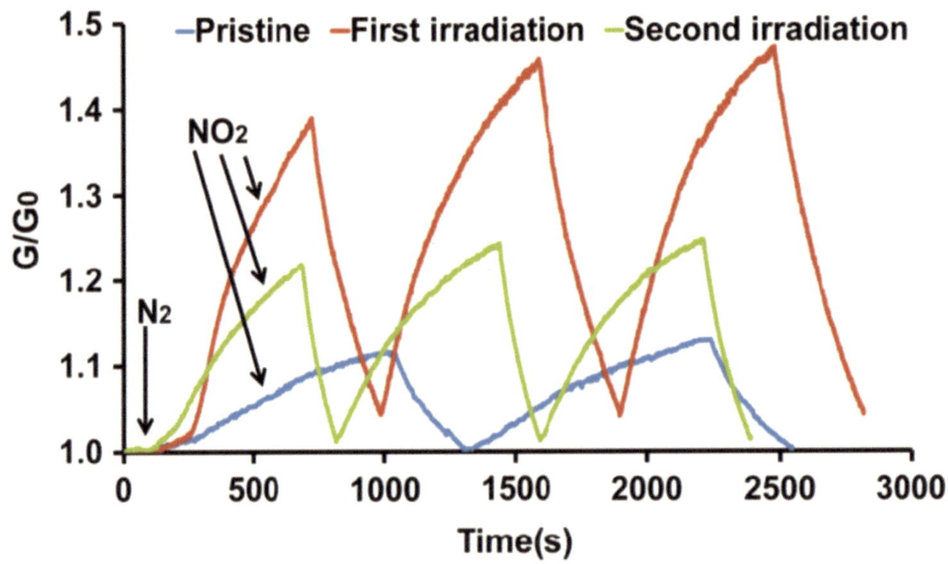

Figure 6. Normalized conductance (G/G_0) response of the graphene gas sensor. The exposure of the NO_2 gas started after 110 s in all three cases. The average rise times for pristine, first, and second defected graphene (during NO_2 exposure) are 500, 328, and 420 s respectively. Reprinted with permission from Hajati et al. [46], copyright (2012) IOP Publishing.

We have also performed *ab initio* density functional calculations in order to understand the interactions between NO_2 and defected graphene. We have used a monolayer graphene for our calculation as most of the defects are same in both monolayer and bilayer graphene. We have studied the binding of NO_2 gas with graphene with different defects, for example, monovacancy, divacancy (585 defect), 686 structure [47], and Stone-Wales (SW) defect. Our analysis shows that the SW vacancy has the highest binding energy with NO_2 molecule (0.72 eV) when compared to the other defects where binding energies are ~0.3 eV.

In **Figure 7**, we have shown the total and molecular NO_2 spin polarized DOS, inverse participation ratio (IPR) [48] for the electronic states of SW + NO_2. The calculated DOS shows that the spin-polarized molecular levels of NO_2 molecules appear near the Fermi energy. These cause 1 μ_B/unit cell magnetic moments. We have also calculated the IPR, which is inversely proportional to the number of atoms contributing to a particular molecular orbital, and hence, IPR gives a quantitative characterization of localization of molecular orbitals. In **Figure 7**, the calculated IPR has very small values near the Fermi energy, which shows conducting character of the states.

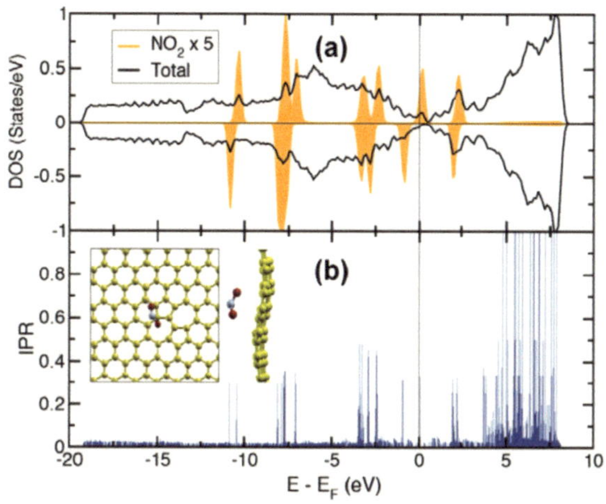

Figure 7. (a) Total and molecular NO_2 spin polarized DOS. (b) Inverse participation ratio (IPR) for the electronic states of SW + NO_2. (b) Optimized geometry of NO_2 at SW-defect site in the graphene lattice is shown as top and side views. Reprinted with permission from Hajati et al. [46]. Copyright (2012) IOP Publishing.

3.3. Fluorination of graphene using defect insertion

Functionalization of graphene has attracted significant attention as it has the potential to make graphene useful for applications. Functionalization of graphene by fluorination is one of the

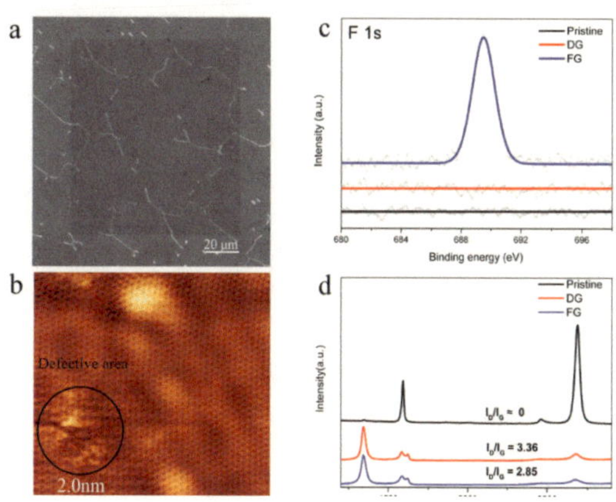

Figure 8. Characterization of pristine graphene, defected graphene (DG) and fluorinated graphene (FG). (a) Scanning electron microscope (SEM) image of local functionalization of graphene (100 μm × 100 μm) with ion doses of 10^{13} ions/cm^2 and simultaneous 167 s gas exposure. (b) Scanning tunneling microscopy image of DG under the same ion dosage. (c) X-ray photoelectron spectroscopy spectra of F 1 s peak of pristine graphene, DG and FG. FG reveals a distinguished F 1 s peak, and the F 1 s spectrum of pristine graphene as well as DG is given as a reference. (d) Raman comparison of pristine graphene, DG and FG. Lower ID/IG in FG in contrast to DG indicates lower degree of defects density and larger crystalline size. Reprinted with permission from Li et al. [49].

ways. In general, the functionalization of graphene is a challenging process. Local functionalization is a promising tool to keep the desired properties of graphene intact after the modification. In this section, we discuss an interesting technique, which allows precise site-selective fluorination [49].

We have used a focused ion beam irradiation under XeF_2 environment in high vacuum to design the site-selective fluorination. In this method, the graphene surface is locally radicalized using high-energy ion irradiation under fluorine contained precursor molecule environment. We have used X-ray photoelectron spectroscopy (XPS), Raman spectroscopy, scanning tunnelling microscopy (STM), and density functional theory (DFT) calculations to verify the fluorination process and explain the mechanism.

The defected structures shown in **Figure 8a**, **b** are obtained by irradiating graphene locally with high-energy (30 kV) Ga^+ ions with an irradiation dose of 10^{13} ions cm^{-2}. Under this amount of irradiation dose, graphene retains most of its lattice structure. However, the damaged part shows significant defect formations, which are mainly vacancies. The formation of fluorinated graphene can be seen from **Figure 8c**, where XPS shows a clear signal of F 1 s peak. We have also used Raman spectroscopy to find out the structural information. From the Raman spectroscopy figure (**Figure 8d**), we can see that the intensity of the D-peak (at 1350 cm^{-1}) increases after irradiation and the intensity of the 2D-peak decreases sharply. It means that the translational symmetry of sp^2 bond is broken. Compared to the defected graphene, in fluorinated graphene, the ratio of D and G peak (I_D/I_G) is lower. This implies that fluorinated graphene contains less structural disorder.

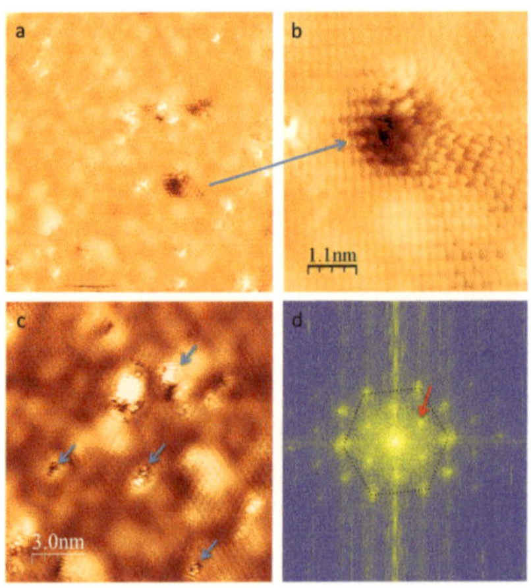

Figure 9. STM images of fluorinated graphene. (a) 20 × 20 nm^2 area. (b) Zoom in image of a hole defect showing standing waves pattern. (c) Other area 15 × 15 nm^2 showing bright feature decorating holes (blue arrows) attributed to fluorine atoms. (d) FFT of (a). It reveals the first Brillouin zone with hexagonal lattice and K points (red arrows) associated to the standing waves pattern due to intervalley scattering. Reprinted with permission from Li et al. [49].

STM experiments were carried out on the fluorinated graphene for a better understanding of structures. **Figure 9** shows these images, which are taken at low bias voltage of −75 mV at Fermi level. In **Figure 9a**, a 20 × 20 nm^2 area of fluorinated graphene is shown. The surface is covered by various defects, corrugations, etc. Standing waves with different structures near the defect area form these corrugations. The Fast Fourier transformation of **Figure 9a** is shown in **Figure 9d**, which clearly shows the first Brillouin zone of the hexagonal lattice and the K points. These K points are related to the standing waves pattern from the intervalley scattering [50]. All the defects in the surface are connected with the standing waves. Larger defects (zoomed in **Figure 9b**) show standing waves as straight lines similar to as observed in step edges. Thus, it can be concluded that the fluorinated graphene remains metallic [51]. In **Figure 9c**, the bright features are associated with the fluorine atoms. Combining the observation of standing waves related to delocalized electron at the conjugated *sp*^{2} bonds from **Figure 9b**, it could be concluded that the fluorination happens near the defect sites created by the ion irradiation.

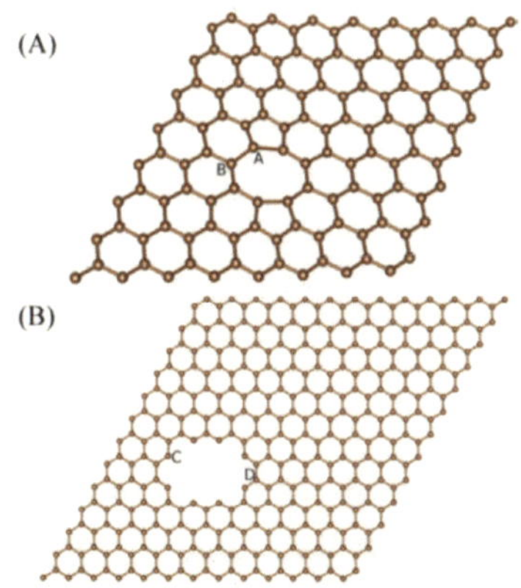

Figure 10. *Ab initio* density functional theory (DFT) calculation models of fluorinated graphene. Di-vacancy model (a) and hole-defect model (b), 0.95 nm in length, are based on the STM observation. Binding energies are shown in **Table 1**.

Structure	**E_{abs} (eV)**	**Hybridization**
Pristine	−1.91	sp^3
Di-vacancy at site A	−2.86	sp^3
Di-vacancy at site B	−2.25	sp^3
Hole-defect site C	−5.64	sp^2
Hole-defect site D	−2.18	sp^3

Reprinted with permission from Li et al. [49].

Table 1. Adsorption energies of fluorine adatom on pristine graphene as well as the edge carbon atoms surrounding the two defects.

We have also performed *ab initio* density functional-based calculations to find out fluorine adsorption characteristics on defected graphene. We have used two models of defected graphene for our calculations as shown in **Figure 10**. These models are: (i) divacancy model and (ii) hole-defect model. These are the two types of models that can be seen in the STM images. In these two types of defects, there are only four possible places for single fluorine adatoms to be adsorbed. These four places are marked as site A–D. In **Table 1**, we have tabulated the energetics. Our calculations show that the adsorption energy of the fluorine adatom on pristine graphene is very high compared to the di-vacancy and the hole defects. It implies that the fluorine atoms are prone to react with the carbon atom surrounding the defect sites. At site C, the carbon atom is radicalized due to the presence of dangling bonds and hence has very low adsorption energy for fluorine adsorption. In this case, the C–F bond is planar with a bond length of 0.136 nm, typical for sp^2 hybridization. The C–F bonds in other sites are all out of plane (perpendicular to the graphene lattice) and have sp^3 hybridization. This strong bond between dangling bond and fluorine atoms implies that different gases could be utilized to functionalize graphene.

In conclusion, we have showed an experimental technique to design site-selective local fluorination using high kinetic energy ion irradiation and simultaneous XeF_2 gas injection. Our method opens up a possibility of functionalize graphene locally with a wide range of other functional materials.

4. Conclusion

From the discussions of Fe adatom adsorption at the partially hydrogenated channels, we conclude that the magnetic adatoms in the zigzag channel interact quite substantially as compared to the armchair channel. The response of the two channels in the presence of magnetic impurities is quite different, viz., localized (delocalized) in the armchair (zigzag) channel. In the semiconducting armchair channel, the magnetic coupling is weakly antiferromagnetic. However, in the delocalized zigzag channel, a relatively stronger ferromagnetic coupling can be observed. Hence, it may be possible to design magnetic graphene lattice by depositing suitable magnetic impurities by means of scanning tunneling microscopy tips, which can lead to the possibility of designing ultrathin magnetic devices.

We have also studied how defects in graphene affect the gas sensing properties. The defects are created using the Ga^+ ion irradiation. The defected graphene shows higher conductivity changes in the presence of NO_2 gas when compared to the pristine graphene. Hence, one can conclude that the defected graphene has higher sensitivity in gas detection. The NO_2 gas molecules bind strongly with SW defects in graphene, which changes the local electronic structure and enhance the transport properties.

We have also demonstrated how defects in graphene can be used for various important applications, for example, spintronics and gas sensing. The presence of defects modifies the structural and electronic properties of the 2D material as well as the binding entities. The understanding of these phenomena can be achieved by materials specific theoretical methods. From the experimental side, the controlled nanoengineering of defects may lead to novel applications and should be pursued seriously in near future.

Author details

Soumyajyoti Haldar and Biplab Sanyal*

*Address all correspondence to: Biplab.Sanyal@physics.uu.se

Department of Physics and Astronomy, Uppsala University, Uppsala, Sweden

References

[1] May JW: Platinum surface LEED rings. Surf Sci. 1969;17:267–270.

[2] Novoselov KS, Geim AK, Morozov SV, Jiang D, Zhang Y, Dubonos SV, et al.: Electric field effect in atomically thin carbon films. Science. 2004;306:666–669. doi:10.1126/science.1102896

[3] Novoselov KS, Jiang D, Schedin F, Booth TJ, Khotkevich VV, Morozov SV, et al.: Two-dimensional atomic crystals. Proc Natl Acad Sci USA. 2005;102:10451–10453. doi:10.1073/pnas.0502848102

[4] Butler SZ, Hollen SM, Cao L, Cui Y, Gupta JA, Gutiérrez HR, et al.: Progress, challenges, and opportunities in two-dimensional materials beyond graphene. ACS Nano. 2013;7:2898–2926. doi:10.1021/nn400280c

[5] Wilson JA, Yoffe AD: The transition metal dichalcogenides discussion and interpretation of the observed optical, electrical and structural properties. Adv Phys. 1969;18:193–335. doi:10.1080/00018736900101307

[6] Bednorz JG, Müller KA: Possible high Tc superconductivity in the Ba–La–Cu–O system. Zeitschrift für Physik B Condensed Matter. 1986;64:189–193. doi:10.1007/BF01303701

[7] Gibney E: 2D or not 2D. Nature. 2015;522:274–276. doi:10.1038/522274a

[8] Watanabe K, Taniguchi T, Kanda H: Direct-bandgap properties and evidence for ultraviolet lasing of hexagonal boron nitride single crystal. Nat Mater. 2004;3:404–409. doi:10.1038/nmat1134

[9] Georgiou T, Jalil R, Belle BD, Britnell L, Gorbachev RV, Morozov SV, et al.: Vertical field-effect transistor based on graphene-WS 2 heterostructures for flexible and transparent electronics. Nat Nanotechnol. 2013;8:100–103. doi:10.1038/nnano.2012.224

[10] Feliciano GT, Sanz-Navarro C, Coutinho-Neto MD, Ordejon P, Scheicher RH, Rocha AR: Capacitive DNA detection driven by electronic charge fluctuations in a graphene nanopore. Phys Rev Appl. 2015;3:034003–034007. doi:10.1103/PhysRevApplied.3.034003

[11] Le Lay G: 2D materials: silicene transistors. Nat Nanotechnol. 2015;10:202–203. doi: 10.1038/nnano.2015.10

[12] Prasongkit J, Amorim RG, Chakraborty S, Ahuja R, Scheicher RH, Amornkitbamrung V: Highly sensitive and selective gas detection based on silicene. J Phys Chem C. 2015;119:16934–16940. doi:10.1021/acs.jpcc.5b03635

[13] Song HS, Li SL, Miyazaki H, Sato S, Hayashi K, Yamada A, et al.: Origin of the relatively low transport mobility of graphene grown through chemical vapor deposition. Sci Rep. 2012;2:337. doi:10.1038/srep00337

[14] Niu T, Zhou M, Zhang J, Feng Y, Chen W: Growth intermediates for CVD graphene on Cu(111): carbon clusters and defective graphene. J Am Chem Soc. 2013;135:8409–8414. doi:10.1021/ja403583s

[15] Banhart F, Kotakoski J, Krasheninnikov AV: Structural defects in graphene. ACS Nano. 2011;5:26–41. doi:10.1021/nn102598m

[16] Meyer JC, Kisielowski C, Erni R, Rossell MD, Crommie MF, Zettl A: Direct imaging of lattice atoms and topological defects in graphene membranes. Nano Lett. 2008;8:3582–3586. doi:10.1021/nl801386m

[17] Ugeda MM, Brihuega I, Guinea F, Gómez-Rodríguez JM: Missing atom as a source of carbon magnetism. Phys Rev Lett. 2010;104:096804. doi:10.1103/PhysRevLett.104.096804

[18] Rocha AR, Martins TB, Fazzio A, Da Silva AJR: Disorder-based graphene spintronics. Nanotechnology. 2010;21:345202. doi:10.1088/0957-4484/21/34/345202

[19] Xu C, Luo G, Liu Q, Zheng J, Zhang Z, Nagase S, et al.: Giant magnetoresistance in silicene nanoribbons. Nanoscale. 2012;4:3111–3117. doi:10.1039/c2nr00037g

[20] Hohenberg P, Kohn W: Inhomogeneous electron gas. Phys Rev. 1964;136:B864–B871. doi:10.1103/PhysRev.136.B864

[21] Kohn W, Sham LJ: Self-consistent equations including exchange and correlation effects. Phys Rev. 1965;140:A1133–A1138. doi:10.1103/physrev.140.a1133

[22] Perdew JP, Zunger A: Self-interaction correction to density-functional approximations for many-electron systems. Phys Rev B. 1981;23:5048–5079. doi:10.1103/PhysRevB.23.5048

[23] Ceperley DM, Alder BJ: Ground state of the electron gas by a stochastic method. Phys Rev Lett. 1980;45:566–569. doi:10.1103/PhysRevLett.45.566

[24] Perdew JP, Chevary JA, Vosko SH, Jackson KA, Pederson MR, Singh DJ, et al.: Atoms, molecules, solids, and surfaces: applications of the generalized gradient approximation for exchange and correlation. Phys Rev B. 1992;46:6671–6687. doi:10.1103/PhysRevB.46.6671

[25] Perdew JP, Burke K, Ernzerhof M: Generalized gradient approximation made simple. Phys Rev Lett. 1996;77:3865–3868. doi:10.1103/PhysRevLett.77.3865

[26] Perdew JP, Burke K, Ernzerhof M: Generalized gradient approximation made simple [Phys. Rev. Lett. 77, 3865 (1996)]. Phys Rev Lett. 1997;78:1396–1396. doi:10.1103/PhysRevLett.78.1396

[27] Geim AK, Novoselov KS: The rise of graphene. Nat Mater. 2007;6:183–191. doi:10.1038/nmat1849

[28] Castro Neto AH, Guinea F, Peres NMR, Novoselov KS, Geim AK: The electronic properties of graphene. Rev Mod Phys. 2009;81:109–162. doi:10.1103/RevModPhys.81.109

[29] Sofo JO, Chaudhari AS, Barber GD: Graphane: a two-dimensional hydrocarbon. Phys Rev B. 2007;75:153401-153404. doi:10.1103/PhysRevB.75.153401

[30] Elias DC, Nair RR, Mohiuddin TMG, Morozov SV, Blake P, Halsall MP, et al.: Control of graphene's properties by reversible hydrogenation: evidence for graphane. Science. 2009;323:610–613. doi:10.1126/science.1167130

[31] Chandrachud P, Pujari BS, Haldar S, Sanyal B, Sanyal B, Kanhere DG: A systematic study of electronic structure from graphene to graphane. J Phys: Condens Matter. 2010;22:465502. doi:10.1088/0953-8984/22/46/465502

[32] Singh AK, Yakobson BI: Electronics and magnetism of patterned graphene nanoroads. Nano Lett. 2009;9:1540–1543. doi:10.1021/nl803622c

[33] Sessi P, Guest JR, Bode M, Guisinger NP: Patterning graphene at the nanometer scale via hydrogen desorption. Nano Lett. 2009;9:4343–4347. doi:10.1021/nl902605t

[34] Lu YH, Feng YP: Band-gap engineering with hybrid graphane–graphene nanoribbons. J Phys Chem C. 2009;113:20841–20844. doi:10.1021/jp9067284

[35] Balog R, Jørgensen B, Nilsson L, Andersen M, Rienks E, Bianchi M, et al.: Bandgap opening in graphene induced by patterned hydrogen adsorption. Nat Mater. 2010;9:315–319. doi:10.1038/nmat2710

[36] Ao ZM, Hernández-Nieves AD, Peeters FM, Li S: Enhanced stability of hydrogen atoms at the graphene/graphane interface of nanoribbons. Appl Phys Lett. 2010;97:233109. doi:10.1063/1.3525377

[37] Nakada K, Fujita M, Dresselhaus G, Dresselhaus MS: Edge state in graphene ribbons: nanometer size effect and edge shape dependence. Phys Rev B. 1996;54:17954–17961. doi:10.1103/PhysRevB.54.17954

[38] Koskinen P, Malola S, Häkkinen H: Evidence for graphene edges beyond zigzag and armchair. Phys Rev B. 2009;80:073401. doi:10.1103/PhysRevB.80.073401

[39] Jia X, Hofmann M, Meunier V, Sumpter BG, Campos-Delgado J, Romo-Herrera JM, et al.: Controlled formation of sharp zigzag and armchair edges in graphitic nanoribbons. Science. 2009;323:1701–1705. doi:10.1126/science.1166862

[40] Girit ÇÖ, Meyer JC, Erni R, Rossell MD, Kisielowski C, Yang L, et al.: Graphene at the edge: stability and dynamics. Science. 2009;323:1705–1708. doi:10.1126/science.1166999

[41] Koskinen P, Malola S, Häkkinen H: Self-passivating edge reconstructions of graphene. Phys Rev Lett. 2008;101:115502. doi:10.1103/PhysRevLett.101.115502

[42] Haldar S, Kanhere DG, Sanyal B: Magnetic impurities in graphane with dehydrogenated channels. Phys Rev B: Condensed Matter and Mater Phys. 2012;85:155426. doi: 10.1103/PhysRevB.85.155426

[43] Valencia H, Gil A, Frapper G: Trends in the adsorption of 3d transition metal atoms onto graphene and nanotube surfaces: a DFT study and molecular orbital analysis. J Phys Chem C. 2010;114:14141–14153. doi:10.1021/jp103445v

[44] Schedin F, Geim AK, Morozov SV, Hill EW, Blake P, Katsnelson MI, et al.: Detection of individual gas molecules adsorbed on graphene. Nat Mater. 2007;6:652–655. doi: 10.1038/nmat1967

[45] Zhu Y, Murali S, Cai W, Li X, Suk JW, Potts JR, et al.: Graphene and graphene oxide: synthesis, properties, and applications. Adv Mater. 2010;22:3906–3924. doi:10.1002/adma.201001068

[46] Hajati Y, Blom T, Jafri SHM, Haldar S, Bhandary S, Shoushtari MZ, et al.: Improved gas sensing activity in structurally defected bilayer graphene. Nanotechnology. 2012;23:505501. doi:10.1088/0957-4484/23/50/505501

[47] Appelhans DJ, Carr LD, Lusk MT: Embedded ribbons of graphene allotropes: an extended defect perspective. New J Phys. 2010;12. doi:10.1088/1367-2630/12/12/125006

[48] Atta-Fynn R, Biswas P, Ordejon P, Drabold D: Systematic study of electron localization in an amorphous semiconductor. Phys Rev B. 2004;69:085207–085210. doi:10.1103/PhysRevB.69.085207

[49] Li H, Daukiya L, Haldar S, Lindblad A, Sanyal B, Eriksson O, et al.: Site-selective local fluorination of graphene induced by focused ion beam irradiation. Sci Rep. 2016;6:19719. doi:10.1038/srep19719

[50] Simon L, Bena C, Vonau F, Cranney M, Aubel D: Fourier-transform scanning tunnelling spectroscopy: the possibility to obtain constant-energy maps and band dispersion using a local measurement. J Phys D: Appl Phys. 2011;44:464010. doi: 10.1088/0022-3727/44/46/464010

[51] Yang H, Mayne AJ, Boucherit M, Comtet G, Dujardin G, Kuk Y: Quantum interference channeling at graphene edges. Nano Lett. 2010;10:943–947. doi:10.1021/nl9038778

Chapter 11

Harvesting Plasmonic Excitations in Graphene for Tunable Terahertz/Infrared Metamaterials

Yuancheng Fan, Fuli Zhang, Quanhong Fu and Hongqiang Li

Additional information is available at the end of the chapter

http://dx.doi.org/10.5772/64361

Abstract

In this chapter, we focus on the development on tunable terahertz/infrared metamaterials enabled with plasmonic excitations in graphene micro-/nanostructures. We aimed the issue that high loss in the plasmonic excitations of graphene limits the performance of graphene's ability in manipulating light. We show the enhancement of light-graphene interactions by employing plasmonic metamaterial design for proper plasmonic excitations, and coherent modulation on optical fields to further increase the bonding of light field for boosted plasmonic excitations. The enhanced plasmonic excitations in graphene provide the possibility of practical applications for terahertz and infrared band graphene photonics and optoelectronics.

Keywords: graphene, plasmonics excitations, tunable metamaterials, terahertz, infrared, surface conductivity

1. Introduction

There have been numerous reports on scientific advances in graphene, a first realistic two-dimensional (2D) material with carbon atoms arranged in a hexagonal lattice. Since its first exfoliation from graphite by Geim and Novoselov [1, 2], graphene stimulated and led the research upsurge in two-dimensional materials [3, 4]. It is attractive for myriad applications that profits from its high electronic mobility (25,000 $cm^2/V^{-1}\ s^{-1}$) [5], exceptional mechanical strength (~1.0 TPa) [6], and thermal conductivity [7]. It has also been widely investigated for potential applications in photonics and optoelectronics [8–10]. Graphene exhibits much stronger binding of surface plasmon polaritons [11–16], and the Dirac fermions in graphene provide ultra wideband tunability in optical response through electrostatic field, magnetic field, or chemical doping [15–20]. All these are good for dynamical control on optical signals. However, there are still some challenges in developing graphene photonics for practical applications. One of them is that graphene is almost transparent to optical waves due to the relatively low carrier concentrations in the monolayer atomic sheet; this might be useful for some cases, for example, for the transparent electrodes, but strong light-matter interactions is the most crucial part for practical optical applications. Boosting the light-matter interaction in graphene is one important issue to address to take further advantage of graphene in optical devices or systems, for example, to realize functionality such as optical insulator similar to gapped graphene [21–23] for nanoelectronics, which is essential for myriad applications in all-optical systems and components of much miniaturized optical circuits [24–26]. It is important to take effectual strategies to improve the light-graphene interactions.

1.1. Plasmonic metamaterials

Artificially engineered microstructures, that is, the plasmonic metamaterial and photonic crystals, are well-known platforms for the enhancement of light-matter interactions [27–29]. Optical absorption enhancement in graphene had been demonstrated in a photonic crystal nanocavity for high-contrast electro-optic modulation, at the regime of critical coupling with photonic crystal-guided resonance, and in photonic crystals for broadband response [30–34]. And in the meanwhile, plasmonic metamaterials [35–42] with even miniaturized elements are promising for the manipulation of light at the deep subwavelength scale by making use of the plasmonic excitations. Especially a kind of metamaterials with a single 2D function layer, named as metasurface [43–49], has been intensively studied in recent years for various possibilities to manipulate light peculiarly. The metasurface is naturally connected to the 2D graphene for the following: (i) changing conversional plasmonic material-metal with graphene will provide frequency-agile responses and make the metasurface even more subwavelength and (ii) the light-graphene interactions can be significantly enhanced in an atomically thin graphene metasurface. Optical absorption enhancement has been studied in graphene nanodisks, in which periodic graphene disks are placed on a substrate or a dielectric layer with metallic ground, the plasmonic excitations in the structure resulting in the complete absorption of incident light. Graphene micro-/nanoribbons, split ring resonators, mantles, nanocrosses, and photonic crystals have also been exploited for controlling terahertz and infrared light.

1.2. Coherently modulated light-matter interactions

Another strategy that was proposed recently to remarkably improve the light-matter interaction is the coherent modulation technique, which is based on the coherence of optical fields. The coherence of laser made it unique in modern optics and photonics. A coherent perfect absorber (CPA), also called anti-laser, was recently proposed [50] and demonstrated [51] in a Si-resonator. The coherent absorption comes from the phase modulation on light fields. Since the first proposal, relevant coherent modulation-assisted processes have attracted considerable research interests with various photonic structures [52–58], for example, laser absorber and symmetry breaking in PT-symmetric optical potentials and strongly scattering systems, unidirectional invisibility in PT-symmetric periodic structures, and perfect mode (polarization or morphology) conversions. It has been proven that coherently modulated optical field provides additional localization of the light within artificially engineered microstructures, including both the photonic crystals and plasmonic metamaterials, to further boost light-matter interactions.

In this chapter, we summarize our recent studies on the excitation of electric/magnetic plasmonic modes in graphene structures [59, 60], and their synergic action with the coherently modulated optical fields that provide strong interaction between graphene and light for practical and tunable terahertz/infrared metamaterials or metasurfaces [61, 62].

2. Tunable plasmonic excitations in graphene metamaterials

2.1. Optical conductivity of graphene

The optical response of a monolayer graphene can be described with the complex surface conductivity in the local-random phase approximation (RPA) approximation as

$$\begin{aligned}\sigma(\omega)=&\frac{2e^2T}{\pi\hbar}\frac{i}{\omega+i\tau^{-1}}\log\left[2\cosh\left(E_F/2k_BT\right)\right]\\&+\frac{e^2}{4\hbar}\left[H(\omega/2)+\frac{4i\omega}{\pi}\int_0^\infty\frac{H(\varepsilon)-H(\omega/2)}{\omega^2-4\varepsilon^2}d\varepsilon\right]\Bigg\},\end{aligned} \tag{1}$$

with $H(\varepsilon)=\frac{\sinh(\hbar\varepsilon/k_BT)}{\cosh(E_F/k_BT)+\cosh(\hbar\varepsilon/k_BT)}$, ω is the angular frequency of the incoming light wave, e is the electron charge, $\hbar$ is the reduced Planck constant, k_B is the Boltzmann constant, and T is the temperature. τ is a phenomenological relaxation time representing the scatterings in graphene or the quality of graphene sample, E_F is the Fermi energy away from the Fermi surface.

Figure 1 shows the surface conductivity of a monolayer graphene (Fermi level: 0.1 eV) at 0 K. We can see that the total conductivity of the graphene includes two parts: (i) contribution from the intraband transition (blue), the response is similar to Drude dispersive metals; and (ii) contribution from the interband transition (green), with near nondispersive real conductivity at frequencies higher than double Fermi level. For that graphene is interesting for dynamically controlled photonic applications, we plot the surface conductivities under different Fermi levels (from 0.06 to 0.14 eV) in **Figure 2**; we can see from the figure that the conductivity of graphene can be s-tuned by changing the Fermi level, especially the low-frequency Drude-like response. The tunable Drude-like metallic behavior has received intensive attention in the past years, and this chapter focuses on the tunable plasmonic excitations in graphene at terahertz/infrared frequencies.

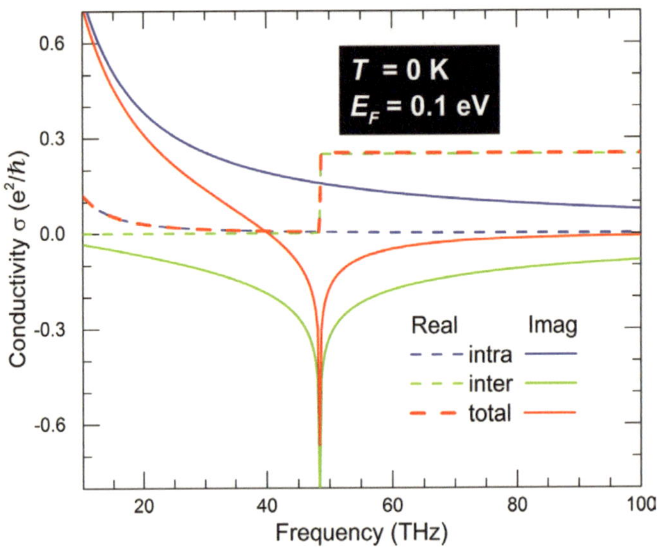

Figure 1. Complex surface conductivity of a monolayer graphene (Fermi level: 0.1 eV) at 0 K.

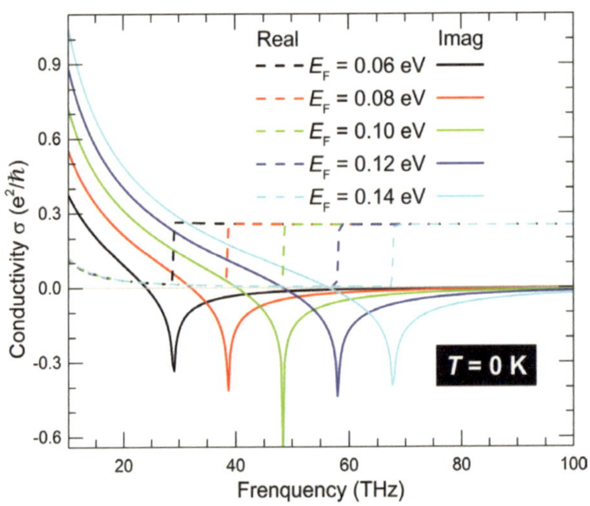

Figure 2. Tunable surface conductivity of a monolayer graphene (Fermi level: from 0.06 to 0.14 eV) at 0 K.

2.2. A comparative study on the plasmonic excitations in graphene split ring resonators (SRRs)

We proposed to enhance infrared extinction and absorption in a monolayer graphene sheet by patterning split ring resonators, a kind of basic structure in the design of plasmonic metamaterials. By introducing asymmetric split ring resonators (ASRRs) into the monolayer graphene sheet, we excited both fundamental magnetic mode and electric mode, and the

http://dx.doi.org/10.5772/64361

contributions on the enhancement of infrared extinction and the absorption of these two modes are comparatively studied. The designed periodic SRR-patterned graphene metamaterial is shown in **Figure 3**, the rings are with width a=2.5 μm and the lattice constant of the square lattice is P=3.0 μm, the gap size of SRR is g=0.6 μm. We note that the line width w of the ring and the gap center position ($\delta$with respect to the center of the ring) are parameters that can significantly influence absorption/extinction properties of the graphene metamaterial.

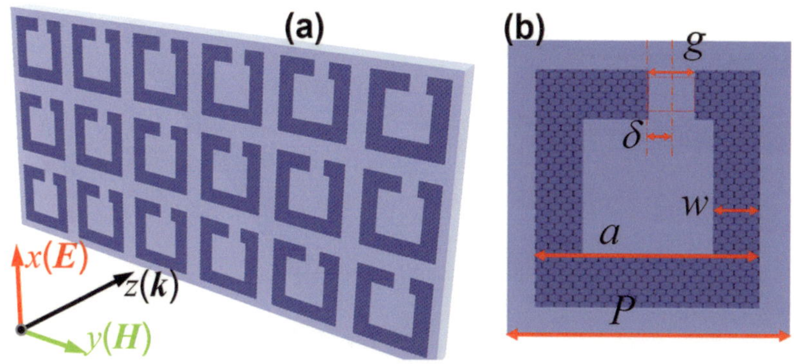

Figure 3. (a) Schematic of the SRR-patterned monolayer graphene sheet and corresponding electromagnetic excitation configuration (the polarization direction is along the x-axis). (b) A unit cell of the SRR-patterned graphene sheet. Geometric parameters are denoted by black letters.

In calculations, the graphene sheet was approximately treated as optical interface with complex surface conductivity, since a one-atom-thick graphene sheet is sufficiently thin compared with the concerned wavelength; the complex surface conductivity can be well described by a Drude model as $\sigma = e^2 E_F / (\pi\hbar^2) * i / (\omega + i\tau^{-1})$, especially at low frequencies and heavily doped region, where E_F=0.5 eV is the Fermi energy away from Dirac point, and $\tau = uE_F / ev_F^2$ is the relaxation rate with $\mu = 10^4 \mathrm{cm^2V^{-1}s^{-1}}$ and $\upsilon \approx 10^6$m/sbeing the mobility and Fermi velocity, respectively. The Fermi level can be easily controlled by electrostatic doping via tuning charge-carrier density.

First, we set δ=0.45 μm≠0, which introduces symmetry breaking in SRRs. It is known that both the electric mode and the fundamental magnetic mode can be excited for normal incidence in the ASRR metamaterial [63–65]. It can be seen from **Figure 4** that there are two transmission dips on the curve. The transmission and reflection coefficients of the electric-field magnitude of the x-polarized incident wave are defined as $t_{ix} = |E_i^{\mathrm{Tran}} / E_x^{\mathrm{Inc}}| (i = x, y)$, and $r_{ix} = |E_i^{\mathrm{Ref}} / E_x^{\mathrm{Inc}}|$), in which E_x^{Inc} is the field of incident wave, and E_i^{Tran} and E_i^{Ref} are the field of transmitted or reflected waves. We confirmed that (through numerical simulation) these two dips correspond to fundamental plasmonic resonances: magnetic mode and electric mode. The electric resonant mode at 3.96 THz is well excited. The magnetic mode at 1.99 THz is with a very shallow dip in the transmission curve, which indicates that the graphene ASRR is weakly excited. When comparing the resonant strength with resonant dips, it is clear that the electric resonant mode is stronger in responding the incident field, and it shows better extinction and absorption compared with the magnetic mode. We believe that this is because the graphene's low-carrier concentration leads to the weak capturing on optical fields. The electric dipolar mode is easier to excite, because to excite the electric mode capturing the incident field into a current loop is not required. We also notice that the polarization transformation $t_{yx}=r_{yx}\neq 0$, which is the radiation from the induced surface current on the graphene ASRRs. This polarization conversion will also influence the optical extinction and absorption of the x-polarized incident waves. The optical extinction and absorption with respect to x-polarized incident wave are defined as $E=1-|t_{xx}|^2$ and $A=1-|t_{xx}|^2-|r_{xx}|^2-|t_{yx}|^2-|r_{yx}|^2$for considering the polarization conversion.

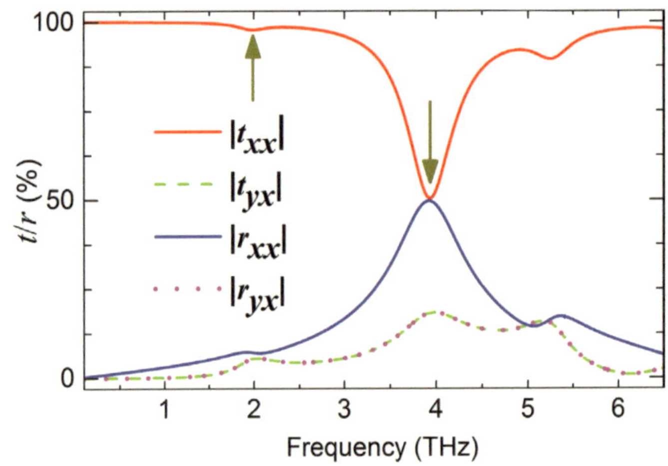

Figure 4. Transmission (t) and reflection (r) coefficients with respect to the x-polarized incident wave of ASRRs-patterned graphene sheet (gap position $\delta=0.45$ μm and line width of graphene ASRR $w=0.5$ μm).

We investigate the influences of geometric parameters on the optical extinction and absorption of the ASRR graphene metamaterial. **Figure 5** shows resonant frequencies of the electric mode and the magnetic mode, and the extinction and absorption at resonant frequencies for graphene metamaterial with different line widths. The extinction on the resonance for both the electric mode and the magnetic mode rises as the line width becomes wider, and the extinction of the electric resonance is nearly one order higher than that of the magnetic resonance. We can see that the extinction can be efficiently boosted at the frequency of electric resonance, for example, the optical extinction of about 87% at a wavelength of 58.5 μm is achieved.

Then, we investigate the influence of symmetry (of the SRR structure) on the optical extinction and absorption. We find that the resonant frequency of the electric mode almost did not shift when changing the asymmetric parameter, and the enhanced extinction and absorption of the symmetric SRR ($E=81\%$, $A=49\%$) is higher than that of the ASRR ($E=75\%$, $A=43\%$). Then, we study comparatively the extinction and absorption in symmetric ($\delta=0$) and asymmetric ($\delta=0.45$ μm) graphene SRRs. We find that the symmetric SRR is better than ASRR for

http://dx.doi.org/10.5772/64361

the enhancement of extinction and absorption in a monolayer of graphene. A maximum extinction of about 90% (w=0.8 μm) and a maximum absorption of 50% (w=0.4 μm) can be achieved in a symmetric SRR graphene metamaterial. Since only an electric resonance is excited in the symmetric SRR structure, and there exists no polarization conversion, the enhancement on the extinction and absorption can be understood simply as follows: First, the SRRs trap optical field that shines on graphene according to their receiving ability. Then, the trapped field is redistributed through an absorbing channel and a radiating channel (including both the forward and backward radiation). We note that the receiving or radiating ability of the graphene SRRs and trapped ratio with respect to the incoming energy are two key characters in this process, and these two aspects are both strongly connected to the line width of the graphene SRRs. On the one hand, the radiating property can be significantly influenced by changing the line width since we know that a thin dipole possesses better radiation. On the other hand, the line width determines directly the graphene area ratio to the whole unit cell and thus remarkably modulates the trapped ratio to the incident light as well as the absorption in the graphene metamaterial. The electric resonance and dissipation of the graphene will compete for an optimized absorption, while for the extinction the incident light is more likely to be scattered for SRRs with wider line widths, which contributes to an incremental optical extinction.

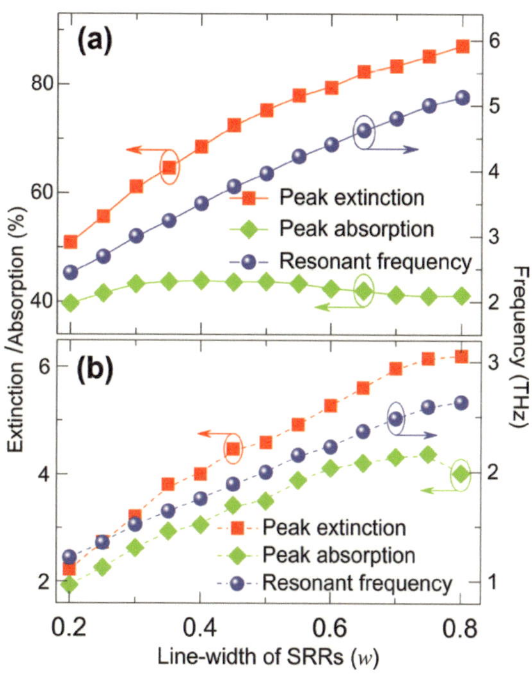

Figure 5. Resonant frequency (blue sphere), extinction (red square), and absorption (olive diamond) at resonant frequencies of the ASRR (with δ=0.45 μm)-patterned monolayer graphene sheet with respect to different line width w for electric mode (solid) (a) and magnetic mode (dashed) (b).

2.3. Electric plasmonic excitation in graphene cut-wires and physics of a maximum 50% absorption in graphene metamaterials

We have found that the electric dipolar mode is stronger in enhancing light-graphene interactions at terahertz frequencies compared to the magnetic mode and other higher-order modes. As the simplest structure that supports electric dipolar excitation, cut-wire is essential in designing plasmonic metamaterials. It has been widely adopted for exploring fundamental physics as well as novel functionalities, such as plasmon-induced transparency, polarization manipulations, and optical antennas. We suggested a tunable metasurface by exploiting a monolayer graphene patterned in a cut-wire array. We mainly focused on the strengthening of graphene's terahertz response by the electric dipolar excitation of the basic cut-wire structure and the influences of the graphene qualities. A 50% maximum absorption at the electric dipolar mode is confirmed by the extraction of effective surface conductivities of a cut-wire array of the theoretical and experimental graphene. Systematic investigations to the graphene metasurface by changing values of graphene sample between two sets of well-known experimental data, that is, Li et al. data [18] and Yan et al. data [10], respectively, show that optical response of the graphene cut-wire-based metasurface can be tuned substantially in terahertz frequencies.

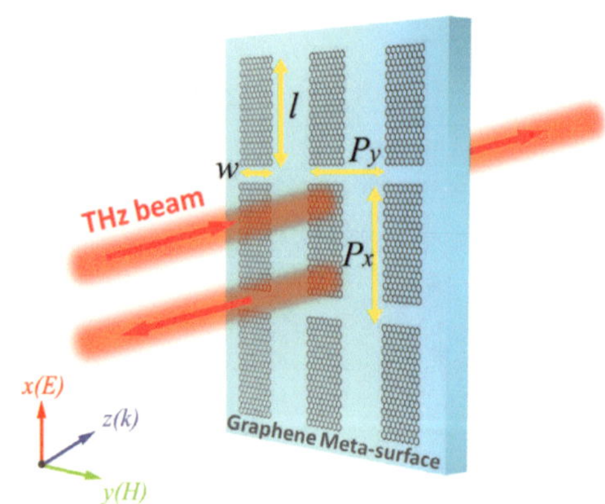

Figure 6. Schematic of the single-layer graphene cut-wire structure together with the corresponding excitation configuration: x-polarized incident light is propagating along the z-axis. Geometric parameters are denoted by black letters.

Figure 6 shows the schematic of the proposed tunable graphene metasurface. The meta-atoms, that is, graphene cut-wires, are periodically arranged in xy-plane with lattice constants P_x=6 μm and P_y=3 μm. The length and width of the cut-wire meta-atoms are l and w, respectively. In the setup, the THz beam, with electric field polarized along the x-direction, normally illuminates graphene metasurface along the z-axis.

We first took E_F=0.5 eV for the graphene, and the width of the wire w=0.7 μm. The changes of the resonant frequency and the peak absorption for the fundamental resonance caused by the increase of the cut-wire length (l) are plotted in the inset of **Figure 7(a)**. We see that the resonant frequency dropped dramatically to a lower frequency while the peak absorption found its maximum of 50% around 5.0 μm. **Figure 7(a)** presents the spectra of transmission (T), reflection (R), and absorption (A) spectra for the case of graphene cut-wire metasurface with the length of the wire l=5.0 μm. We can see an obvious resonance around 3.30 THz. From our investigation on the local field and current distributions, the resonance was confirmed to be an electric dipolar mode. It is well known that the resonance frequency of cut-wires changes with the length of the corresponding meta-atom. However, it is found that the width of cut-wires also plays an important role in determining the electromagnetic (EM) properties of graphene metasurface. Then, we specifically focus on studying the influence of w to the optical response of the cut-wire-array metasurface, by setting the length l at a constant value 5.0 μm. The absorption spectra of cut-wires with different line width show the influence of line width to the terahertz performance of graphene cut-wire array; we find that the resonant frequency of the electric mode shows a monotonous blue shift as we increase the line width w. On the other hand, we also notice that there exists an inflection for the absorption enhancement, which runs up to the maximum under w=0.7 μm at around 3.30 THz. The simple cut-wire structure confirms that the 50% maximum absorption enhancement still holds for the excitation of electric dipolar mode in a graphene sheet [59].

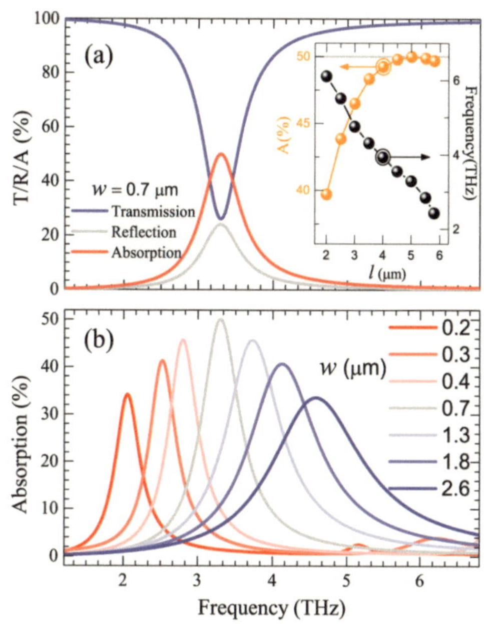

Figure 7. (a) Transmission, reflection, and the absorption of a graphene cut-wire array; the inset shows resonant frequencies and the peak absorptions for cut-wires with different lengths. (b) The absorption spectra for graphene cut-wire arrays with different cut-wire widths: the resonant spectra show a blue shift (from red to blue) regarding the increase of the widths.

Actually, the 50% maximum absorption of graphene metasurface can be understood simply with a transfer matrix study on a conductive sheet: since the graphene cut-wires are of deep subwavelength, we can neglect high-order scatterings of graphene metasurface, for that the unit cells of graphene metasurface are all of deep subwavelength, then we have the absorption (A) of a free-standing metasurface (described as a conductive sheet with effective complex conductivity σ)

$$
\begin{aligned}
A &= 1-|r|^2-|t|^2 \\
&= 1-\left|\frac{\sigma\eta_0}{2+\sigma\eta_0}\right|^2-\left|\frac{2}{2+\sigma\eta_0}\right|^2 \\
&= \frac{4}{\left(\left[4+(\sigma_i\eta_0)^2\right]/\sigma_r\eta_0\right)+\sigma_r\eta_0+4} \\
&\leq \frac{1}{2}
\end{aligned}
\tag{2}
$$

where η_0 represents the characteristic impedance of vacuum, r and t are the complex scattering coefficients, and σ_r and σ_i are the real and imaginary parts of σ, respectively. We can find that the absorption receives the maximum, that is, 50%, when $\sigma_r\eta_0$=2 and σ_i=0 (or simply $\sigma\eta_0$=2).

The retrieval method [66, 67] for the calculation of effective EM parameters from measured r and t coefficients provides a very intuitive route to understand the EM properties of metamaterials. Here, we applied a recently proposed retrieval method for sheet materials [68] to understand the THz response of the metamaterial designs; the sheet retrieval method is suitable for our 2D graphene structure. The effective electric surface conductivity $\sigma_{\parallel}^{e}$ can be obtained from the complex scattering coefficients

$$
\sigma_{\parallel}^{e} = \frac{2}{\zeta}\left(\frac{1-r-t}{1+r+t}\right) \tag{3}
$$

with ζ representing the wave impedance. The extracted complex surface conductivity of the graphene metasurface for the case of w=0.7 μm is shown in **Figure 8**. The resonant response around 3.30 THz can be seen on the effective electric conductivity spectrum, which is the proof of the excitation of electric dipolar mode. So far, we have investigated the graphene metasurface with the theoretical data, that is, the Drude weight and the collision frequency are α=58.86 GHz/Ω and γ=2 THz. We then consider the metasurface with graphene data from experimental measurements, that is, the Yan et al. graphene with α=76.0 GHz/Ω and γ=9.8 THz THz. The same procedure for studying the influence of w was performed and the 50% maximum absorption occurs when w=2.6 μm. The retrieved electric conductivity for Yan et al. graphene metasurface is presented in **Figure 8**, the electric resonance is shifted to 5.06 THz for this case. It is interesting that the amplitudes of the effective conductivity of the metasurfaces are nearly the same (as indicated in **Figure 8**). We also see from the figure that $\sigma\eta_0$=2 is almost fulfilled for the maximum 50% absorption at the resonant frequencies.

http://dx.doi.org/10.5772/64361

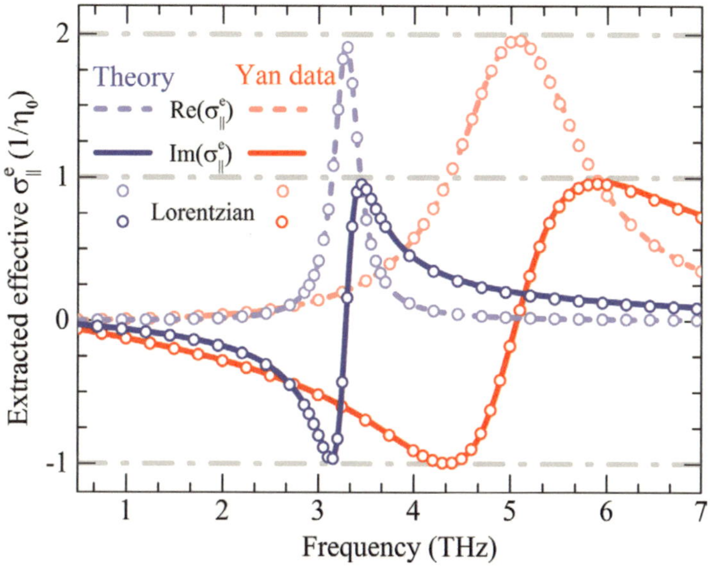

Figure 8. Extracted effective surface conductivities for graphene cut-wire arrays with theoretical graphene data and Yan et al. graphene data, respectively. Lorentzian fittings of the resonant conductivity responses are plotted on the extracted curves.

Since the effective electric surface conductivity of the graphene metasurface shows a Lorentz response, we used a Lorentzian function to fit the conductivity for quantitative descriptions of the electric resonances

$$\sigma_{\parallel}^{e} = \frac{i\kappa f 2}{f^2 + i\Gamma f - f_r^2} - i\beta f \quad (4)$$

where f_r is the frequency of resonance, Γ represents the damping, κ is a constant measuring coupling, and β characterizes the background polarization. The fitted Lorentz factors of the theoretical graphene data case are f_r=3.30 THz, Γ=315.0 GHz, and κ=606.4 GHz, while for the Yan et al. graphene data case these parameters are f_r=5.06 THz, Γ=1554.0 GHz, and κ=3048.1 GHz. We found two connections from the fitted Lorentz parameters: $\Gamma_2/\Gamma_1 \approx \gamma_2/\gamma_1$ and $\kappa_1/\Gamma_1 \approx \kappa_2/\Gamma_2$ (in which subscripts 1 and 2 denote the cases with theoretical and Yan et al. graphene data). These two relations imply the following: (i) the damping factor of plasmonic excitation of the graphene metasurface is determined by the collision frequency in graphene; and (ii) graphene metasurfaces with the same absorption (50% maximum absorption in our study) have nearly the same resonant amplitudes—κ/Γ.

To further investigate the graphene metasurface with different surface conductivities, we comparatively studied the graphene metasurface by changing the values of α and γ between that of the two sets of well-known experimental data, that is, Li et al. graphene (α=19.9 GHz/Ω, γ=29.4 THz) and Yan et al. graphene. The line width of cut-wires w is set as 2.6 μm for the discussions in this section. The false-color maps of the extracted electric surface conductivity are plotted in **Figure 9(a)** and **(b)**, respectively. We change the parameters of graphene from the Li et al. data (denoted as 1) to the Yan et al. data (denoted as 3) in two steps: decreasing collision frequency and increasing Drude weight (2 represents a hypothetic graphene data with α=19.9 GHz/Ω and γ=9.8 THz). From 1→3, it can be seen that the electric dipolar resonance changed considerably as we improved the quality of graphene (from the Li et al. data in 2008 to Yan et al. data in 2012), which can be confirmed quantitatively from **Figure 9(c)**, in which we plotted the fitted resonant frequencies and Q-factors (f_r/Γ) of the studied graphene metamaterial: (i) the resonant frequency maintains nearly the same when decreasing the collision frequency; however, the changing of the damping frequency makes the Q-factor to increase linearly; (ii) the damping frequency maintains when increasing the Drude weight; however, the blue shift of the resonant frequency makes the Q-factor to increase continuously during the increase in the Drude weight.

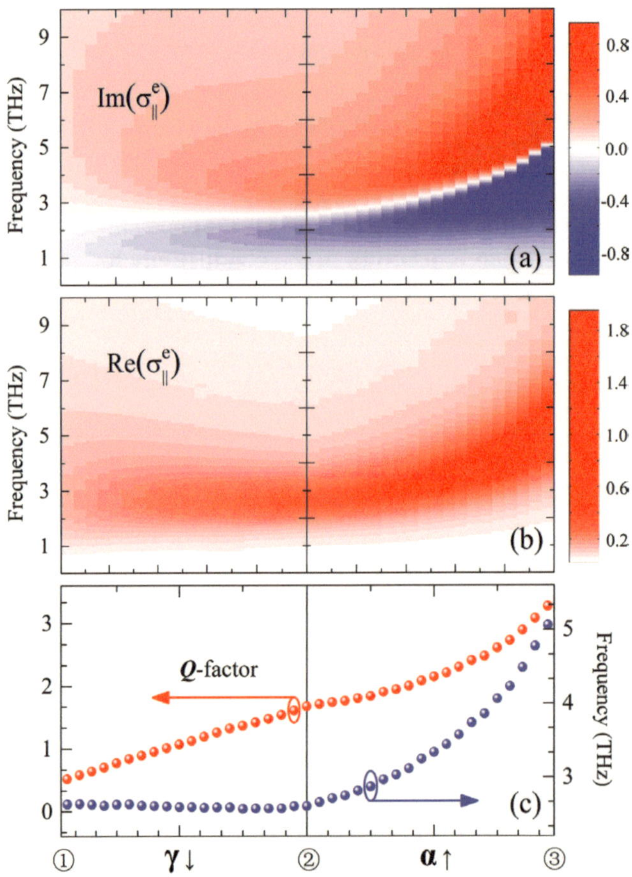

Figure 9. Spectra of imaginary (a) and real (b) parts of extracted effective surface conductivities as a function of the graphene Drude weight α and the collision frequency γ. (c) Q-factors and resonant frequencies of graphene cut-wire arrays with different Drude weights and collision frequencies.

3. Graphene plasmonic excitations in coherently modulated optical fields

3.1. A monolayer graphene as a tunable terahertz CPA

We suggest enhancing the terahertz absorption with the technique of coherent modulation in an unstructured and nonresonant monolayer graphene. We found that the quasi-CPA frequency, at which the formation condition of CPA is fulfilled, does exist in the terahertz band for suspending graphene. The scattering of coherent beams can be perfectly suppressed with proper coherent modulation on the input beams. In our theoretical study, a layer of graphene is free standing in vacuum, and it is illuminated by two counter-propagating and coherently modulated input beams ($I_{\pm}$), $O_{\pm}$ are the respective output magnitudes.

In the monolayer graphene system, the complex scattering coefficients ($O_{\pm}$) can be related to the two input beams ($I_{\pm}$) through a scattering matrix, S_g, which is defined as

$$\begin{pmatrix} O_+ \\ O_- \end{pmatrix} = S_g \begin{pmatrix} I_+ \\ I_- \end{pmatrix} = \begin{pmatrix} t_+ & r_- \\ r_+ & t_- \end{pmatrix} \begin{pmatrix} Ie^{i\phi_+} \\ Ie^{i\phi_-} \end{pmatrix} \tag{5}$$

where $t_{\pm}/r_{\pm}$ are the scattering elements of the forward (I_+) and backward (I_-) beams; the scattering matrix can be simplified for that the system is of reciprocity and spatial symmetry. We consider only phase modulation on the coherent input beams for simplicity, then the two input beams are of equal amplitude I; the amplitude of the output beams would be

$$|O_+| = |O_-| = \left|tIe^{i\phi_+} + rIe^{i\phi_-}\right| \tag{6}$$

In a terahertz coherent perfect absorber, the coherent modulation of the input beams performance is required to inhibit the scatterings and thus stimulate the complete absorption of coherent terahertz beams, which requires $tIe^{i\phi_+}=rIe^{i\phi_-}$, we get the necessary condition for acquiring CPA performance: $|t| = |r|$.

In calculations, the graphene sheet can be considered as an optical interface described by complex surface conductivity (σ). Reflection and transmission coefficients of forward- and backward-propagating light through the graphene can be driven with the assistance of Ohm's law. The scattering elements t and r at the normal incidence are given by

$$t = \frac{2}{2+\sigma\eta_0}, \tag{7}$$

$$r = \frac{\sigma\eta_0}{2+\sigma\eta_0}, \tag{8}$$

where $\eta_0$is the wave impedance of free space. From these equations, we can give the condition for coherent perfect absorption:

$$|\sigma\eta_0| = 2 \tag{9}$$

In the work, we also found that the CPA based on a monolayer graphene is of angularly sensitivity, which is good for wide angular tunability. For oblique incidence, we should consider both *s*- and *p*-polarized modes; it is found with a transfer matrix formalism that the quasi-CPA frequency splits into two frequency branches for the two polarizations: the CPA of *s*-polarization has a blue shift compared to the normal incident case, while the CPA of *p*-polarization has a red shift compared to the normal incident situation. The two bands together covered a wide frequency range.

The charge-carrier density and thus the Fermi level can be easily changed through electrostatic doping, which makes graphene promising for wide-tunable and broadband optoelectronic and photonic applications. With the increase in the electrostatic doping, we get higher charge-carrier concentration and thus higher Fermi energies, and we found a blue shift of the quasi-CPA frequency. This process can be understood as follows: the Drude weight of graphene (with higher carrier concentration) becomes higher, or we can say the graphene is with reenforced metallicity, which will have more scattering, then the reflection will be increased and the transmission will be decreased, and the quasi-CPA point will show a blue shift.

3.2. Graphene metamaterial interaction with coherent-modulated optical field as a tunable infrared CPA

The discussed suspending monolayer graphene CPA is physically based on the intrinsic Drude response of graphene, which is realizable only in the few-terahertz range with achievable graphene samples. And then we exploited the possibility of coherent perfect absorption at infrared frequencies. We designed a graphene nanoribbon-based metasurface and found that quasi-CPA frequencies, which is the necessary formation condition of coherent absorption, do exist in the mid-infrared regime for properly designed graphene nanoribbon arrays. The scattering of coherent beams can be perfectly suppressed at the quasi-CPA frequencies with proper phase modulations on the input beams. For the case with two crosses on the transmission and reflection spectra, we can achieve coherent perfect absorption at the two quasi-CPA frequencies, simultaneously. The flexible tunabilities of the graphene metasurface-based CPA are of interests for tunable infrared detections and signal modulations.

Figure 10 shows the schematic of the proposed graphene nanoribbons-based metasurface and the corresponding excitation configuration with two counter-propagating and coherently modulated optical beams ($I_{\pm}$), $O_{\pm}$ being the respective output magnitudes; the graphene sheet is illuminated with perpendicularly polarized light. Both lattice constant and width of the graphene nanoribbon meta-atoms play important roles in determining the optical resonant behaviors of graphene metasurface, the lattice constant is set to be $P = 0.7$ μm for the study of CPA in the mid-infrared regime.

We took $E_F = 0.5$ eV for the graphene. **Figure 11(a)** presents the spectra of the reflection coefficients (r) and transmission coefficients (t) for the case of graphene nanoribbon metasurface with the width of $w = 0.33$ μm. We can see an obvious resonance around 23.0 THz. The resonance was confirmed to be electric dipolar mode (which will be verified from the effective surface conductivity below), similar to the low-frequency resonance of split ring resonators as that in reference [59]. The excitation of electric dipolar mode leads to the enhancement of absorption in the graphene sheet; however, the maximum limit is 50%.

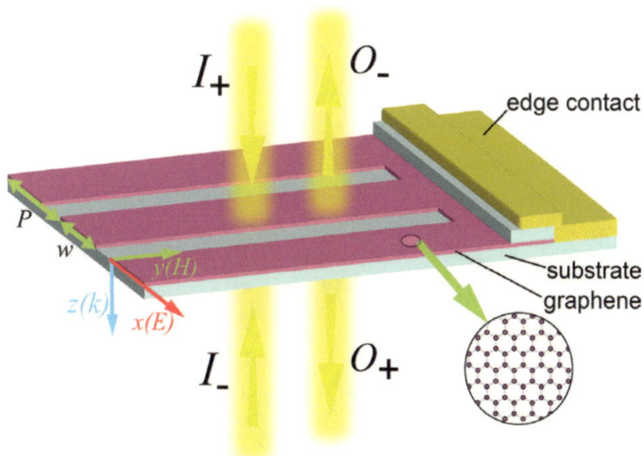

Figure 10. Schematic of a graphene ribbon metasurface illustrated by two counter-propagating and coherently modulated input beams.

Since the high-order scatterings of the deep subwavelength graphene nanoribbons are negligible, the graphene metasurface can be formalized with effective surface conductivities, then its interactions with the coherent modulated optical fields are the same as previously discussed suspending graphene case. The scatterings of the input beams are required to be inhibited to demonstrate a CPA, the necessary condition for CPA performance is $|t| = |r|$.

It can be seen from **Figure 11(a)** that there exists two frequencies (22.65 and 23.33 THz), which we call *quasi*-CPA points, which implies the necessary condition for suppressing the scattering fields to completely absorb coherent input beams of equal intensity. In the view of experiments, graphene generally needs to be transferred onto some substrate; we studied the scattering responses of a nanoribbon array (with the same geometry as that in **Figure 11(a)**) sandwiched in between two 45-nm-thick hexagonal boron nitride (h-BN) substrates, which was suggested as an exceptionally clean environment for achieving high confinement and low levels of plasmon damping in graphene [69] and is suitable for the one-dimensional high-quality electrical contact [70] (see the illustration in **Figure 10**). The dielectric function was taken from experimental studies [69, 71]. As can be seen from the inset of **Figure 11(a)**, the resonant frequency shifts to lower frequency as expected because of the introduction of the substrate, and there exist two quasi-CPA frequencies. For simplicity and without loss of generality, we will consider free-standing graphene metasurfaces.

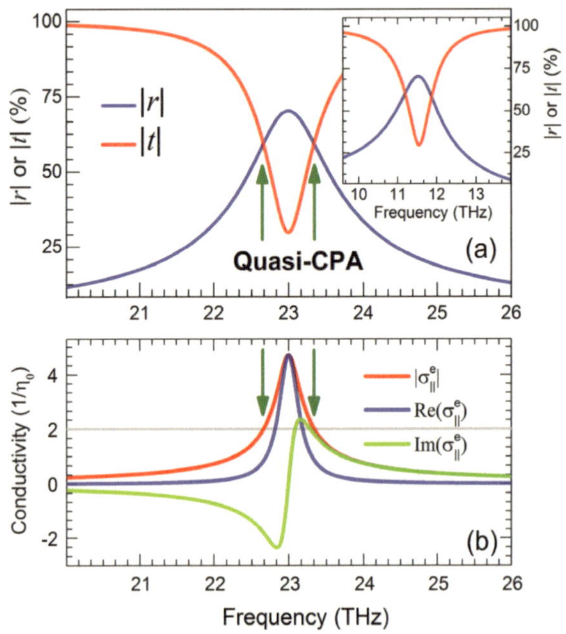

Figure 11. (a) Spectra of the reflection and transmission coefficients of a graphene nanoribbon metasurface; the quasi-CPA points at the crosses of the scattering spectra are indicated with arrows. (b) Absolute values, real and imaginary parts of the effective surface conductivities of the corresponding graphene nanoribbon metasurface.

Equation (**9**) gives the formation condition for CPA in an effective medium scheme. To verify this, we used a recently proposed sheet retrieval method (see Eq. (**3**)) to extract the effective surface conductivity $\sigma_{\|}^{e}$of the graphene nanoribbon metasurface. **Figure 11(b)** shows the absolute, real, and imaginary parts of the effective surface conductivities corresponding to **Figure 11(a)**. It is obvious that there is a Lorentz resonance around 23.0 THz on the effective electric conductivity spectrum, which confirms the excitation of electric dipolar mode. The extracted magnetic conductivity does not show any resonant feature around this electric resonance, so we have not included the corresponding result here. And it is also seen that the condition $|\sigma_{\|}^{e}\eta_0|=2$ is fulfilled at the two quasi-CPA frequencies, which indicate the validity of describing the graphene nanoribbon metasurface with the effective surface conductivity.

To implement the perfect absorption with the graphene metasurface, we set a chirped phase modulation $\Delta\phi(f)=\phi_+-\phi_-=\phi_0+kf$ on the beams $I_\pm$, with f being the frequency, and $k=-1.87\times10^{-12}\pi$ being the chirped factor to compensate the frequency dispersion. The plotted false-color map of the normalized spectra of total absorptions in **Figure 12(a)** shows the detailed dependence on ϕ_0. We see that a proper phase modulation (ϕ_0=1.03π) of the input coherent beams leads to significant suppression of the scattering outputs at the quasi-CPA frequencies. The normalized total absorption as a function of the frequency for the phase modulation ϕ_0=1.03π is plotted in **Figure 12(b)**. We can see total absorption at both the two quasi-CPA frequencies. The significant boosting of the absorption implies destructive interference, which prevents the coherent beams from escaping the absorbing channel of the graphene metasurface, demonstrating completely absorption of the coherent input beams.

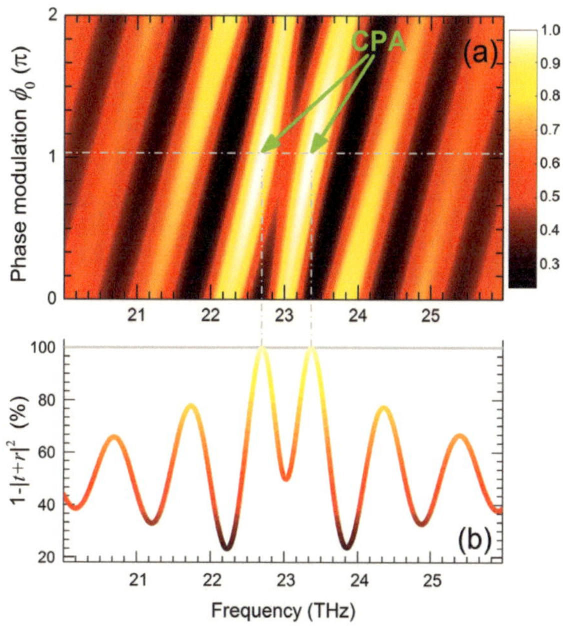

Figure 12. A two-dimensional false-color plot of the normalized total absorptions as a function of frequency and phase modulation, the exact CPA points are denoted with green arrows. (b) Normalized total absorption as a function of frequency for the phase modulation ϕ_0=1.03π.

The metasurface structures together with the electrically controlled graphene will provide more wide tunable space for the design of mid-infrared CPA; we first consider the geometric tunability of the graphene nanoribbon-based CPA. **Figure 13** shows the dependence of the difference ($|r|-|t|$) of the scattering coefficients of the graphene metasurface on the widths of nanoribbons. We can see that the resonant frequency of the electric dipolar mode shows a monotonous red shift with the increase of w, which is similar to the cut-wire case [60] (actually, the ribbon structure is the special situation of cut-wire with graphene covering all the lattice range along the x-axis). Increasing the width or the graphene-filling factor in the unit cell of the metasurface leads to stronger light-graphene interaction, that is, high resonant strength of the electric dipolar resonance, and thus higher r and lower t around the resonance that introduce a regime where $|r|-|t|=0$ starting from w=0.138 μm, which has its boundary (as the solid line indicated) being the quasi-CPA points. The discrete spheres on top of the solid curve, representing the extracted surface conductivities with $|\sigma_{\parallel}^{e}\eta_0|=2$, also imply that the formation condition of CPA is fulfilled at the boundary. At these quasi-CPA points, we can completely suppress the scatterings of the graphene metasurface with proper phase modulations as that shown in **Figure 12**.

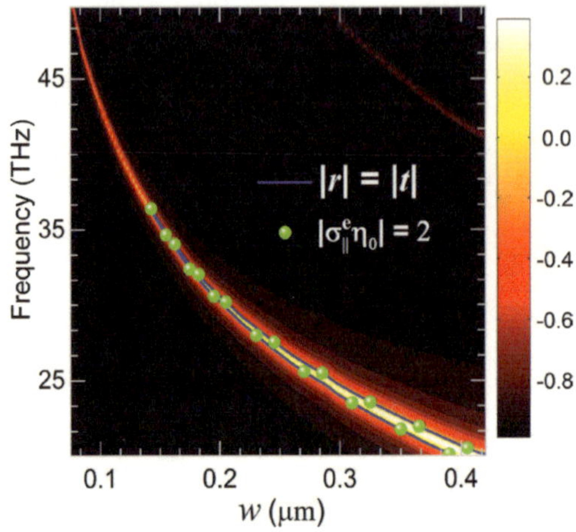

Figure 13. Geometric tunability of the graphene metasurface CPA.

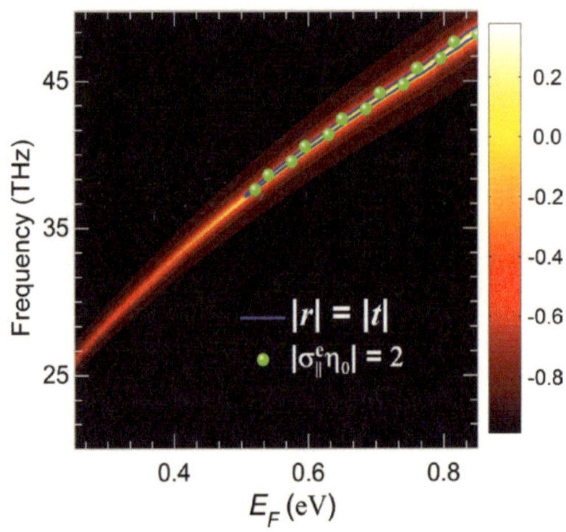

Figure 14. Electric tunability of the graphene metasurface CPA.

On the other hand, the graphene metasurface is also expected to have higher resonant strength for graphene with larger Fermi level. The dependence on Fermi energy of the difference of the scattering coefficients is plotted in **Figure 14** (the width of the graphene nanoribbon was set to be 0.138 μm, the left edge of the solid boundary in **Figure 13**); it can be seen that the resonant frequency shifts to higher frequencies and the resonant strength becomes higher while the Fermi level changes from0.25 to 0.85 eV. Similar to the influence of the width of nanoribbons, the light-graphene interaction is enhanced for better resonant behaviors, and thus a regime

http://dx.doi.org/10.5772/64361

starting from 0.5 eV where with its boundary $|t|-|r|=0$ being the quasi-CPA frequencies. Combining the two functional bands for CPA, we see that it is free to achieve CPA at desired frequency in an ultra-wide range by merging the geometric and electrical tunabilities. And the discrete spheres representing the extracted surface conductivities in **Figure 14** again confirm that $|\sigma_{||}^{e}\eta_0|=2$ is equivalent to the quasi-CPA condition $|t|=|r|$ for graphene nanoribbons-based metasurface.

4. Conclusion

In summary, this chapter summarized the recent progresses in the subfield of graphene plasmonics. Aimed to the issue in practical applications based on graphene: high loss in the plasmonic excitations of graphene limits the performance of graphene's ability in manipulating light. We show some reported results on the enhancement of light-graphene interactions by employing the new strategies including plasmonic metamaterial design and coherent modulation on optical fields. We found that the terahertz/ infrared extinction and absorption can be enhanced in a single graphene sheet by patterning plasmonic metamaterial structures, such as SRRs and cut-wires. It is found that we can significantly control the plasmonic excitations by manipulating geometric symmetry. It is shown that the electric plasmonic mode is stronger in enhancing infrared extinction and absorption compared to the magnetic mode and higher-order modes. We prove that the condition for maximum 50% absorption is $\sigma_r\eta_0=2$. It is also found that graphene metamaterials interacting with coherent-modulated optical fields can be employed to further take advantages from plasmonic excitations in infrared frequencies to perfectly suppress scattering for CPA. We prove that the necessary condition for realizing CPA for both suspending graphene and graphene metamaterial cases is $|\sigma\eta_0|=2$. Furthermore, the CPA can be tuned in an ultra-wide frequency band by considering both the geometric tunability and electrically controlled charge-carrier density in graphene. Our results on enhancing optical response in graphene with plasmonic metamaterial design, coherent modulations, and synergic action of them have potential applications for terahertz and infrared band graphene photonics and optoelectronics. We also expect potential applications of plasmonic metamaterial design and coherent modulations in two-dimensional materials beyond graphene.

Acknowledgements

Y. Fan would like to acknowledge Prof. C. M. Soukoulis, Prof. Q. Zhao, Dr. N.-H. Shen, and Mr. P. Zhang for collaboration and helpful discussion. This work was supported by the NSFC (Grant Nos. 11674266, 61505164 and 11372248), the Program for Scientific Activities of Selected Returned Overseas Professionals in Shaanxi Province, the Fundamental Research Funds for the Central Universities (Grant Nos. 3102015ZY079 and 3102015ZY058), and the Shaanxi Project for Young New Star in Science and Technology (Grant No. 2015KJXX-11).

Author details

Yuancheng Fan[1*], Fuli Zhang[1], Quanhong Fu[1] and Hongqiang Li[2]

*Address all correspondence to: phyfan@nwpu.edu.cn

1 Department of Applied Physics, School of Science, Northwestern Polytechnical University, Xi'an, China

2 The Institute of Dongguan–Tongji University, Dongguan, Guangdong, China

References

[1] Novoselov KS, Geim AK, Morozov SV, Jiang D, Zhang Y, Dubonos SV, Grigorieva IV, Firsov AA. Electric field effect in atomically thin carbon films. Science. 2004;306(5696): 666–669. DOI: 10.1126/science.1102896.

[2] Geim AK, Novoselov KS. The rise of graphene. Nat. Mater. 2007;6(3):183–191. DOI: 10.1038/nmat1849.

[3] Butler SZ, Hollen SM, Cao LY, Cui Y, Gupta JA, Gutierrez HR, Heinz TF, Hong SS, Huang JX, Ismach AF, Johnston-Halperin E, Kuno M, Plashnitsa VV, Robinson RD, Ruoff RS, Salahuddin S, Shan J, Shi L, Spencer MG, Terrones M, Windl W, Goldberger JE. Progress, challenges, and opportunities in two-dimensional materials beyond graphene. ACS Nano. 2013;7(4):2898–2926. DOI: 10.1021/nn400280c.

[4] Fiori G, Bonaccorso F, Iannaccone G, Palacios T, Neumaier D, Seabaugh A, Banerjee SK, Colombo L. Electronics based on two-dimensional materials. Nat. Nanotechnol. 2014;9(12):1063–1063. DOI: 10.1038/nnano.2014.283.

[5] Morozov SV, Novoselov KS, Katsnelson MI, Schedin F, Elias DC, Jaszczak JA, Geim AK. Giant intrinsic carrier mobilities in graphene and its bilayer. Phys. Rev. Lett. 2008;100(1). DOI: 10.1103/PhysRevLett.100.016602[1–4].

[6] Lee C, Wei XD, Kysar JW, Hone J. Measurement of the elastic properties and intrinsic strength of monolayer graphene. Science. 2008;321(5887):385–388. DOI: 10.1126/science.1157996.

[7] Balandin AA, Ghosh S, Bao WZ, Calizo I, Teweldebrhan D, Miao F, Lau CN. Superior thermal conductivity of single-layer graphene. Nano Lett. 2008;8(3):902–907. DOI: 10.1021/nl0731872.

[8] Bonaccorso F, Sun Z, Hasan T, Ferrari AC. Graphene photonics and optoelectronics. Nat. Photonics. 2010;4(9):611–622. DOI: 10.1038/nphoton.2010.186.

[9] de Abajo FJG. Graphene plasmonics: challenges and opportunities. ACS Photonics. 2014;1(3):135–152. DOI: 10.1021/ph400147y.

[10] Yan H, Li X, Chandra B, Tulevski G, Wu Y, Freitag M, Zhu W, Avouris P, Xia F. Tunable infrared plasmonic devices using graphene/insulator stacks. Nat. Nanotechnol. 2012;7:330–334. DOI: 10.1038/nnano.2012.59.

[11] Jablan M, Buljan H, Soljacic M. Plasmonics in graphene at infrared frequencies. Phys. Rev. B. 2009;80(24). DOI: 10.1103/PhysRevB.80.245435[1–7].

[12] Chen J, Badioli M, Alonso-Gonzalez P, Thongrattanasiri S, Huth F, Osmond J, Spasenovic M, Centeno A, Pesquera A, Godignon P, Zurutuza Elorza A, Camara N, Javier Garcia de Abajo F, Hillenbrand R, Koppens FHL. Optical nano-imaging of gate-tunable graphene plasmons. Nature. 2012;487(7405):77–81. DOI: 10.1038/nature11254.

[13] Fei Z, Rodin AS, Andreev GO, Bao W, McLeod AS, Wagner M, Zhang LM, Zhao Z, Thiemens M, Dominguez G, Fogler MM, Castro Neto AH, Lau CN, Keilmann F, Basov DN. Gate-tuning of graphene plasmons revealed by infrared nano-imaging. Nature. 2012;487(7405):82–85. DOI: 10.1038/nature11253.

[14] Gerber JA, Berweger S, O'Callahan BT, Raschke MB. Phase-resolved surface plasmon interferometry of graphene. Phys. Rev. Lett. 2014;113(5). DOI: 10.1103/PhysRevLett.113.055502[1–5].

[15] Vakil A, Engheta N. Transformation optics using graphene. Science. 2011;332(6035): 1291–1294. DOI: 10.1126/science.1202691.

[16] Koppens FHL, Chang DE, de Abajo FJG. Graphene plasmonics: a platform for strong light-matter interactions. Nano Lett. 2011;11(8):3370–3377. DOI: 10.1021/nl201771h.

[17] Wang F, Zhang YB, Tian CS, Girit C, Zettl A, Crommie M, Shen YR. Gate-variable optical transitions in graphene. Science. 2008;320(5873):206–209. DOI: 10.1126/science.1152793.

[18] Li ZQ, Henriksen EA, Jiang Z, Hao Z, Martin MC, Kim P, Stormer HL, Basov DN. Dirac charge dynamics in graphene by infrared spectroscopy. Nat. Phys. 2008;4(7):532–535. DOI: 10.1038/nphys989.

[19] Horng J, Chen CF, Geng BS, Girit C, Zhang YB, Hao Z, Bechtel HA, Martin M, Zettl A, Crommie MF, Shen YR, Wang F. Drude conductivity of Dirac fermions in graphene. Phys. Rev. B. 2011;83(16). DOI: 10.1103/PhysRevB.83.165113[1–5].

[20] Tassin P, Koschny T, Soukoulis CM. Graphene for terahertz applications. Science. 2013;341(6146):620–621. DOI: 10.1126/science.1242253.

[21] Kane CL, Mele EJ. Quantum spin Hall effect in graphene. Phys. Rev. Lett. 2005;95(22). DOI: 10.1103/PhysRevLett.95.226801[1–4].

[22] Yao YG, Ye F, Qi XL, Zhang SC, Fang Z. Spin-orbit gap of graphene: First-principles calculations. Phys. Rev. B. 2007;75(4). DOI: 10.1103/PhysRevB.75.041401[1–4].

[23] Zhang ZR, Li HQ, Gong ZJ, Fan YC, Zhang TQ, Chen H. Extend the omnidirectional electronic gap of Thue-Morse aperiodic gapped graphene superlattices. Appl. Phys. Lett. 2012;101(25). DOI: 10.1063/1.252104[1–4].

[24] Engheta N. Circuits with light at nanoscales: Optical nanocircuits inspired by metamaterials. Science. 2007;317(5845):1698–1702. DOI: 10.1126/science.1133268.

[25] Fan YC, Han J, Wei ZY, Wu C, Cao Y, Yu X, Li HQ. Subwavelength electromagnetic diode: One-way response of cascading nonlinear meta-atoms. Appl. Phys. Lett. 2011;98(15). DOI: 10.1063/1.151903[1–3].

[26] Fan YC, Wei ZY, Han J, Liu XG, Li HQ. Nonlinear properties of meta-dimer comprised of coupled ring resonators. J. Phys. D-Appl. Phys. 2011;44(42). DOI: 10.1088/0022-3727/44/42/425303[1–4].

[27] Yablonovitch E. Inhibited spontaneous emission in solid-state physics and electronics. Phys. Rev. Lett. 1987;58(20):2059–2062. DOI: 10.1103/PhysRevLett.58.2059.

[28] Smith DR, Pendry JB, Wiltshire MCK. Metamaterials and negative refractive index. Science. 2004;305(5685):788–792. DOI: 10.1126/science.1096796.

[29] Soukoulis CM, Wegener M. Past achievements and future challenges in the development of three-dimensional photonic metamaterials. Nat. Photonics. 2011;5(9):523–530. DOI: 10.1038/nphoton.2011.154.

[30] Gan X, Shiue R-J, Gao Y, Mak KF, Yao X, Li L, Szep A, Walker D, Hone J, Heinz TF, Englund D. High-contrast electrooptic modulation of a photonic crystal nanocavity by electrical gating of graphene. Nano Lett. 2013;13:691–696. DOI: 10.1021/nl304357u.

[31] Piper JR, Fan S. Total absorption in a graphene monolayer in the optical regime by critical coupling with a photonic crystal guided resonance. ACS Photonics. 2014;1:347–353. DOI: 10.1021/ph400090p.

[32] Gu T, Andryieuski A, Hao Y, Li Y, Hone J, Wong CW, Lavrinenko A, Low T, Heinz TF. Photonic and plasmonic guided modes in graphene–silicon photonic crystals. ACS Photonics. 2015;2:1552–1558. DOI: 10.1021/acsphotonics.5b00209.

[33] Zheng J, Barton RA, Englund D. Broadband coherent absorption in chirped-planar-dielectric cavities for 2D-material-based photovoltaics and photodetectors. ACS Photonics. 2014;1:768–774. DOI: 10.1021/ph500107b.

[34] Fan YC, Wei ZY, Li HQ, Chen H, Soukoulis CM. Photonic band gap of a graphene-embedded quarter-wave stack. Phys. Rev. B. 2013;88(24). DOI: 10.1103/PhysRevB.88.241403[1–5].

[35] Liu N, Guo HC, Fu LW, Kaiser S, Schweizer H, Giessen H. Three-dimensional photonic metamaterials at optical frequencies. Nat. Mater. 2008;7(1):31–37. DOI: 10.1038/nmat2072.

[36] Zhang FL, Zhao Q, Kang L, Zhou J, Lippens D. Experimental verification of isotropic and polarization properties of high permittivity-based metamaterial. Phys. Rev. B. 2009;80(19). DOI: 10.1103/PhysRevB.80.195119[1–6].

[37] Wei ZY, Cao Y, Han J, Wu C, Fan YC, Li HQ. Broadband negative refraction in stacked fishnet metamaterial. Appl. Phys. Lett. 2010;97(14). DOI: 10.1063/1.141901[1–3].

[38] Han J, Li HQ, Fan YC, Wei ZY, Wu C, Cao Y, Yu X, Li F, Wang ZS. An ultrathin twist-structure polarization transformer based on fish-scale metallic wires. Appl. Phys. Lett. 2011;98(15). DOI: 10.1063/1.151908[1–3].

[39] Wei ZY, Cao Y, Fan YC, Yu X, Li HQ. Broadband polarization transformation via enhanced asymmetric transmission through arrays of twisted complementary split-ring resonators. Appl. Phys. Lett. 2011;99(22). DOI: 10.1063/1.221907[1–3].

[40] Wei ZY, Cao Y, Fan YC, Yu X, Li HQ. Broadband transparency achieved with the stacked metallic multi-layers perforated with coaxial annular apertures. Opt. Express. 2011;19(22):21425–21431. DOI: 10.1364/oe.19.021425.

[41] Zhang FL, Zhao Q, Zhou J, Wang SX. Polarization and incidence insensitive dielectric electromagnetically induced transparency metamaterial. Opt. Express. 2013;21(17): 19675–19680. DOI: 10.1364/oe.21.019675.

[42] Zhang FL, Feng SQ, Qiu KP, Liu ZJ, Fan YC, Zhang WH, Zhao Q, Zhou J. Mechanically stretchable and tunable metamaterial absorber. Appl. Phys. Lett. 2015;106(9). DOI: 10.1063/1.091907[1–5].

[43] Yu N, Capasso F. Flat optics with designer metasurfaces. Nat. Mater. 2014;13(2):139–150. DOI: 10.1038/nmat3839.

[44] Kildishev AV, Boltasseva A, Shalaev VM. Planar photonics with metasurfaces. Science. 2013;339(6125). DOI: 10.1126/science.1232009[1–6].

[45] Yu NF, Genevet P, Kats MA, Aieta F, Tetienne JP, Capasso F, Gaburro Z. Light propagation with phase discontinuities: generalized laws of reflection and refraction. Science. 2011;334(6054):333–337. DOI: 10.1126/science.1210713.

[46] Ni XJ, Emani NK, Kildishev AV, Boltasseva A, Shalaev VM. Broadband light bending with plasmonic nanoantennas. Science. 2012;335(6067):427–427. DOI: 10.1126/science.1214686.

[47] Sun SL, He Q, Xiao SY, Xu Q, Li X, Zhou L. Gradient-index meta-surfaces as a bridge linking propagating waves and surface waves. Nat. Mater. 2012;11(5):426–431. DOI: 10.1038/nmat3292.

[48] Yin XB, Ye ZL, Rho J, Wang Y, Zhang X. Photonic spin Hall effect at metasurfaces. Science. 2013;339(6126):1405–1407. DOI: 10.1126/science.1231758.

[49] Shitrit N, Yulevich I, Maguid E, Ozeri D, Veksler D, Kleiner V, Hasman E. Spin-optical metamaterial route to spin-controlled photonics. Science. 2013;340(6133):724–726. DOI: 10.1126/science.1234892.

[50] Chong YD, Ge L, Cao H, Stone AD. Coherent perfect absorbers: Time-reversed lasers. Phys. Rev. Lett. 2010;105(5). DOI: 10.1103/PhysRevLett.105.053901[1–4].

[51] Wan WJ, Chong YD, Ge L, Noh H, Stone AD, Cao H. Time-reversed lasing and interferometric control of absorption. Science. 2011;331(6019):889–892. DOI: 10.1126/science.1200735.

[52] Longhi S. PT-symmetric laser absorber. Phys. Rev. A. 2010;82(3). DOI: 10.1103/PhysRevA.82.031801[1 –4].

[53] Lin Z, Ramezani H, Eichelkraut T, Kottos T, Cao H, Christodoulides DN. Unidirectional invisibility induced by PT-symmetric periodic structures. Phys. Rev. Lett. 2011;106(21). DOI: 10.1103/PhysRevLett.106.213901[1–4].

[54] Noh H, Chong YD, Stone AD, Cao H. Perfect coupling of light to surface plasmons by coherent absorption. Phys. Rev. Lett. 2012;108(18). DOI: 10.1103/PhysRevLett.108.186805[1–5].

[55] Crescimanno M, Dawson NJ, Andrews JH. Coherent perfect rotation. Phys. Rev. A. 2012;86(3). DOI: 10.1103/PhysRevA.86.031807[1–4].

[56] Zhang J, MacDonald KF, Zheludev NI. Controlling light-with-light without nonlinearity. Light-Sci. Appl. 2012;1. DOI: 10.1038/lsa.2012.18[1–5].

[57] Hao JM, Zhou L, Qiu M. Nearly total absorption of light and heat generation by plasmonic metamaterials. Phys. Rev. B. 2011;83(16). DOI: 10.1103/PhysRevB.83.165107[1–12].

[58] Pu MB, Feng Q, Hu CG, Luo XG. Perfect absorption of light by coherently induced plasmon hybridization in ultrathin metamaterial film. Plasmonics. 2012;7(4):733–738. DOI: 10.1007/s11468-012-9365-1.

[59] Fan YC, Wei ZY, Zhang ZR, Li HQ. Enhancing infrared extinction and absorption in a monolayer graphene sheet by harvesting the electric dipolar mode of split ring resonators. Opt. Lett. 2013;38(24):5410–5413. DOI: 10.1364/ol.38.005410.

[60] Fan YC, Shen NH, Koschny T, Soukoulis CM. Tunable terahertz meta-surface with graphene cut-wires. ACS Photonics. 2015;2(1):151–156. DOI: 10.1021/ph500366z.

[61] Fan YC, Zhang FL, Zhao Q, Wei ZY, Li HQ. Tunable terahertz coherent perfect absorption in a monolayer graphene. Opt. Lett. 2014;39(21):6269–6272. DOI: 10.1364/ol.39.006269.

[62] Fan Y, Liu Z, Zhang F, Zhao Q, Wei Z, Fu Q, Li J, Gu C, Li H. Tunable mid-infrared coherent perfect absorption in a graphene meta-surface. Sci. Reports. 2015;5. DOI: 10.1038/srep13956[1–8].

[63] Fedotov VA, Rose M, Prosvirnin SL, Papasimakis N, Zheludev NI. Sharp trapped-mode resonances in planar metamaterials with a broken structural symmetry. Phys. Rev. Lett. 2007;99(14). DOI: 10.1103/PhysRevLett.99.147401[1–4].

[64] Al-Naib IAI, Jansen C, Koch M. High Q-factor metasurfaces based on miniaturized asymmetric single split resonators. Appl. Phys. Lett. 2009;94(15). DOI: 10.1063/1.153505[1–3].

[65] Fan YC, Wei ZY, Li HQ, Chen H, Soukoulis CM. Low-loss and high-Q planar metamaterial with toroidal moment. Phys. Rev. B. 2013;87(11). DOI: 10.1103/PhysRevB.87.115471[1–5].

[66] Smith DR, Schultz S, Markos P, Soukoulis CM. Determination of effective permittivity and permeability of metamaterials from reflection and transmission coefficients. Phys. Rev. B. 2002;65(19). DOI: 10.1103/PhysRevB.65.195104[1–5].

[67] Chen XD, Grzegorczyk TM, Wu BI, Pacheco J, Kong JA. Robust method to retrieve the constitutive effective parameters of metamaterials. Phys. Rev. E. 2004;70(1). DOI: 10.1103/PhysRevE.70.016608[1–7].

[68] Tassin P, Koschny T, Soukoulis C. Effective material parameter retrieval for thin sheets: Theory and application to graphene, thin silver films, and single-layer metamaterials. Phys. B. 2012;407(20):4062–4065. DOI: 10.1016/j.physb.2012.01.119.

[69] Woessner A, Lundeberg MB, Gao Y, Principi A, Alonso-Gonzaolez P, Carrega M, Watanabe K, Taniguchi T, Vignale G, Polini M, Hone J, Hillenbrand R, Koppens FHL. Highly confined low-loss plasmons in graphene-boron nitride heterostructures. Nat. Mater. 2015;14(4):421–425. DOI: 10.1038/nmat4169.

[70] Wang L, Meric I, Huang PY, Gao Q, Gao Y, Tran H, Taniguchi T, Watanabe K, Campos LM, Muller DA, Guo J, Kim P, Hone J, Shepard KL, Dean CR. One-dimensional electrical contact to a two-dimensional material. Science. 2013;342(6158):614–617. DOI: 10.1126/science.1244358.

[71] Caldwell JD, Kretinin AV, Chen Y, Giannini V, Fogler MM, Francescato Y, Ellis CT, Tischler JG, Woods CR, Giles AJ, Hong M, Watanabe K, Taniguchi T, Maier SA, Novoselov KS. Sub-diffractional volume-confined polaritons in the natural hyperbolic material hexagonal boron nitride. Nat. Commun. 2014;5. DOI: 10.1038/ncomms5221[1–9].

Chapter 12

Graphene-Based Materials Functionalization with Natural Polymeric Biomolecules

Edgar Jimenez-Cervantes Amieva,
Juventino López-Barroso,
Ana Laura Martínez-Hernández and
Carlos Velasco-Santos

Additional information is available at the end of the chapter

http://dx.doi.org/10.5772/64001

Abstract

The use of 2D nanocarbon materials as scaffolds for the functionalization with different molecules has been rising as a result of their outstanding properties. This chapter describes the synthesis of graphene and its derivatives, particularly graphene oxide (GO) and reduced graphene oxide (rGO). Both GO and rGO represent a tunable alternative for applications with biomolecules due to the oxygenated moieties, which allow interactions in a either covalent or non-covalent way. From here, other discussed topics are the biofunctionalization with keratin (KE) and chitosan (CS). The non-covalent functionalization is based primarily on secondary interactions such as van der Waals forces, electrostatics interactions, or π–π stacking formed between KE or CS with graphenic materials. On the other hand, covalent functionalization with KE and CS is mainly based on the reaction among the functional groups present in those biomolecules and the graphenic materials. As a result of the functionalization, different applications have been proposed for these novel materials, which are reviewed in order to offer an overview about the possible fields of application of 2D nanocarbon materials. In a nutshell, the objective of this work is as follows: first, overhaul different aspects about the synthesis of graphene chemically obtained, and second, make a review of different approaches in the functionalization of 2D carbon materials with specific biomolecules.

Keywords: graphene oxide, reduced graphene oxide, reduction, functionalization, covalent, non-covalent, chitosan, keratin

1. Introduction

Carbon is one of the most interesting elements of nature and is one of the primary constituents for the formation of all organic matter. Its capacity to bind to itself in different ways makes carbon a very versatile element, giving rise to a series of structures called allotropes. Some of these carbon forms have attracted great interest in the past few decades due to their small size, in the scale of nanometers, and their very particular shape or dimensionality, which directly affect their chemical and physical properties. Among these carbon nanostructures, graphene (GE) has become an outstanding material presenting a unique set of "superlative" properties such as mechanical, electrical, thermal and electronic characteristics, enlisted in many reports [1–4].

Graphene is the name of a single layer of carbon atoms arranged in a two-dimensional (2D) crystalline hexagonal lattice, due to the sp^2 hybridization of carbon. Thus, graphene has strong in-plane σ bonds, responsible for its high mechanical strength and flexibility, and it also has weak out-of-plane π bonds responsible for its thermal carrying, electrical charge, and transparency, and graphene is also impermeable. Nevertheless, all these properties are only observed in a single defect-free graphene layer, which is costly to produce in a scalable degree. Alternatively, there are other ways to produce graphene with relative ease, such as the chemical phase exfoliation of graphite oxide. This method yields the synthesis of graphene oxide (GO), a highly oxidized version of graphene. The subsequent reduction in the oxygen content brings a partial restoration to graphenic state, producing reduced graphene oxide (rGO) or chemically converted graphene (CCG) (**Figure 1a**). These graphene-based materials

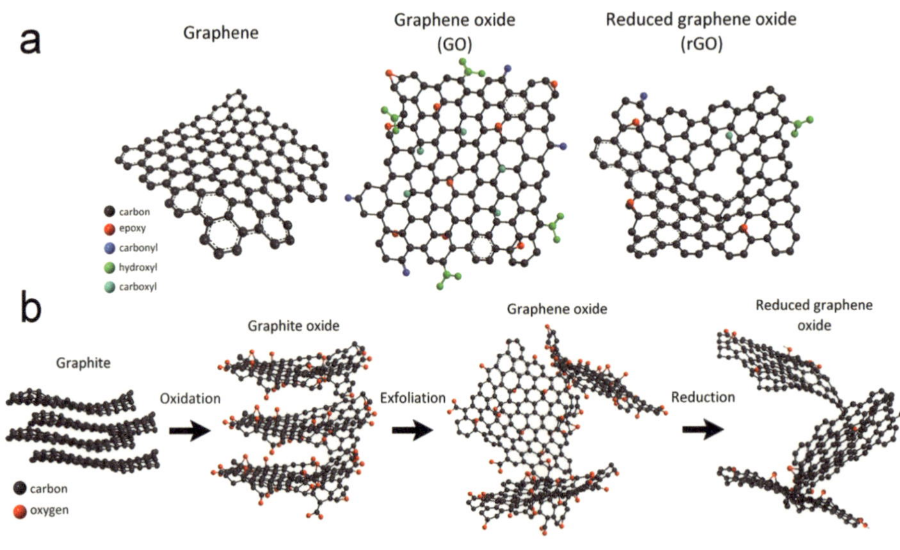

Figure 1. (a) Schematic chemical structures of graphene, graphene oxide, and reduced graphene oxide. (b) Route of graphite to reduce graphene oxide.

are considered as a functionalization of a graphene sheet, because of the presence of oxygen species [5].

Even when GO and rGO have a lower set of properties compared with those of GE, they can be improved through an additional functionalization among others with organic and different bio-molecules. Since the presence of oxygen groups is greater in GO, its reactivity is higher compared with rGO, and thus, graphene oxide is more suitable to be functionalized through covalent interactions. On the other hand, even when rGO retain some oxygen sites, its partial graphitic surface makes it adequate for a non-covalent functionalization. In this manner, tailored-made properties can be successfully achieved giving rise to a wide range of potential applications.

In this manner, GO and rGO have been modified with all kind of biomolecule groups, such as nucleic acids, proteins and peptides, antibodies, enzymes, polysaccharides and amino acids, for applications related to biological/biomedical fields which require a good degree of compatibility, adherence, or biocide characteristics. Long-chain biopolymers containing amino groups, such as keratin (a protein) and chitosan (a polysaccharide), show an excellent covalent linkage with GO, and a good adsorption on rGO (keratin), and they could function either increasing the bacterial adhesion or inhibiting the bacterial growth. This chapter focuses in the study and insight of the interactions between amine-containing biopolymers through covalent and non-covalent interactions with GO and rGO. The details of synthesis of graphene oxide and reduction are reviewed and discussed; a brief discussion involving the different types of functionalization is mentioned; finally, some researches related to functionalization of biomolecules on graphenic materials along with their diverse potential uses is also discussed.

2. Graphene oxide and reduced graphene oxide

As a result of the exceptional properties of graphene, different efforts have been made in order to scale up its mass production. In a brief historically account, one of the first approaches were conducted by Lang in 1975, where few layer graphite was obtained by chemical deposition method. Nevertheless, in those days, the characterization techniques were unable to show what Lang has achieved. More recently in 1999, was reported the mechanical cleavage of highly ordered pyrolytic graphite (HOPG) with atomic force microscopy (AFM) tips, in an attempt to exfoliate a single layer of graphene [1]. Finally, in 2004, Novoselov reported the isolate one layer of graphene through "peeling" many times natural graphite with "Scotch" tape. In this method, the graphite layers are sliced down by mechanical exfoliation until one layer is deposited in a substrate [2]. Despite the effectiveness of this method and the high-quality obtained graphene, this process requires a great deal of time and the amount of the as-produced graphene is not enough for practical applications [3, 4].

A possible solution, in order to obtain graphenic materials in larger amounts, has been found on the graphite oxide route, starting from the modified Hummers method [6], which is based on the introduction of functional groups in graphite layers using a mixture of sulfuric acid and potassium permanganate. The versatility of the method, the excellent dispersion acquired in different solvents [5], and the possibility of a high yield production, makes the graphite oxide route one of the most promising scalable methods. This process mainly consists of three stages (**Figure 1b**).

First, it is necessary to produce graphite oxide by the oxidative intercalation reaction. In this step, the sp^2 carbon arrangement is disrupted by the introduction of oxygenated functional groups, such as hydroxyl and epoxy in the basal plane and carbonyl and carboxyl at the edges (surroundings). During this process, interlaminar space between graphene sheets is increased two or three times than that of pristine graphite, from 3.34 Å to 5.62 Å [1, 7]. Second, graphene oxide (GO) is produced through exfoliation and dispersion of oxidized graphite. It has been reported that cavitation produced by ultrasonic waves within a fluid (commonly distilled water) produces hot spots with temperatures approximately to 5000°C, high local pressures about 500 atm and rates of heating–cooling of 10^9 K/s [8]. Therefore, water molecules (or polar solvents) can be intercalated in graphite oxide, producing an interlaminar spacing of roughly 1.2 nm [9]. Finally, the third stage consists on reducing or diminishing the oxidation degree of GO, in other words the production of reduced graphene oxide. In terms of recovery thermal, electrical, and mechanical properties, GO has been subjected to a variety of treatments to diminish the oxidized state or the number of oxidized moieties on it. Treatments such as chemical, thermal, UV radiation, and electrochemical reduction have been applied in either individual or multi-steps, partially recovering the sp^2 arrangement found in pristine graphene [6, 7].

2.1. Synthesis of graphene oxide

GO is a covalent carbon structure where at least 60% of its atoms show sp^3 hybridization connected by σ bonds to oxygen atoms, whereas the structure of basal plane is preserved even though it has substantial deformations. Therefore, GO is a hybrid material considered as an insulator exhibiting a resistance about 10^{12} Ω/sq depending on sp^2 and sp^3 formed during oxidation [10]. GO thickness is theoretically within 0.7–0.8 nm because of the oxygenated moieties present on its surface, that is, roughly twice as the thickness of single layer of pristine graphene; however, GO could be larger considering that the occasional presence of bulkier functionalities, organic adsorbates, even some contaminants and so on [11].

Inasmuch as GO is not an stoichiometric material, its structure depends on the oxidizing process, this produces an aromatic disrupted lattice rich in functional groups maintaining arranged zones sp^2 (2–3 nm) within a matrix C–O sp^3 [12], and, for this reason, its chemical structure is difficult to determine. Lerf–Klinowski model based on the evidence obtained through nuclear magnetic resonance (NMR) of graphite oxide describes this material such as randomized of oxidized areas. As a result of oxidation sp^2/sp^3 regions are conformed of aromatic or aliphatic six-membered rings [13]. This model is one of the most accepted in order to describe GO structure. From Lerf–Klinowski model, Gao et al. [14], proposed a new model including the presence of lactole rings of five and six members on the edges besides of carboxyl and carbonyl and hydroxyl, while in the basal plane are found tertiary alcohol esters as well as epoxy and hydroxyl groups.

On the other hand, graphite is considering an inexpensive available material for scalable production of GO; additionally, it can be obtained from either natural or synthetic sources [15], which can be influential over the final properties of GO, as has been reported. Undoubtedly for GO is really important, though not unexpected that the final dimension of the GO sheets obtained is strongly influenced for both the incipient size of graphite flakes and its inherent defects in π-structure in conjunction with the oxidation protocols and exfoliation procedure [16, 17].

The fundamental procedure involved in the synthesis of graphene oxide had originally been developed to oxidize graphite by Brodie in 1859 which used fuming nitric acid (HNO_3) with potassium chlorate $KClO_3$, evolving to the Staudenmaier method where oxidation of graphite was carried out with the addition of sulfuric acid (H_2SO_4) and $KClO_3$ in different parts of the oxidation reaction [18]. Finally, Hummers and Offeman developed the most commonly method derived from Staudenmaier's work, where GO is extensively produced from the oxidation of graphite and the process is completed by a subsequent exfoliation of graphite oxide generally made by ultrasound wave or rapid heating [19]. The reaction takes place when a mixture of potassium permanganate ($KMnO_4$) as strong oxidizing agent is combined with a concentrated H_2SO_4 (Eq. (1)). In terms of the permanganate used as an oxidizing agent, it is generally accepted that the main reactive specie in oxidation of graphite is the dimanganese heptoxide (Mn_2O_7) as can been seen in Eq. (2), not taking into account the changes in the method. Mn_2O_7 can react explosively as long as the temperature raise above 55°C or to come into contact with organic compounds [3, 6].

$$KMnO_4 + H_2SO_4 \rightarrow K^+ + MnO_3^+ + H_3O^+ + 3HSO_4^- \tag{1}$$

$$MnO_3^+ + MnO_4^- \rightarrow Mn_2O_7 \tag{2}$$

Chemical species involved in the synthesis of graphene oxide.

Few years ago, Stankovich et al. [20] reported to obtain chemically modified graphene sheets from complete exfoliation of graphite by Hummers method, due to the van der Waals forces are weakened by the oxygenated moieties formed during the oxidation and the change of graphite to hydrophilic character, the hydration and dispersion in aqueous media can be done to obtain stable colloidal dispersions of GO on water. Additionally, the oxygenated moieties on GO have a negative charge, which grant first a good dispersion as well as stability in some organic solvents, alcohol, and water throughout electrostatic repulsion [17]. Consequently, GO have been used in applications such as sensing, composites, electronic devices, it can not only get well dispersion but also offers a platform for functionalization through chemical reactions such as amidation, esterification, and so on [11, 21]. Additionally, another important feature of GO is its amphiphilicity caused by the heterogeneous distribution of functional groups producing both hydrophobic and hydrophilic domains, having great results in order to interact with other materials such as polymers [11] or creating Pickering emulsions with organic solvents acting as molecular dispersing agent of materials such as carbon nanotubes (CNT) tuning only the solution pH [22].

Generally, the dispersion of GO is made by ultrasonic waves because is an easy, quick and efficient technique in comparison with the stirring and rapid heating process; however, long periods of sonication can be damaged and reduced the size of GO layers impacting on its properties [6, 7]. Sonication can be followed by centrifugation in order to obtain different lateral sizes on graphene oxide sheets based on density-gradient [17]. Furthermore, salts and ions formed, while the oxidation reaction was carried out are removed [3]. Finally, GO can be reduced in order to recover the initial structure present in graphite, after that, properties such as conductivity can be restored until four times as much as in GO.

2.2. Synthesis of reduced graphene oxide

The sp^2 hybridization loss in GO can be partially restored through the removal of oxygen moieties present on its surface. In this regard, GO has been subjected to the different process of reduction resulting in reduced graphene oxide (rGO). In other words, in rGO, there occurs the formation of percolation pathways between nanometric sp^2 domains disrupted in oxidation reaction of graphite [17]. Therefore, rGO is an important material which has captivated scientific attention for its properties alike to those of pristine graphene that has permitted its use in many potentials applications.

The chemical structure of rGO does not have a specific arrangement regarding oxygenated functionalities distribution and aromatic/aliphatic domains. The remaining oxygen atoms exist as a result of the formation of stable carbonyl and ether groups that cannot be removed without making damage to the basal plane. In addition, another kind of defects found in rGO is the so called Stone–Wales defects (heptagons and pentagons pairs) as well as holes caused by losses of carbon in the form of CO and CO_2 in the reduction process [12]. Nonetheless, properties such as resistivity have important changes after the reduction; values of $\approx$28.6 kΩ/sq have been reported for rGO [3]. On the other hand, even with an incomplete reduction, rGO produced has some advantages in view, it can be both electrically better than GO and still keeps some functional group reactive sites where further functionalization can be made [23]. Therefore, it is possible that the properties on rGO can be tuned, according with the degree of reduction. Hu et al. highlighted the close relationship between oxidation degree of graphite oxide and defects of rGO. Experimentally, they found that it is possible to obtain rGO with a few amount of defects, first with a relative high oxidation degree and second with low addition of $KMnO_4$ used like oxidizing agent (chemical route) in terms of low addition, defects appeared on edges of graphite [24].

As a result of the reduction process visible changes can be observed, for instance, the brown dispersion of GO turns into a black precipitate due to aromatic restoration and re-agglomeration of the rGO sheets, respectively, as a result of the change from hydrophilic to hydrophobic character after removal of functional groups [16]. Additionally, remarkable differences between GO and rGO have been reported under optical microscopy, while GO is mostly transparent rGO has a very light contrast with the substrate (SiO_2/Si wafer), this reflects the change from insulator GO to rGO an electrical conductive material [25]. In this regard, for UV–vis spectroscopy, GO shows an absorption peak near 230 nm due to π–π* plasmon, and a shoulder around 300 nm associated with π–π* transitions in C=O, while rGO exhibits only an absorption peak between 225 and 275 nm, attributed to a better structural order and more C=C bonds [17]. Another important parameter to take into account is the C/O atomic ratio which is obtained in most cases by X-ray photoelectron spectroscopy, elemental analysis and/or energy dispersive spectroscopy [26]. In this regard, different values have been reported, from 2.2 to 2.7 typically for GO to 10–12 for rGO by chemical reduction, even C/O ratio of 50.2 was reported for reduction under acetylene (C_2H_2) atmosphere [10, 17, 27]. Besides, C/O ratio, electrical conductivity also demonstrates the restoration of sp^2 domains through the formation of percolation pathways after reduction. For GO, values of 10^{-6} S cm^{-1}, characteristic of an insulator material, have been reported, [17], whereas for rGO sheets functionalized with pyrene, a conductivity of 1314 S cm^{-1} has been reported [28]. In addition to these parameters, other characterization techniques, such as NMR, atomic force microscopy (AFM), Raman spectroscopy, transmission electron microscopy (TEM), are also used to reveal the structure and properties of rGO.

One method of reduction of GO is through heating of the samples, also called thermal annealing. This approach is generally employed to exfoliate graphite oxide using gradients of temperature higher than 2000°C/min [29]; for example, it has been reported the use of arc discharge method under hydrogen atmosphere yielding C/O molar ratio of 15–18 and a conduction 10 times higher than the traditional arc discharge method, this can be attributable to the *in situ* elimination and healing during the exfoliation [30]. The exfoliation–reduction can be explained given the CO and CO_2 found within layers of graphite oxide, suffering an abrupt expansion increasing the interlayer distance of GO sheets; moreover, the decomposition of oxygenated moieties also produces high pressure and both are able to overcome the van der Waals forces [10, 29]. McAllister et al. [31] reported the generation of pressures of 40 MPa and 130 MPa at 300 and 1000°C, respectively, based on state equation, this exceeds the value of 2.5 MPa, calculated from Hamaker constant as the pressure necessary to exfoliate GO platelets with interlayer distance approximately of 0.7 nm according to the X-ray Diffraction (XRD) made during the oxidation [31]. However, the rGO sheets obtained are small and wrinkled as a result of both, the release of CO_2 and the removal of functional groups; this can be the main difficulty to scale this method [10].

On the other hand, another important method to reduce GO has been the electrochemical reduction mainly caused by the electron exchange among GO and an electrolyte; this has been performed within a typical electrochemical cell through two different routes [10]. First, a one-step electrochemical approach consisting in a reduction from aqueous colloidal suspension of GO in the presence of buffer electrolyte to produce rGO on the electrode surface. The second, a one- or two-step electrochemical approach performed with thin films of GO deposited on a substrate (electrode) in order to form a GO coated-electrode, and subsequently, this is subjected to electrochemical reduction in a standard three electrode system [32]. C/O ratios from 3.57 to 5.57 have been reported with this technique in addition to be an ecofriendly and controllable method; however, it cannot produce rGO in mass [32, 33]. Other emerging methods to produce rGO such as solvothermal, photocatalyst, and multistep reduction have been reported [10];notwithstanding, they will not be discussed in this chapter.

In spite of the successful reduction with thermal and electrochemical routes, chemical reduction of GO is the most promising method to scale the production of rGO due to the availability of graphite, the relating ease of processing and the possibility to make modifications to the method in order to obtain high-quality graphene. In this regard, recently Chua et al. [26] make a possible classification of chemical reduction approaches in two categories. The first one called "well-supported" mechanism which consist in chemical reduction through widely known agents used in synthetic chemistry such as aluminum hydride and sodium borohydride, which have a determined mechanism of reaction over specific oxygenated moieties. The second one or "proposed" mechanism consists in reducing agents that have not been used in synthetic chemistry in order to reduce GO; consequently, reaction mechanisms are not well determined.

One of the agents used in "proposed" mechanisms is hydrazine (N_2H_4). Hydroxyl, epoxide, and carbonyl groups are the dominant functionalities. Hydrazine is known to form hydrazone and hydrazides in the combination of carbonyl moieties; however, in GO reduction, only hydrazone is formed by the removal of oxygen. Even, when the mechanism of how reduction is carried out by hydrazine has been widely studied, one of the most accepted is the reduction via nucleophilic substitution proposed by Ruoff and co-workers. In this mechanism, the initial resulting derivative from epoxide opening and hydrazine, an alcohol functional group that releases water, is aminoaziridine, which is finally thermally eliminated in form of diimide, forming at the end a double carbon bond on the graphene surface [20]. In spite of the success of hydrazine in order to yield rGO, its toxicity and dangerous handling have limited its use. This has originated different studies about new reduction agents which not only should be safe but also they need to be ecofriendly.

Owing to the extensive search to scale rGO in a safe and nontoxic manner, "green" and safe reduction agents, take ascorbic acid or plant extracts for instance, have been used. L-ascorbic acid (L-AA) or Vitamin C, as it is commonly known, is an essential nutrient which exhibits antioxidant properties. Gao et al. [34] produced rGO through L-AA as reducing agent, while L-tryptophan was added to stabilize the produced rGO aqueous dispersion by electrostatic repulsion. They proposed a two-step reduction mechanism, first S_N2 nucleophilic reaction and second thermal elimination. Due to electron withdrawing, five-membered ring makes the hydroxyls more acidic that result in dissociation of two protons to form an oxygen anion of L-AA. Subsequently, the reduction could be continued with a back side S_N2 nucleophilic as well as releasing H_2O and formation of intermediate species. Finally, thermal elimination of those precedes the yield of rGO. During the reduction, the L-AA pass through oxidation to dehydroascorbic acid. Later was reported that dehydroascorbic acid can be converted into oxalic and gluluronic acid which can interact with the remaining carboxylic moieties in the periphery of rGO, disrupting π–π interactions between rGO sheets avoiding the restacking and consequent agglomerates [35].

Table 1 (Adapted from Ref. [26]) shows a list of "green" agents for the reduction of GO. Additionally, some recent approaches made with antioxidant agent from plant extracts, high H_2-rich water, potassium carbonate (K_2CO_3), and caffeic acid are included.

Reducing agents	C/O ratio	Conditions	Refs.
Reduction methods with "proposed" mechanisms Nitrogen-containing reducing agents			
Hydrazine	10.3[b]	100°C, 24 h	[26]
Oxygen-containing reducing agents			
Methanol	4.0[b]	100°C, 5 days	[26]
Ethanol	6.0[b]	100°C, 5 days	[26]
Isopropyl alcohol	6.9[b]	100°C, 5 days	[26]
Benzyl alcohol	30[b]	100°C, 5 days	[26]
Hydroquinone	–	RT, 20 h	[26]
L-Ascorbic acid/L-tryptophan/NaOH	–	80°C, 24 h	[26]
L-Ascorbic acid	–	RT, 48 h	[26]
L-Ascorbic acid/NH_3	12.5[a]	95°C, 15 min	[26]
Glucose/NH_3	–	95°C, 1 h	[26]
Dextran/NH_3	–	95°C, 3 h	[26]
Gallic acid	5.3[a]	95°C, 6 h	[26]
Caffeic acid[*]	7.5[a]	95°C, 24 h	[26]
Amino acid			
L-Cysteine	–	RT, 72 h	[26]
Glycine			[26]
L-Lysine	–	50°C, 9 h	[26]
L-Glutathione	–	50°C, 6 h	[26]
Plant extracts			
Green tea	–	90°C, 2.5 h	[26]
C. esculenta leaf	7.1[b]	RT	[26]
M. ferrea Linn. leaf	6.1[b]	RT	[26]
C. sinensis peel	6.0[b]	RT	[26]
R. damascena	–	95°C, 5.5 h	[26]
Rose water/$NaAuCl_4$	–	95°C, 5 h	[36]
Syzygium aromaticum	–	100°C, 30 min	[37]
Spinacea oleracea	–	100°C, 30 min	[38]
Cinnamomum zeylanicum		Refluxed 45 min	[39]
Asian Red Ginseng		80°C, 10 min	[40]
Microorganisms			
Shewanella	–	Anaerobic, 72 h	[26]

Reducing agents	C/O ratio	Conditions	Refs.
Shewanella	3.1[a]	Aerobic, 60 h	[26]
E. coli culture	–	37°C, 48 h	[26]
E. coli biomass	–	37°C, 72 h	[26]
Baker's yeast/NADPH	5.9[a]	35–40°C, 72 h	[26]
Wild carrots roots	11.9[a]	26°C, 72 h	[26]
Proteins			
Bovine serum albumin/NaOH	–	55–90°C, 3–24 h	[26]
Hormones			
Melatonin/NH_3	–	80°C, 3 h	[26]
Others Green reducing agents			
K_2CO_3	–	90°C, 2 h	[41]
H_2-rich water	6.26[a]	90°C, 3 h	[42]

C/O ratio obtained by [a]X-ray Photoelectron spectroscopy [b]Elemental analysis [c]Energy dispersive spectroscopy

Table 1. Reducing agents for GO toward a green reduction.

3. Functionalization

In spite of, the great potential related to pristine graphene, the fact that it possesses no band gap, it is practically chemically unreactive and it has a poor water dispersibility, have a severe effect limiting its applications in certain fields compared to other well-established materials. Nevertheless, functionalization of graphene-based materials has become one of the main alternatives to address these problems. Graphene derivatives, specifically GO and rGO, could be modified with a wide range of organic or inorganic molecules through chemical or physical functionalization [43]. The availability of different oxygen-containing groups and the presence of sp^2 domains enable these materials to interact with a covalent, non-covalent, and the combination of both interactions with other molecules. This produces hybrids or composite materials with a particular set of properties and potential applications [12], such as the increment in their dispersibility, processability, purification, device fabrication, biocompatibility, band gap modification [44].

As Georgakilas emphasized [43, 45], the presence of several oxygen reactive sites in GO yields a higher covalent reactivity compared with rGO, and thus, GO is frequently chosen as a starting material where molecules will be attached to oxygen atoms. On the other hand, non-covalent functionalization makes use of hydrophobic or π-interactions on the surface or basal plane of the graphenic layer, giving rise to the preference of rGO as a starting material for this approach. Furthermore, the remnant oxygen moieties offer the possibility to form hydrogen bonding, electrostatic interactions, or the combination of all of them. Although these preferences are not a rule, it could be useful in some cases depending on the type of materials to immobilize [46].

3.1. Covalent functionalization of GO

Covalent modification of GO yields chemical derivatives produced by the next routes, analogous to those obtained with other materials [47]:

1. **Click chemistry**. This type of functionalization is related to a quick joining of small organic units with relative ease and a high efficiency under mild conditions, and in some cases even as a semi-quantitative methodology. One of the best-known reactions is the cyclo-addition between an azide and alkyne group catalyzed with Cu. However, there are other reactions different from the azide/alkyne that fits with this click approach [48, 49].

2. **Linker reaction**. In some cases, the use of small functional molecules which aid as a bridge or linkage between the surface of the GO and other materials is required, due to the lack of affinity, or to preserve a functional property specially with some biomolecules, where a direct contact with the surface of the carbon material may produce denaturation as in the case of proteins (enzymes, antibodies) [45, 50].

3. **Direct chemical attachment**. In this route, the oxygen functionalities in GO are directly covalently bound to other molecules with or without the aid of a catalyst. This approach yields a stale and reproducible immobilization of molecules.

3.1.1. Chemical reactions by oxygen moiety

Both, click chemistry and liker reaction are considered as a post-functionalization of graphene oxide. Direct chemical attachment could be briefly explained in terms of the organic chemistry of the oxygen moieties in GO (**Figure 2**), according to Georgakilas [48].

i. Carboxylic acids could react mainly through amidation or esterification.

ii. Hydroxyl groups could react mainly through silanization, silylation, or etherification.

iii. Epoxy groups react mainly through a nucleophilic addition with an amine-containing compound, from small molecules to large polymeric chains, with the subsequent ring opening.

iv. Miscellaneous reactions, frequently in a random way, where two or more oxygen moieties could react with one or more functional groups, as in the case of peptides and proteins which present a rich variety of available reactive groups.

After the functionalization, the remaining unreacted oxygen groups on GO could be removed with a post-reduction chemical reaction to recover some of the graphenic character. This is a very common approach that exploits the dispersibility and chemical properties of GO, while the final material resembles in certain degree to graphene. In fact, in some cases, a partial reduction takes place through the functionalization reaction itself due to the presence of a reducing group (like amines) in the attached molecule or because of the conditions/catalyst employed. In either case, a covalently functionalized rGO is produced with improved properties such as conductivity, thermal stability, or mechanical properties, although the hydrophilic character is diminished [48].

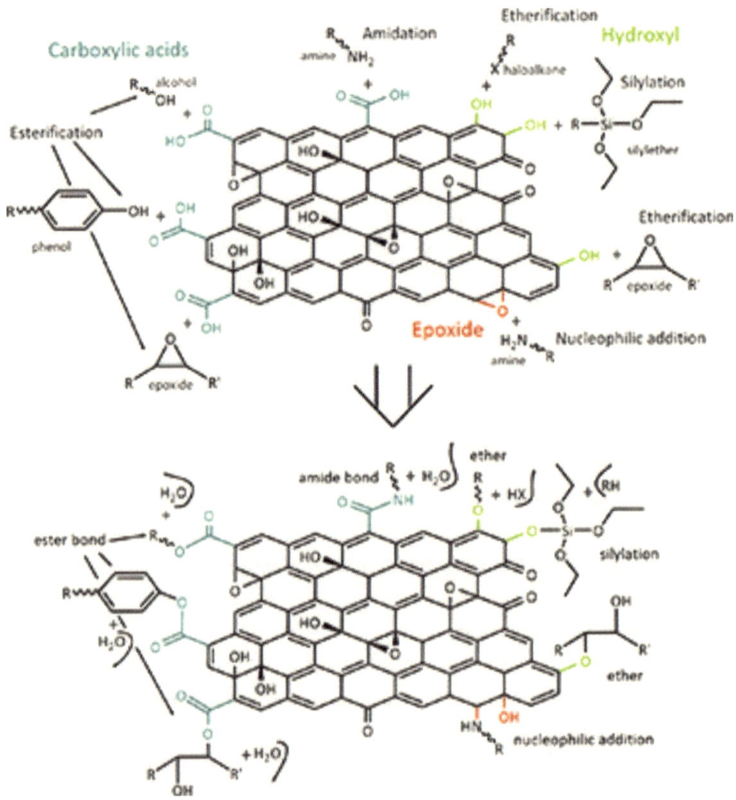

Figure 2. Schematic representation of the covalent functionalization of GO through direct chemical attachment to its oxygen functionalities.

3.2. Non-covalent functionalization of rGO

Unlike the strong covalent functionalization, non-covalent functionalization is an attractive modification route because it offers the possibility to immobilize molecules on both sides of the graphenic basal plane without any further chemical modification of the carbon lattice, avoiding the generation of additional defects, and thus lowering the loss of desired properties, while new properties are introduced [43, 46, 51].

Non-covalent functionalization consists of π-stacking interactions, hydrophobic effects, van der Waals forces, electrostatic interactions, and hydrogen bonding, and in particular situations, even the geometry of the materials plays an important role [52]. In many cases, non-covalent functionalization is accomplished through simple mixing of the materials in an adequate medium [48, 53, 54] or following an incubation protocol for specific biomolecules and cells [55].

As previously mentioned, even though rGO is frequently preferred for this approach, there are many works on the non-covalent modification of GO, followed by a chemical reduction similar to the description above. These non-covalent modifications are usually governed by electrostatic interactions, and thus they depend on the pH of the medium [56–60].

3.2.1. Non-covalent interactions

π interactions occur at the aromatic domains on the surface of both GO and particularly on rGO. π-stacking interactions are fundamental for the stabilization of aromatic systems as in the case of dyes, pyrene functionalized molecules, etc.; aromatic or π-electron-based polymers such as polystyrene, polypyrrol, polyaniline; and biomolecules such as nucleic acids, aromatic residues in polypeptides and some drugs. Such stability is a result of the strong π–π stacking interaction ruled by dispersion forces, not by electrostatic forces. There are other types of π complexes that can be formed with polar gases, cations, and anions. Hydrophobic and van der Waals interactions are exploited by molecules that lack of aromatics or charged groups, such as many polymers, surfactants, quantum dots, and predominantly hydrophobic polypeptides, among others [43, 46, 51].

Oxygen groups such as epoxy and hydroxyl on the surface, and carboxyl and carbonyl at the edges of the layers, allow the adsorption of polar and/or charged molecules through electrostatic interactions and/or the formation of strong hydrogen bonding. While the surface charge of materials is a pH-dependent property, hydrogen bonds is only presented with materials that possesses amine or hydroxyl moieties. Polycationic polymers, polysaccharides, proteins, and enzymes are some examples of molecules, which can present this type of interactions. Even when certain molecules present electrostatic interactions, it can occur that some biomolecules are pH independent, and thus, they are mainly adsorbed due to the hydrophobic or π-interactions [45, 52].

Thus, the combination of the previously discussed non-covalent forces depends on the type of materials and the reaction conditions, and they are very important for the fabrication of new devices [46, 51]. Finally, a common approach consist on the synthesis of systems through the combination of covalent/non-covalent interactions in multiple consecutive stages, yielding very complex hybrid materials or in a multilayer arranged, highly functional, selective, and efficient [46, 61].

Although functionalization of graphene is applicable to very diverse research fields, systems focus on biological applications which represent an attractive and rapidly growing area of research.

4. Functionalization with biomolecules

Even when graphene-based materials have been studied intensively in different fields, their potential in biotechnology and biomedicine applications is still in development, but their research is growing quickly [46]. Graphenic surfaces are ideal to interact with several biomolecules as it has been seen with carbon nanotubes [47, 62]. The combination between graphene materials and biotechnology gives rise to new nano-/bio-interfaces [63] through different biofunctionalization process. Biofunctionalization is defined as the modification of a material by the attachment of biomolecules, ranging from organic groups to very large proteins or even cells. However, biofunctionalization can also be understood in terms of a temporal or permanent biological function as a result of the materials modification [43, 47, 64, 65]. This modification enhances graphene biocompatibility, solubility, immobilization of other molecules, and/or molecular recognition [47, 63].

The different sort of biomolecules could be attached to graphene materials by means of the covalent and non-covalent interactions previously discussed. Notwithstanding, a couple of descriptive examples include the incorporation of nucleic acids and aptamers through π interactions owing to the aromatic character of the nucleobases; the easy assembly of phospholipid chains onto GO and rGO layers through hydrophobic interactions; the immobilization of proteins and enzymes through a combination of hydrophobic and π interactions, and in some cases aided with the contribution of electrostatic forces, in accordance with the amino acid residue composition [46, 51].

Covalently, functionalized graphene biosystems are mainly produced by amidation or esterification reactions of the carboxyl groups, with the aid of coupling reagents or by means of a specific chemical reaction mechanism, although functionalization also includes the epoxy ring opening and hydroxyl modification [46].

Proteins and polysaccharides have the advantage of possessing a rich chemical structure and/or a large amount of functional groups.

4.1. Amine and other functional groups Interactions with graphenic surfaces

Amine functional groups are nucleophiles that possess a basic nitrogen atom with a lone pair. Basically, amines can be classified in primary, secondary, and tertiary, although some cyclic/aromatic amines could be present in certain molecules (tryptophan and histidine in proteins). Nevertheless, amines for chemical graphene modification are frequently found in their primary form, as a pendant or terminated group, whether they are intrinsically present or introduced by chemical modification (amination process).

Amines can react with carboxylic acids of graphene through a condensation reaction forming a stable amide bond. In addition, aminated compounds react through nucleophilic substitution to epoxide groups, yielding an amine addition and a hydroxyl with the ring opening. Chitosan is an excellent example for a macromolecule that can be amide bonded to GO. Also, the N-terminal or amino acid side chain from a protein is commonly reacted through an amide bonding, as well as the lysine residue, or through nucleophilic addition to epoxides. Nonetheless, if the nucleophile is an alcohol (forming an ester with carboxyl groups), a thiol or a carboxylate anion, the variety of reactions becomes larger, just as the non-covalent interactions, as in the case of proteins, such as keratin (KE) among others.

http://dx.doi.org/10.5772/64001

4.2. Keratin functionalization of GO and rGO

Although there are several reports on the usage of proteins, particularly enzymes, for the functionalization of graphene-based materials [46], few works have been done on the surface modification of graphene through the attachment of structural fibrous proteins,[1] and, to the best of our knowledge, only a couple of reports have used keratin [66, 67], specifically, chicken feather keratin, for this purpose.

Fibrous proteins give resistance and flexibility to several biological structures and they possess a high concentration of hydrophobic residues. Keratin is a distinctive fibrous protein compared to collagen, elastin, and myofibril proteins, due to its high degree of disulfide bonds due to the presence of cysteine in its sequence [68]. Keratins are chemically stable and durable against hard environmental conditions; thus, it is found in hair, wool, horns, claws, and feathers.

Chicken feather keratins are small polypeptide chains with a molecular mass of about 10 kDa, and a predominantly hydrophobic character, although it also has a high amount of serine and some other polar or charged groups. It possesses seven cysteine residues, but lacks of lysine, tryptophan, histidine, and methionine. Keratin is therefore adequate for covalent and non-covalent binding with graphene materials.

4.2.1. *Covalent attachment of keratin onto GO through a $KMnO_4$ redox system*

Redox reaction systems with Mn(III) have been successfully used for the polymerization of vinyl monomers and the grafting of certain macromolecules [66, 69–73]. Mn(III) is obtained from the combination of $KMnO_4$ in acid media with the aid of a reducing reagents, like malic acid, as an electron donor for the reduction of Mn(VII) into Mn(IV) as manganese dioxide. Mn(IV) reacts with the reducing agent to produce the highly reactive Mn(III) ions along with free radicals. This specie can generate active free radicals (primary radicals R) with the reducing reagents, or in the presence of a macromolecule such as a protein, polysaccharide, or graphenic material, yielding a macroradical. These macroradicals may also be formed by the direct attack of primary radicals [74–76].

Keratin and graphene oxide possess different X–H functional groups that may act as electron donors through the abstraction of a hydrogen, thus becoming a radical. In the case of KE, these groups are –SH from cysteine, –OH from serine, threonine, and tyrosine, –$CONH_2$ from glycine and asparagine, and –COOH from glutamic and aspartic acids. In particular, KE also possesses some disulfide bonds (–S–S–) due to the oxidation of thiols. These groups could also be transformed into free radicals with the abstraction of a hydrogen from a neighbor carbon atom [68, 77, 78].

Keratin polypeptide chains are suitable to be grafted onto graphene oxide through a redox reaction system [67]. In this process, keratin needed to be dissolved and dialyzed in order to be chemically attached. Then, under mild conditions in aqueous media and in the presence of sulfuric acid, malic acid, and potassium permanganate, the functionalization was conducted

[1] We are not considering here the case of nanocomposites where graphene materials act as a reinforcement for the polymer matrix.

at two different conditions varying the amount of H_2SO_4/malic acid employed. Keratin polypeptide chains were successfully attached to graphene oxide as it is confirmed by the infrared spectroscopy in **Figure 3a**. The graph shows well-defined peaks from keratin (numbers in red) and graphene oxide (numbers in blue) characteristic vibrations in the covalent bonded materials, and apparently, there is an amide and/or ester bond formation during the reaction, where KE/GO-1 seems to have a higher degree of polypeptide incorporation. In this case, graphene oxide was preferred over reduced GO, due to its higher degree of reactive oxygen sites thus giving a higher yield of chemical bonding.

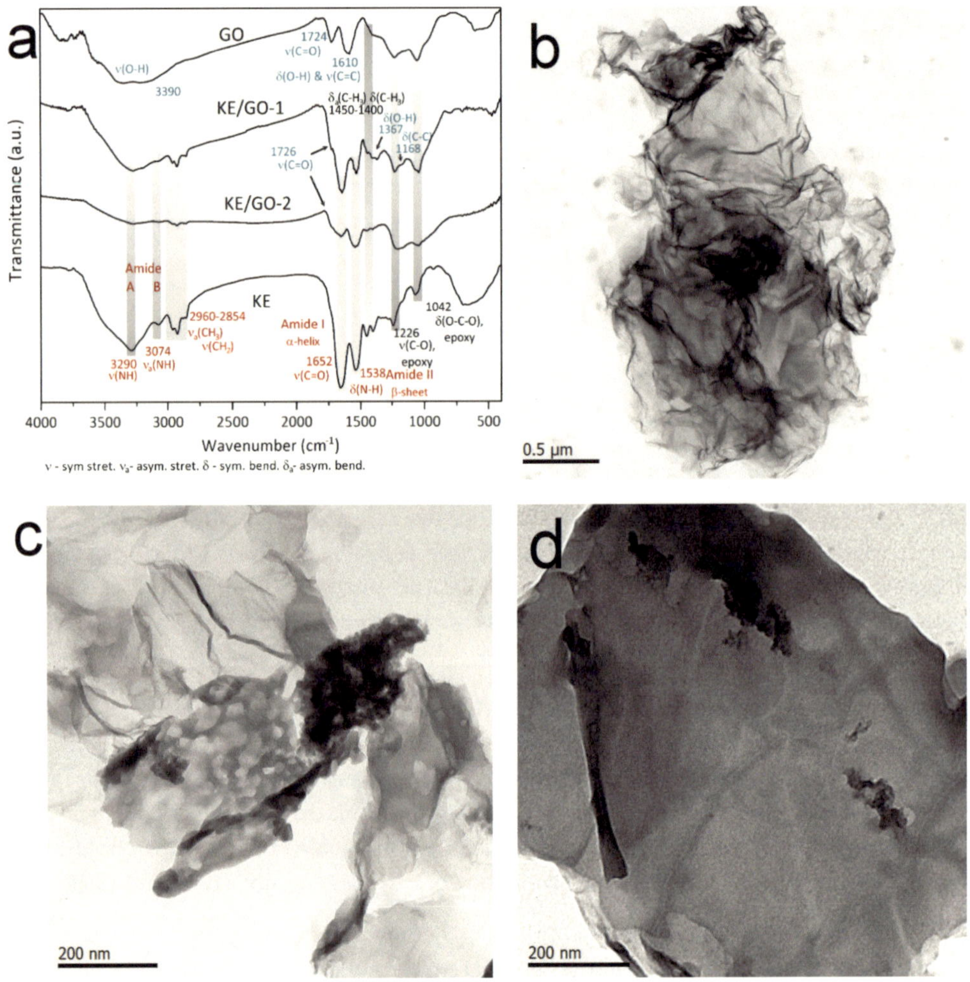

Figure 3. (a) FTIR spectra of graphene oxide, keratin, and the functionalization between them at two conditions (KE/GO-1 and KE/GO-2) and TEM images of (b) GO, (c) KE/GO-1, and (d) KE/GO-2.

Typical TEM image of GO with its intrinsic foldings and wrinkles is shown in **Figure 3b**, where even though the sheets are very large (a couple of microns), they have a very thin structure.

Figure 3c and **d** clearly shows that some morphologies are different than those of the carbon layer sheets, due to the presence of the keratin chains on the surface of the material. To further demonstrate the polypeptide immobilization onto the carbon nanostructure surface, a line-profile chemical composition analysis is also shown in **Figure 4**, confirming the presence of S and N.

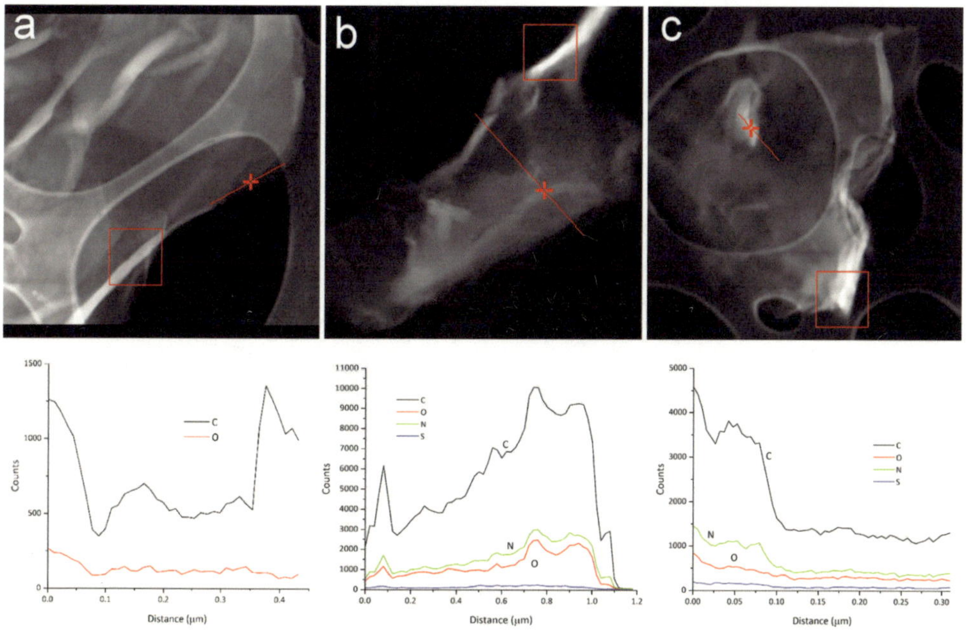

Figure 4. STEM-HAADF images with EDX line scan (red) and chemical profile (below) of (a) GO, (b) KE/GO-1, and (c) KE/GO-2.

4.2.2. Non-covalent attachment of KE onto rGO

Reduced graphene oxide possesses a higher degree of sp^2 hybridization than its graphene oxide counterpart, thus these π-electron domains may interact with hydrophobic or aromatic molecules as well as with aromatic pendant groups, for a non-covalent modification. The oxygen remaining moieties are also useful for an electrostatic- or hydrogen-bonding interactions. In the case of the previously described keratin, the mainly hydrophobic nature of the protein allows its immobilization or adsorption onto the surface of the rGO without the aid of any coupling reagent, while the charged and polar residues could also contribute for the protein tethering forming electrostatic functions depending on the mixing conditions.

Following the procedure for the solubilization of KE, a non-covalent attachment of the protein had been made an aqueous media by simply mixing a KE/rGO mass ratio of 1:1 for 3 h [67]. From the keratin sequence, two amino acids may exhibit:

i. π-stacking: phenylalanine and tyrosine onto the graphene basal plane.

ii. Hydrophobic interaction: proline, glycine valine, cysteine, leucine, isoleucine and alanine, onto both, the basal, and the edges of the layers.

iii. Electrostatic and/or hydrogen bonding: serine, glutamine, threonine, asparagine, glutamic, and aspartic acids.

iv. Strong electrostatic/hydrogen bonding: arginine.

Arginine has a particular strong interaction with negatively charged groups like carboxylic acid, due to its positively charged structure. This residue could be attached to some remaining carboxyl groups in rGO, as it has certainly done in the case of GO [79]. In addition, it is possible that some of the polar residues of the KE will be oriented toward the aqueous media. The effectively non-covalent functionalization of KE onto rGO is indicated by the FTIR spectra of the modified material compared with its precursors (**Figure 5a**).

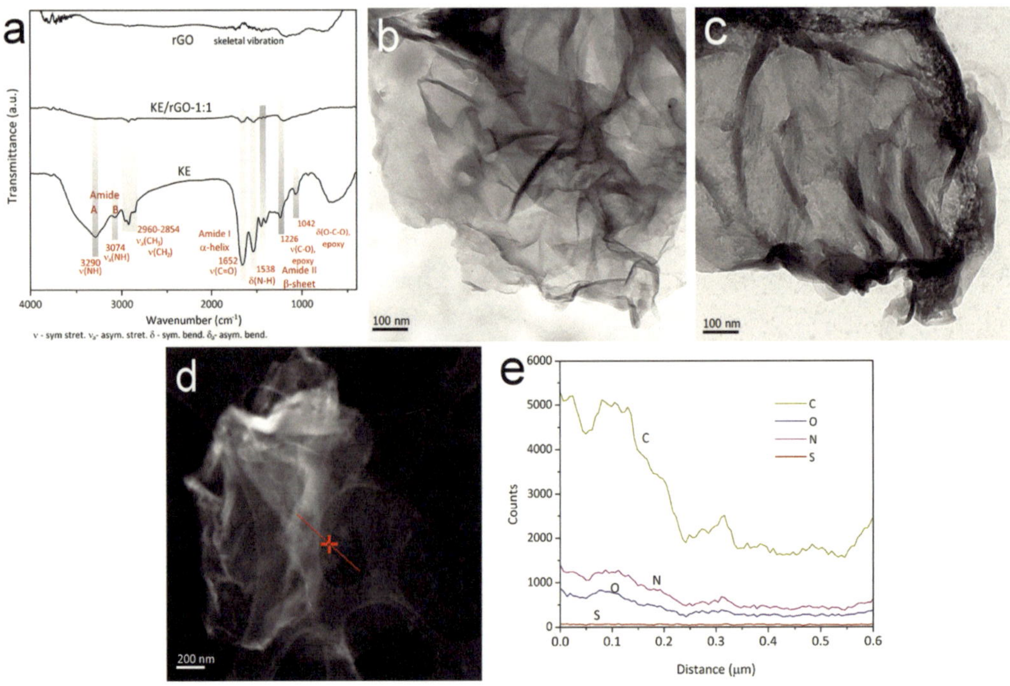

Figure 5. (a) FTIR spectra of reduced graphene oxide, keratin, and the functionalization between them at a mass ratio of 1:1 (KE/rGO-1:1). TEM images of (b) rGO, (c) KE/rGO-1:1, (d) STEM-HAADF images with EDX line scan, and (e) chemical profile of KE/rGO-1:1.

In the same way as with keratin/graphene oxide, transmission electron microscopy and line scan chemical composition had been performed in order to further confirm the functionalization. As can be seen from **Figure 5b** and **c**, a surface roughness besides those intrinsically found in graphene is appreciated when keratin has been immobilized. Also, chemical compositions clearly show the presence of N and S on the surface of the graphenic layers (**Figure 5d** and **e**).

Based on these results, keratin has an affinity to interact with rGO in a strong manner, practically wrapping the carbon material, even when there is no covalent linkage and no coupling reagents present in the mixture. But there still more future research in order to manipulate the degree or sites of interactions between these two systems

4.3. Chitosan functionalization of GO and rGO

GO and rGO have been functionalized through covalent and non-covalent interactions with several biopolymers; one of them with great relevance due to its potential applications such as tissue engineering, cell adhesion, and food delivery is chitosan (CS) [46]. CS exhibits an unusual combination of properties such as biological activities, mechanical, and physical properties that make it the most important derivative of chitin, the second most abundant natural polysaccharide [80]. CS is composed of β-(1,4)-2-amino-2-deoxy-D-glucose, and it is the result of the deacetylation of chitin (β-(1,4)-2-acetoamido-2-deoxy-D-glucose) [21] which

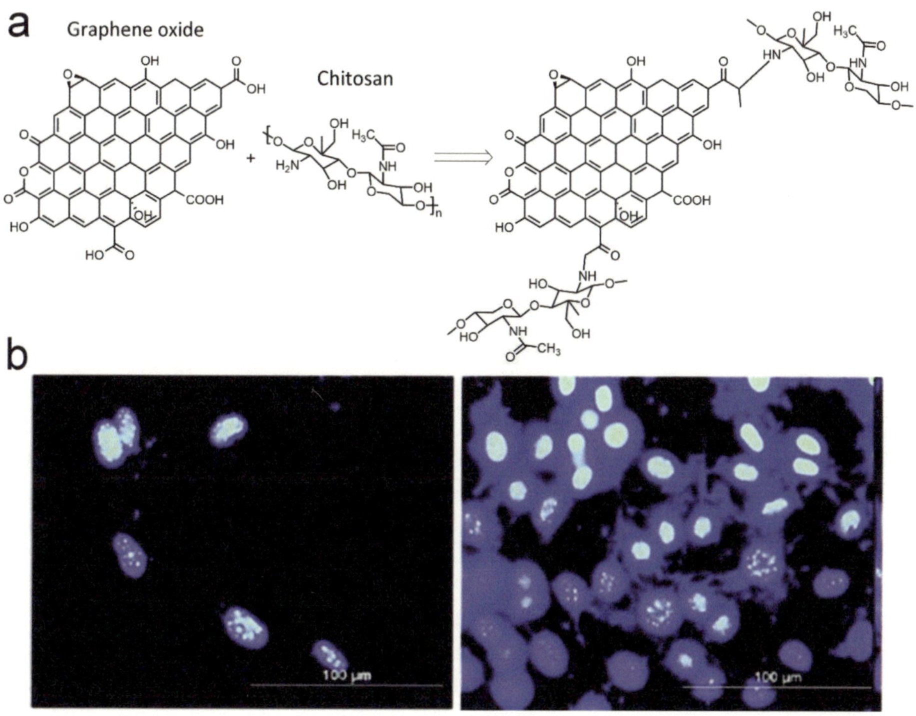

Figure 6. (a) Schematic illustration of covalent interaction between GO and CS trough amide bond formed between –COOH moieties of GO and $-NH_2$ groups, present in CS. (b) Fluorescence micrographs of immunocytochemistry of cell nuclei of pre-osteoblasts after 2 day culture on GO (left) and GO–CS–HAP (right). The graphs shows high fluorescence intensity zones as pre-osteoblasts growth, GO–CS–HAP system has notorious development in comparison to GO. (Adapted from [91] with permission of The Royal Society of Chemistry).

can be found in crustacean shells and shellfish wastes, both of them by-products from marine bioprocessing plants and probably one of the most important sources of CS [81]. CS is a polycationic polymer with two amino and two hydroxyl groups and a carbohydrate backbone similar to cellulose as can be seen in **Figure 6a**. Molecular weight (MW) and degree of deacetylation depend on the alkaline reaction conditions yielding a random distribution of acetylated/deacetylated units coupled in its chains [80]. These two features are chiefly responsible of the physicochemical properties of CS, principally impacting on its biological characteristics [82]. The amino functional groups in CS provide a versatile behavior when CS is in aqueous solution. At low pH (<6), the amino groups are protonated, thus giving CS a polycationic behavior. On the other hand, when pH is higher than 6.5, functional groups in CS are deprotonated, losing its charge and reducing its hydrophilicity [83]. CS is a biocompatible, biodegradable, non-toxic material approved by the FDA; additionally, amino and hydroxyl groups present in its structure offers reactive sites for functionalization; for these reasons, it has been considered together with GO and rGO as a promising material to synthesize new materials for different applications such as biomedical, pharmaceutical, gene therapy, and waste water treatment to mention but a few [80].

4.3.1. Covalent functionalization of GO and rGO with CS

The covalent functionalization of GO and rGO accordingly with reports is made through the interaction between the oxidized moieties on surface of 2D carbon nanomaterial (GO, rGO), and the functional groups present in CS molecule (amino). GO functionalized with CS have been used to developed novel materials via amidation [84], esterification [85], or nucleophilic addition [86, 87]. Esterification was reported by Xu et al., in order to obtain CS grafted onto GO. First GO was dissolved in dimethyl formamide (DMF) in sonication, followed by the addition of thionyl chloride ($SOCl_2$) which reacts with –COOH groups to obtain acyl-chloride functionalized GO (GO–COCl). Finally, –NH_2 groups present in chitosan react with GO–COCl via esterification [85]. On the other hand, abundant epoxy groups in GO can react with –NH_2 at elevated temperatures following the nucleophilic addition mechanism similar to cross-linking mechanism in epoxy resin during curing [86]. In spite of the successful grafting of CS onto GO via esterification and nucleophilic addition, the most common route to functionalize GO with CS is amidation, which is done by the interaction of –COOH and –NH_2 groups present in GO and CS, respectively [84]. GO functionalized with CS (GO–CS) has been used as a scaffold in different fields due to its interesting properties.

Recently, GO–CS system was used as a platform for drug delivery and sensing devices [88–94]. Depan et al. reported a biomimetic mineralization route of conjugated material made up of GO and CS for hydroxyapatite (HAP) biomineralization. HAP as a major component in bones with important features, like great biocompatibility and bioactivity, has been utilized as a part of new multicomponent system (GO–CS–HAP) for bone tissue engineering. The functionalization was completed via covalent modification between highly decorated GO with –COOH moieties and CS with amino groups. Fast Fourier transform infrared spectroscopy (FTIR) in earlier reports of Depan and coworkers showed the formation of amide bond. Prior to functionalization, the characteristic signals of functional groups in CS were observed in 1636 and 1597 cm^{-1} attributed to the stretching of C–O (in –NHCO amide group) and bending of N–H (in $-NH_2$), respectively, whereas for GO grafted with CS, $-NH_2$, signal was shifted to a lower value, while amide group (–NHCO) was shifted to a larger value, this point out that functional groups present in CS and GO interacted to form covalent bonds [95].

On the other hand, HAP nucleation on GO–CS was corroborated by FTIR, X-ray diffraction (XRD) and scanning electron microscopy (SEM). A fine dispersion of HAP was observed, and it was attributed to the electrostatic interactions with GO–CS system which promoted the HAP growth. The adhesion and proliferation of osteoblasts on CS–GO–HAP system was observed in order to evaluate their distribution and attachment, both features are necessary for tissue formation. Samples were analyzed after 1 and 24 h of cells incubation, fluorescence microscopy with DAPI showed a synergistic effect in mineralization of HAP in GO–CS–HAP system as can be seen in **Figure 6b**, attributed to interaction of CS and HAP. Another study about GO functionalized with CS was reported by Mohandes et al., using GO–CS–HAP nanocomposite as a scaffold to grow apatite (AP) due to its high porosity and interconnectivity, both features in nanocomposite are important for the cell attachment and new bone formation. Additionally, thermal stability of GO and GO–CS–HAP was evaluated by termogravimetric analysis which suggests that the loss of mass in GO of approximately 25 wt% is attributed to functional groups, while GO–CS–HAP shows stability at 310°C where the weight loss was important. New AP was formed in GO–CS–HAP nanocomposite after 14 days socked in simulated body fluid (SBF). Grain size of 20–15 nm was observed by scanning electron microscopy (SEM) [96].

Due to its haemostatic properties and safe excretion CS were used to build microneedle arrays as transdermal preloaded drug delivery nanocomposite. Fluorescein sodium (FS) was attached by non-covalent interactions to CS–rGO at the same time reduction of GO was carried out. rGO–CS–FL array shows better mechanical properties and drug release in comparison to pristine CS. Nanocomposite with 2% of rGO achieved maximum release (91%) of available drug in 48 h, while CS–FL system only achieved around 33% according to the report [94]. Other approaches in drug delivery were reported by Rana et al. and Bao et al. The first one reported the use of ibuprofen (IBU) and 5-fluoracil (5-FU) drugs loaded in rGO–CS system via simple physisorption, and additionally, they evaluated the cytotoxicity and cell viability of GO–CS–IBU and GO–CS–5-FU systems over CEM and MCF-7 cancer cells [89]. The second one reported a novel nanocarrier of campthotecin (CPT), an inhibitor of topoisomerase I and anticancer drug. Cell viability of 80% was found in methylthiazoloterazolium (MTT) assay. Additionally, π–π stacking and hydrophobic interactions between GO–CS and CPT allow high load (20 wt %) of drug in nanocarrier [90].

Other approaches regarding removal of contaminants in water have been done. Removal of Cr(IV) from simulated wastewater with magnetic GO–CS ionic liquid (MCGO-IL) was reported with a maximum of 143.35 mg/g of adsorption capacity (Q_{max}), this was described by the Langmuir isotherm. A removal mechanism was proposed. Briefly, first, the electrostatic attraction between Cr(IV) with $-OH_2^+$ and $-NH_3^+$ takes place, second, the cooperation between ionic liquid, functional groups of GO–CS and Cr(IV) occurs and, third, the reduction of Cr(IV) to Cr(III) assisted by π electron of carboxylic six-membered ring in MCGO-IL is achieved [97]. On another report magnetic CS nanoparticles (MCGO) acting as magnetic bioadsorbent shown to have excellent properties. Magnetic biosorbents can be recovered, easily separated and CS adsorption is better in view of its high surface area along functional groups present [98]. According to Fuschine dye adsorption experiments, pH 5.5 is the optimal value to carry out the removal by MCGO. Moreover, the calculated and experimentally values of adsorption capacity were close, which confirms the pseudo-second-order kinetic model of adsorption proposed. Additionally, the studies over recycling of MCGO as adsorbent shows only slightly decay adsorption capacity of MCGO after the fifth cycle. These results suggested that this method can be used to yield graphene-absorbents in commercial scale. In **Figure 7**, the synthesis of MCGO is briefly represented by a schematic diagram [99].

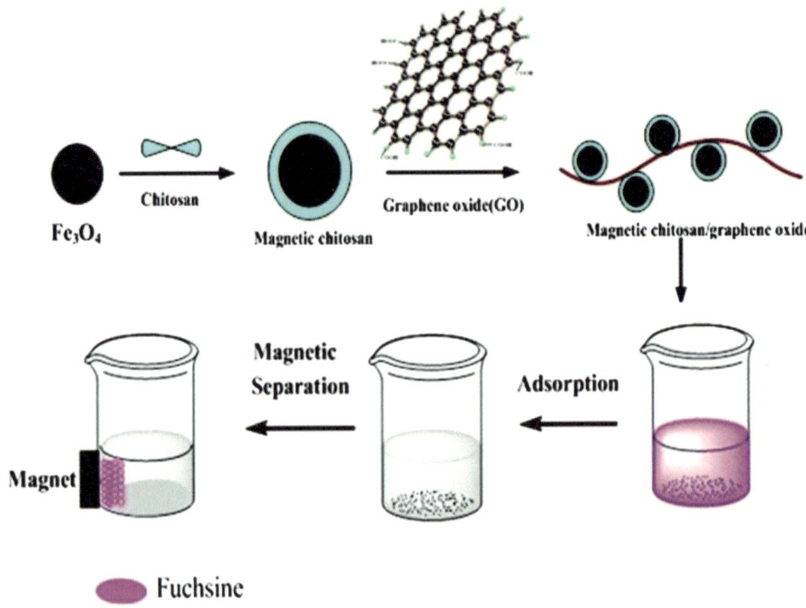

Figure 7. In the synthesis of MCGO, first magnetic particles are attached to CS and second trough amide bond formation GO surface is functionalized. (Reproduced from [99] with permission of The Royal Society of Chemistry).

On the other hand, antifouling membranes were developed by depositing a thin film of GO–CS on polyamide surface to form brackish water thin-film membrane for reverse osmosis (BWTFC-RO). In this approach, the BWTFC-RO with high content of GO shows better antifouling response because of the negative moieties in GO–CS increase hydrophilicity, which can reduced the interfacial energy among membrane surface and water and thus membrane fouling resistance was incremented. A water contact angle of 63.68° was measured for unmodified membrane, while membrane with GO–CS system has a contact angle of 19.13° [100]. In addition, in order to remove Uranium U(vi) released in wastewater, GO–CS systems grafted via covalent were done. Different probes of adsorption and desorption were made having Q_{max} = 225.78 mg/g at pH 4.0 and fitted to a Langmuir model. The removal of U(VI) can

http://dx.doi.org/10.5772/64001

be attributed to interactions between amine groups and the metal; this was corroborated by FTIR where –NHCO– (at 1530 and 1630 cm^{-1}) and $–NH_2$ (at 1420 cm^{-1}) peaks decreased when the U is adsorbed [101].

Owing to the large surface-to-volume ratio and good electrochemical activity of GO and the compatibility of CS, GO–CS nanocomposite enhances transfer electron and DNA immobilization between electrode surface and DNA in electrochemical biosensor for typhoid diagnosis. This was reported using glutaraldehyde (GA) as a bridge between the GO–CS nanocomposite film on indium tin oxide (ITO) electrode and *Salmonella typhi,* specifically its 5′-amine labeled single-stranded DNA (ssDNA). GO–CS–ssDNA–ITO bioelectrode shows improvements in detection limits in comparison to other studies [88], and this is attributed to the excellent conductivity and small band gap from biomolecules; the bioelectrode shows the limits of 100 fM in buffer solution and shelf life of 15 days with a 100% recovery. Otherwise, *Candida rugose* lipase (CRL) immobilization was reported by magnetite particles (Fe_3O_4) assembled in GO–CS system. GO–CS–Fe_3O_4 nanocomposite was synthetized from ferric chloride hexahydrate ($FeCl_3{\cdot}6H_2O$) and 1,6-hexadiamine by solvothermal reaction. The nanocomposite needs only one step to be synthetized which is important when large-scale production can be possible. In this study, three strategies were used to immobilize CRL. Firstly, the electrostatic adsorption by GO–CS–Fe_3O_4, secondly, covalent bonding with GO–CS–Fe_3O_4–GA and thirdly metal-chelate ligand anchorage on GO–CS–Fe_3O_4–IDA-CU. The activity recovery was measured for three systems and the adsorption results showed the best protein CRL immobilization for GO–CS–Fe_3O_4–IDA-CU followed by GO–CS–Fe_3O_4–GA and GO–CS–Fe_3O_4 systems with 65.5, 60.2, and 57.2% of activity recovery, respectively. GO–CS–Fe_3O_4–IDA-CU leads the activity recovery due to exposition of functional groups thus giving low diffuse resistance. On another approach, the biocompatibility of mouse mesenchymalstem C310T1/2 cells adhered to GO–CS covalent attached system was evaluated by DAPI fluorescence, tracing the nucleus of cells and low cytotoxicity was further reported. Additionally, cell viability after 24 h shown low death cell indications. This can be traduced in an acceptable biocompatibility of GO–CS system [93].

Another emergent field of study of GO–CS is catalysis. Green nanocomposite based on GO–CS have demonstrated high thermal stability (165°C) in differential scanning calorimetry (DSC) in comparison with pure CS (118–119°C), both were used as a support for novel catalyst. In this research, average size distribution of 50 nm was measured by atomic force microscopy (AFM) for GO–CS nanocomposite synthesized in free-solvent conditions. Efficient synthesis of 2,4,5-trisubstituied-1H-imidazoles were done, GO–CS in catalytic amount was combined with benzyl 2 or benzoin 1, benzaldehyde 3 and ammonium acetate at 120°C that shows great thermal stability of GO–CS system [102]. Another novel catalyst was developed for reduction of aromatic nitroarenes and azo dye degradation [103]. Silver (AgNPs) and gold (AuNPs) nanoparticles were attached onto GO–CS system in order to avoid aggregations of either Ag or Au nanoparticles which affect its catalytic activity. Both GO–CS-AgNPs and GO–CS-AuNPs catalysts were obtained in similar process using 0.1 M solution of silver nitrate ($AgNO_3$) and tetrachloroauric (III) acid hydrate respectively in a solution of sodium borohydride ($NaBH_4$). Particles of 20 nm for AgNPs and 5 nm for AuNPs were observed by high resolution transmission electron microscopy (HR-TEM) that demonstrates homogeneous deposition of particles. The excellent catalytic activity and selective reduction of nitroarenes was mainly attributed to the high surface area of GO–CS system which improves adsorption of organic substrates, and to the presence of $-NH_2$ and –OH groups in GO–CS system which might assist the stabilization and attraction of particles toward the catalytic sites, because of the combination of hydrophobic–hydrophilic nature present in GO and CS.

4.3.1.1. Covalent attachment of chitosan onto GO through a $KMnO_4$ redox system

Functionalization of GO with CS by redox system with $KMnO_4$, H_2SO_4, and malic acid also has been reported [75]. In this approach, covalent attachment of CS onto GO was done varying temperature conditions, this shows influence over morphology and features of the final material produced. A better dispersion behavior was obtained as a result of chemical modification (grafting) originated from the interaction of GO and CS functional groups. As can be seen in **Figure 8a**, well-exfoliated graphene oxide sheet alike a thin film and wrinkled surface is shown. Additionally, TEM images of CS grafted on GO exhibit the differences on morphology acquired in different conditions **Figure 8b–d**. At the lowest temperature, GO sheet is completely covered by CS. However, as the temperature is increase (75–80°C), CS only partially covers the GO sheets, and finally for the highest temperature, the GO–CS system shows great differences in its morphology, like a scrolled material. This can be attributable to the loss of free water in CS that increases as the temperature does, impacting on the amount of water and hydrogen bonds formed during grafting. Dispersion behavior of different GO–CS produced on water and hexane. A water stable dispersion even 24 h after the sonication process was

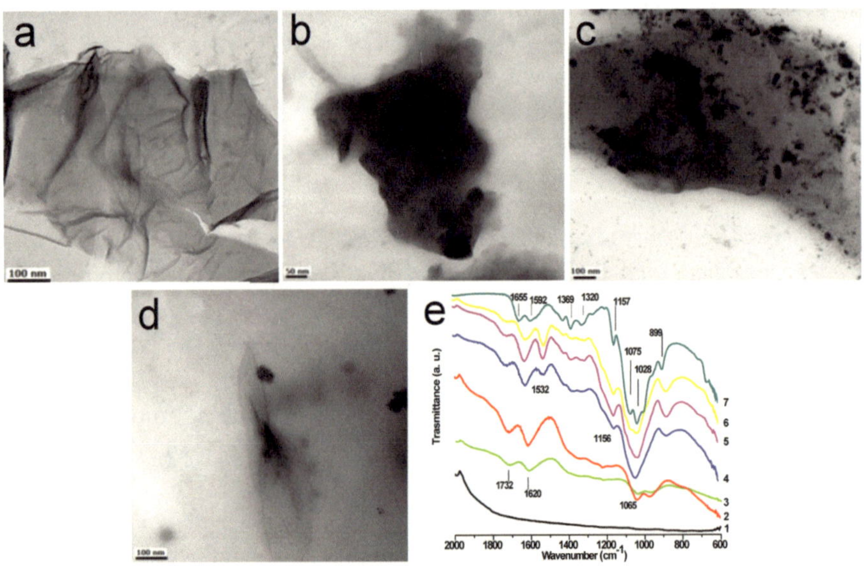

Figure 8. TEM images of (a) GO; (b) GO–CS obtained at 55–60°C; (c) GO–CS obtained at 75–80°C; (d) GO–CS obtained at 95–100°C; (e) FTIR spectra of (1) graphite; (2) graphite oxide; (3)GO; (4) GO–CS obtained at 55–60°C; (5) GO–CS obtained at 75–80°C; (6) GO–CS obtained at 95–100°C and (7) CS.

observed, unlike in hexane non-polar solvent which shows poor dispersion and rapid precipitation of GO–CS. In addition, AFM measurements show a roughness between 3 and 6 nm for the GO–CS which is the result of bundles of CS chains grafted on GO that can be observed as dark dense zones in TEM images.

FTIR analysis was done in order to provide evidence about the chemical modification of graphite, and grafting of CS onto GO. The appearance of new signals respecting the interaction of moieties presents in both GO and CS can be clearly seen in the **Figure 8e**. Signal from new amides or carbamate esters formed during grafting, this band appears in 1535 cm^{-1} attributed to the combination of ν(C–N) and δ(CNH). Both, the band at 1156 cm^{-1} corresponding to ν_a(C–O–C) and signal at 899 cm^{-1} for C–O–C were detected, they are related with glycosidic linkage. The latter are typical in chitosan, whereas after grafting they are shifted to 881 cm^{-1} in GO–CS systems that show the modifications of C–O–C bond vibrations and the interaction of GO–CS hybrids. In addition, Raman spectroscopy and energy dispersive X-ray spectroscopy (EDS) characterization techniques were used in order to corroborate the grafting of CS onto GO.

Finally based on previous works [99, 100] and evidence obtained in FTIR measurements, few reactions can be proposed for the novel bonds as a result of grafting of CS. As can be seen in Eqs. (3)–(5), first, linkage can be done by carbonyl moieties ((3) and (4)) and second with breaking of epoxy groups (5). Despite that other reactions can be carried out but evidence provided by FTIR points out just the aforementioned.

$$GO-COOH+CS-NH_2 \xrightarrow[H^+,\ malic\ ac.]{KMnO_4} GO-CONH-CS+Subproduct \quad (3)$$

$$GO-COOH+CS-OH \xrightarrow[H^+,\ malic\ ac.]{KMnO_4} GO-COO-CS+Subproducts \quad (4)$$

Subproducts = HO, RH, ROH (R = Radical from malic acid)

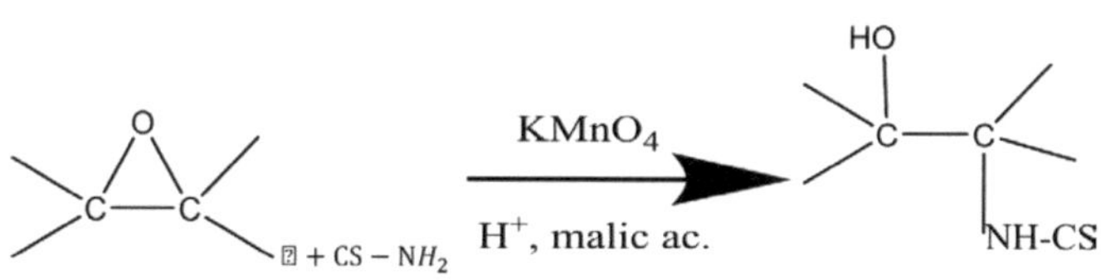

(5)

4.3.2. Non-covalent functionalization of GO and rGO with CS

Non-covalent functionalization of GO and rGO has been done through different ways such as hydrogen bonding, electrostatic interactions, Van der Waals forces, and ionic interactions. Electrostatic forces in functionalized GO have been reported in order to develop applications such as hemolytic activity and nanocomposites [104–107]. Li et al. reported removal of single or multi-system ions based on Cu (II), Pb (II) and Cd (II). GO was functionalized with a sulfhydryl compound and CS resulting in CS–GO–SH adsorbent material, basically due to the affinity of CS, known as a cationic polymer with plenty –OH and $-NH_2$ groups and negative charged in GO–SH complexes. In this work, ultrasonication was applied to yield GO–CS–SH nanocomposite; sulfhydryl groups were previously grafted on GO reacting in stirring with 4-aminothiophenol. Adsorption tests result in Q_{max} of 235, 226, and 117 mg/g for Cu (II), Pb (II), and Cd (II), respectively, from a starting concentration of 250 mg/L, and thus, a higher adsorption was obtained in comparison with other results reported with GO and CS as adsorbent. This suggests a synergistic effect originated by increasing the specific surface area and space between CS and GO–SH. Moreover, operational factors such as pH, adsorbent dosage, and temperature play important roles in results [107]. Other approach about electrostatic interactions to functionalize GO was reported by Liao et al., they discussed the effect of GO size and coated GO over hemolytic activity. GO obtained in different exfoliation parameters were tested by methylthiazolyldiphenyl-tetrazolium bromide (MTT) and water-soluble tetrazolium salt (ST-8) to show blood compatibility. The functionalization of GO with CS reveals lower hemolytic activity in comparison of GO sheets; this can be attributable to disruption of red blood cells membrane (RBC) as a result of the strong electrostatic interaction between oxygenated functional groups of GO sheets with negative charge and positively charged phosphatidylcholine lipids of RBC outer membrane [104]. Unlike a traditional CS polymer with $-NH_2$ groups which only become charged in acidic media, quaternized CS (QCS) is a polymer attached with quaternary ammonium groups which confers a cationic polyelectrolyte permanently charged feature [108]. Consequently, cellulosic paper fibers can be cover with a positively charged coating made of QCS which attract the negatively charged GO sheets. So QCS can be used as a "glue" linking cellulose fibers to GO via electrostatic interactions. This was reported by Ling and co-workers in order to fabricate composite paper in addition to load GO with AuNPs and its subsequent reduction by hydrogen iodide vapor. The final material shows an outstanding conductivity of 831 S/m [105]. Another interesting applications recently reported are Janus GO–CS membrane; in this work; non-covalently assembly GO–CS membrane was used as a support of poly(styrene) (PS) and poly (N,N-dimethyl methacrylate) (PDMAEMA) in upper and lower membrane surface; respectively. Photografting and photopolymerization were used to yield Janus membranes a possible material for applications such as sensing or catalyst as a result of its versatility [109].

On the other hand, another route of non-covalent functionalization of GO is through hydrogen bonding which has been reported to develop novel nanocomposites. Recently; a facile GO/CS conducting biocompatible hydrogel production method by extrusion printing was reported. CS and lactic acid (LA) as matrix and GO as reinforcement material were used, and L929 cells were cultured in diluted GO–CS–LA dispersions and were compared with GO–CS–LA nanocomposites.

Another via of non-covalent functionalization of GO is the formation of hydrogen bond with CS which has been reported to develop novel nanocomposites. As a result of homogeneous distribution as well as excellent mechanical properties achieved in GO–CS nanocomposites originated from non-covalent interactions, high-performance nanocomposites have been prepared [110]. High mechanical properties exhibited in GO–CS nanocomposite can be used for drug delivery, GO–CS nanocomposites offer a platform to yield a tunable control through pH transdermal system for drug release [111]. Recently, another approach was done using CS and lactic acid (LA) as matrix and GO as reinforcement material by extrusion printing. Due to the relevance of the morphology, dimensions, shape, and biocompatibility in materials for tissue engineering, GO–CS was used to yield conducting biocompatible hydrogel. FTIR shows strong interaction by hydrogen bond between GO–CS–LA. Additionally L929 cells were cultured in composites; first diluted GO–CS–LA dispersions were used to note cell growth. Cell growth showed an increment of cell density 10–15 times ($45 \pm 4E^4$ cells cm^{-2} and 1.5% dead cells) in comparison to the original amount. No cell showed inclusion of dark material in either cytoplasm or any organelle; this was observed by bright field microscope. Second for GO–CS–LA films and scaffolds; no toxicity or changes on migration were observed in cells; moreover, mechanical properties were enhanced by the addition of GO having increments in tensile strength of 320% and 162% for dry and wet state for GO–CS–LA nanocomposite with only 3 wt% of GO added [112].

Due to the versatility and interesting capacity of self-healing against damage, polymeric gels have attracted great attention. GO and CS hydrogels were prepared by non-covalent interactions; in this regard, GO can be considered as a 2D cross-linker as a result of functional groups present in both faces of GO sheets. Electrostatic and hydrogen bond are the predominant interaction in this kind of materials according with the report [113]. Water, CS, and GO are the molecules present in hydrogel, interacting mainly by electrostatic forces when mixed at room temperature. Particularly CS shows a compact state owing to hydrogen bond present in its structure, thus restricting the interaction with GO, this is diminished with the increase of temperature of hydrogel preparation according with the methodology. This kind of materials can be applied in fields such as biomaterials, absorbents, and so on. For example, recently heparin/GO–CS hybrid hydrogel (Hep-GO–CS) were synthetized for bilirubin adsorption. Adsorption measurements in phosphate buffer solution (PBS) show for Hep-GO–CS four times adsorption capacity than CS hydrogel; this can be attributable to GO added to hydrogel. Additionally, this material exhibits good hemocompatibility and low degree of hemolysis in the tests realized [114]. On the other hand, other applications for GO–CS system based on both electrostatic interactions and hydrogen bond have been reported in fields such as nanocomposites for electrochemical biosensors [115], pathogen agent detection [116], membranes in microbial fuel cells [117], and cell growth [118]. Additionally, preparation of poly(vinyl alcohol) PVA–CS nanocomposite reinforced by rGO was reported; simultaneously, reduction and functionalization were done and synergistic effect by the interaction of rGO–CS–PVA was demonstrated with high-performance nanocomposite obtained. CS acts as a bridge between rGO and PVA enhancing the quantity of formed hydrogen bonds, which improved mechanical properties taking for instance an increase of 40% in value of tensile strength in comparison of CS–PVA polymer [119].

As can be seen in **Figure 9**, multifunctional rGO magnetic nanosheet functionalized via non-covalent with chitosan (rGOMCS) was developed for *Pseudomonas aeruginosa* (BCRC 10303) and *Staphylococcus aureus* (BCRC 1045) detection in aqueous suspension and mouse blood. Owing to its inherent fluorescence properties and high surface area, rGO as a support to magnetic nanoparticles enhanced detection of bacteria in fluorescence spectroscopy and matrix-assisted laser desorption/ionization mass spectrometry (MALDI-MS). On one hand, fluorescence as bacteria detection method has been used widely, in spite of interference created by biomolecules in biological samples, furthermore external factors, among others make it a complicated task. On the other hand, great influence of CS and complementary effect of functional groups on graphene attached on rGOMCS was observed over bacteria in MALDI-MS. This allowed different non-covalent interactions (hydrogen bond, hydrophobic and electrostatic interactions, acid-base interactions, $\pi-\pi$ interactions and polar functional groups interactions) together with pathogen agents reflecting high adsorption. Additionally, Abdel-hamid et al. [116] reported thermodynamic analysis which reveals better affinity of CS to BCRC 1045 as a result of increment on peptidoglycan content and consequently better sensitivity for this pathogen agent.

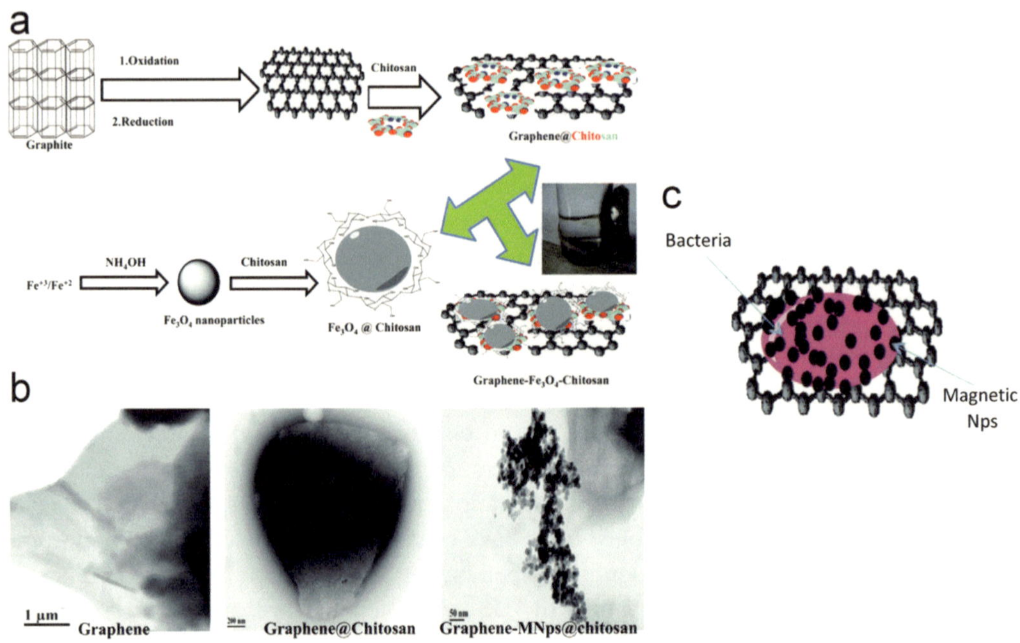

Figure 9. (a) Schematic diagram of the assembly of rGOMCS; (b) TEM images of rGO functionalized with magnetic particles and CS (c) Interaction between bacteria and rGOMCS. (Adapted from [116] with permission of The Royal Society of Chemistry).

ζ potential measurements reveal the influence of ionic interactions between rGO–CS solutions, another via of non-covalent functionalization of rGO. First, notable differences in the responses of GO–CS solutions were observed only changing the order of aggregation of the components.

In the **Figure 10**, as can be seen when GO is added to CS solution the GO sheets are wrapped rapidly by ionic interactions with considerable amount of CS (ζ potential of +40 mV); however, if the GO is added greatly dispersed in solution to CS forms a "bridging" flocculation as a result of attachment of two or more GO sheets to polymer chain (ζ potential of +8 mV). Second, one can take advantage that after GO reduction by L-AA in the presence of CS was done; the remaining functional groups on rGO and $-NH_2$ groups in CS are still available for non-covalent functionalization. This offers the possibility to control and change behavior of rGO–CS solutions inasmuch as pH is changed. ζ potential measurements show both behaviors on pH 6 (+33 mV) and at pH 7 (-9 mV) for a stable suspension and formation of agglomerates, respectively. It can be explained because at low pH, $-NH_2$ groups are protonated creating a good environment for electrostatic repulsion between rGO sheets. In spite of that, if the pH is increased the intermolecular repulsion tend to decrease, originated by the deprotonation of NH_2 groups of CS [60]. Ko et al., performed a nanocomposite for growth of *Escherichia coli* bacteria. They carried out the simultaneous reduction and non-covalent functionalization of GO in acid media solutions, attributing to remaining protonated $-NH_3^+$ groups on CS the contributions for the well-dispersed rGO–CS system. Moreover, enhanced antibacterial activity for *E. coli* was detected [120]. In **Table 2** are summarized some applications for both GO-CS and rGO-CS with covalent and non-covalent interactions.

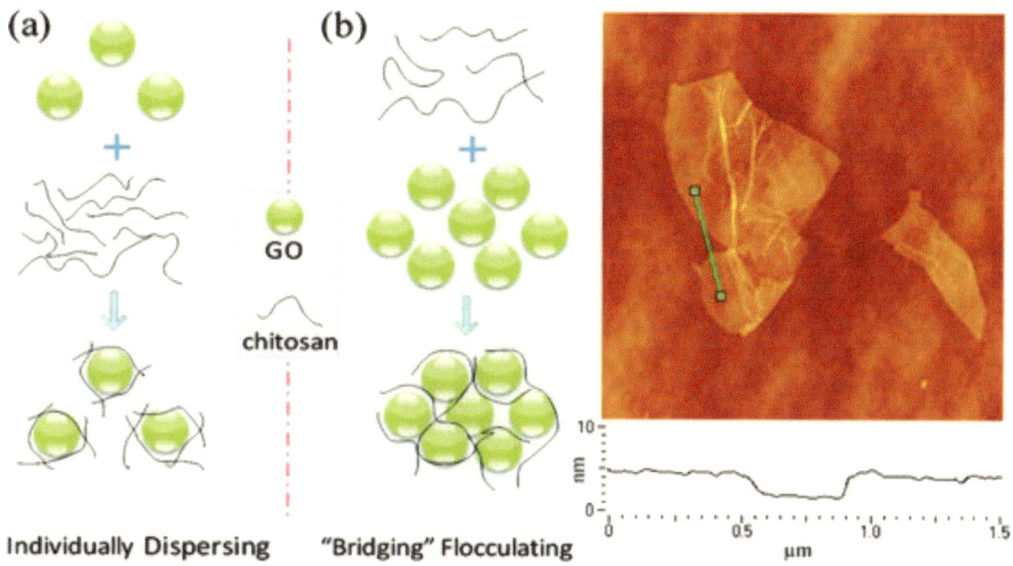

Figure 10. (a) Schematic diagram of GO–CS suspensions behavior as a result of the variation in the order of the addition of components. (b) Measurements taken by atomic force microscopy (AFM) of rGO extracted from rGO–CS suspension and deposited in cleaved mica. (Adapted with permission from [60]. Copyright 2016 American Chemical Society).

Potential application	**Interaction GO–CS or r GO–CS *via covalent***	**Reference**
Water treatment (U(vi) removal)	Amidation	[101]
Water treatment (antifouling)	Amidation	[100]
Water treatment (removal Cr(IV))	Amidation	[97]
Water treatment (desalting)	Nucleophilic addition (Ring opening)	[87]
Water treatment (Fuschine removal)	Amidation	[99]
DNA biosensor	Amidation	[88]
Drug delivery	Amidation	[89]
Drug delivery	Amidation	[90]
Lipase immobilization	Amidation	[92]
Osteoblasts growth	Amidation	[91]
Growth of cells C3H10T1/2	Amidation	[93]
Drug delivery (Fluorescein sodium)	Amidation	[94]
Nanocatalyst	Amidation	[103]
Nanocatalyst	Amidation	[102]
Nanocomposite (GO–CS –HAP, bioactivity)	Amidation	[96]
Nanocomposite	Amidation	[84]
Nanocomposite	Amidation	[121]
Nanocomposite	Amidation/Esterification	[21]
Nanocomposite	Nucleophilic addition (Ring opening)	[86]
Nanocomposite	Esterification	[85]
Potential application	Interaction GO–CS or r GO–CS *via non-covalent*	
Hemolytic activity	Electrostatic adsorption	[104]
Multifunctional paper	Electrostatic interaction	[105]
Glucose and urea detection	Electrostatic interaction	[106]
Removal contaminants	Electrostatic interaction	[107]
Nanocomposite (PVA)	Hydrogen bond/electrostatic interaction	[119]
Supramolecular hydrogel	Hydrogen bond/electrostatic interaction	[113]
L929 cells and MG-63 cells growth	Hydrogen bond/electrostatic interaction	[118]
Bulirubin adsorbent hydrogel	Hydrogen bond/electrostatic interaction	[114]
Electrochemical electrode	Hydrogen bond/electrostatic interaction	[115]
E. Coli growth	Hydrogen bond/electrostatic interaction	[117]
Sensor of *Pseudomonas aeruginosa and Staphylococcus aureus*	Hydrogen bond/electrostatic attractions	[116]

Potential application	Interaction GO–CS or r GO–CS *via covalent*	Reference
Nanocomposite	Hydrogen bond/electrostatic attractions	[122]
Janus membranes	Polycationic interaction	[109]
pH responsive nanocomposite	GO-Ionic interactions (Electrostatic)	[60]
pH responsive nanocomposite	rGO ionic interaction/hydrogen bond	[60]
Nanocomposite	Hydrogen bond	[110]
Fluorescein sodium	Hydrogen bond	[111]
L929 cells growth	Hydrogen bond	[112]
Antibacterial activity	Van der Walls/Ionic/hydrogen bonding	[120]

Table 2. Functionalization of GO and rGO by CS and its progress in different fields of application.

5. Conclusions

During the last years, many efforts have been done in order to exploit the outstanding properties of graphenic materials in fields such as catalysis, nanocomposites, cell batteries, and so on. As a result of its versatility, the chemically route to produce graphene oxide and reduced graphene oxide offers a possible solution for requirements about scalability, cost-effective, and easy way to obtain suited amounts of graphene for its applicability. Both GO and rGO can take some advantages over pristine graphene because of the rich chemistry of their functional groups present on the surface. These groups can be used as points to anchor specific molecules according the necessities, which granted the shown multifunctionality. Specifically, in rGO, the restoration of sp^2 domains through the reduction reaction offers a conductive material which can be applied in fields where the conductivity is overriding. Additionally, emerging routes to achieve non-covalent and covalent functionalization of graphenic materials have been reported. In this chapter, we outline recent progress in covalent and non-covalent functionalization of graphenic materials with biopolymers such as keratin and chitosan. Both are aimed on the development of a multifunctional platform capable of being used in different fields.

Functionalization of graphene-based materials with biomolecules offers an additional advantage to orient these materials for biological applications, increasing, for example, their compatibility. Biofunctionalization with small or medium size biopolymer chains, as in the case of some proteins or polysaccharides, enables the formation of hybrid or conjugated systems through a series of several interaction or reaction points. This, as a result of the presence of different chemical functionalities, as in the case of keratin, or a repeated functional group, as in the case of chitosan, in each polymeric unit. This aspect allows to fully cover the nanocarbon surface with along with a strong biomolecule immobilization.

To the best of our knowledge, as we have discussed in this chapter, there are just a couple of works related to the functionalization of graphenic materials through a redox reaction system. This approach has been commonly used to achieve high degree of attachments in natural and synthetic polymeric materials. A redox reaction brings the benefit to yield a random covalent functionalization practically using all the available chemical groups in both materials, at relatively low reaction temperature. Therefore, the resulting surface will possess few functional groups, if any at all. In the case of the non-covalent functionalization of reduced graphene oxide, the opposite behavior will take place. The functional groups of the biomolecule will be exposed on the conjugated system surface. This proves its utility if an additional functionalization, specific group reactions, or a further adsorption process is desired.

We want to awake the interest of the reader in the study of these approaches and to compare their advantages in disadvantages for other systems.

Author details

Edgar Jimenez-Cervantes Amieva[1,2], Juventino López-Barroso[1], Ana Laura Martínez-Hernández[1*] and Carlos Velasco-Santos[1]

*Address all correspondence to: almh72@gmail.com

1 Division of Postgraduate Studies and Research, Technological Institute of Queretaro, Santiago de Querétaro, Querétaro, Mexico

2 Centre of Applied Physics and Advanced Technology, National Autonomous University of Mexico, Campus Juriquilla, Querétaro, Mexico

References

[1] Das S., Choi W. Graphene synthesis. In: Choi W., Lee J.W., editors. Graphene: synthesis and applications. Boca Raton, FL: CRC Press, Taylor and Francis group; 2012. p. 27–63.

[2] Novoselov K.S., Geim A.K., Morozov S.V., Jiang D., Zhang Y., Dubonos S.V., Grigorieva I.V., Firsov A.A. Electric field effect in atomically thin carbon films. Science. 2004;306(5696):666–669. doi:10.1126/science.1102896

[3] Sundaram R.S. Chemically derived graphene. In: Skákalová V., Schäffel F., Bachmatiuk A., Rümmeli M.H., editors. Graphene: properties, preparation, characterisation and devices. Sawston, Cambridge: Woodhead Publishing; 2014. p. 50–80. doi: 10.1533/9780857099334.1.50

[4] Ibrahim I., Rümmeli M.H. Mechanical exfoliation. In: Warner J.H., Schäffel F., Bachmatiuk A., Rümmeli M.H., editors. Graphene: fundamentals and emergent applications. Kidlington, Oxford: Elsevier; 2013. p. 129–137. doi:10.1016/B978-0-12-394593-8.00004-7

[5] Zhong Y.L., Tian Z., Simon G.P., Li D. Scalable production of graphene via wet chemistry: progress and challenges. Materials Today. 2015;18(2):73–78. doi:10.1016/j.mattod.2014.08.019

[6] Hummers W., Offeman R.E. Preparation of graphitic oxide. Journal of the American Chemical Society. 1958;80(6):1339. doi:10.1021/ja01539a017

[7] Bonaccorso F., Lombardo A., Hasan T., Sun Z., Colombo L., Ferrari A.C. Production and processing of graphene and 2d crystals. Materials Today. 2012;15(12):564–589. doi:10.1016/S1369-7021(13)70014-2

[8] Chen G., Weng W., Wu D., Wu C., Lu J., Wang P., Chen X. Preparation and characterization of graphite nanosheets from ultrasonic powdering technique. Carbon. 2004;42(4):753–759. doi:10.1016/j.carbon.2003.12.074

[9] Warner J.H. Reduced graphene oxide. In: Warner J.H., Schäffel F., Bachmatiuk A., Rümmeli M.H., editors. Graphene: fundamentals and emergent applications. Kidlington, Oxford: Elsevier; 2013. p. 155–162. doi:10.1016/B978-0-12-394593-8.00004-7

[10] Pei S., Cheng H.-M. The reduction of graphene oxide. Carbon. 2012;50(9):3210–3228. doi:10.1016/j.carbon.2011.11.010

[11] Hu K., Kulkarni D.D., Choi I., Tsukruk V.V. Graphene-polymer nanocomposites for structural and functional applications. Progress in Polymer Science. 2014;39(11):1934–1972. doi:10.1016/j.progpolymsci.2014.03.001

[12] Loh K.P., Bao Q., Eda G., Chhowalla M. Graphene oxide as a chemically tunable platform for optical applications. Nature Chemistry. 2010;2(12):1015–1024. doi:10.1038/nchem.907

[13] Lerf A., He H., Forster M., Klinowski J. Structure of graphite oxide revisited. The Journal of Physical Chemistry B. 1998;102(23):4477–4482. doi:10.1021/jp9731821

[14] Gao W., Alemany L.B., Ci L., Ajayan P.M. New insights into the structure and reduction of graphite oxide. Nature Chemistry. 2009;1(5):403–408. doi:10.1038/nchem.281

[15] Li D., Mueller M.B., Gilje S., Kaner R.B., Wallace G.G. Processable aqueous dispersions of graphene nanosheets. Nature Nanotechnology. 2008;3(2):101–105. doi:10.1038/nnano.2007.451

[16] Dreyer D.R., Park S., Bielawski C.W., Ruoff R.S. The chemistry of graphene oxide. Chemical Society Reviews. 2010;39(1):228–240. doi:10.1039/B917103G

[17] Hernandez Y., Pang S., Feng X., Müllen K. Graphene and its synthesis. In: Matyjaszewski K., Möller M., editors. Polymer science: a comprehensive reference. Amsterdam, NL: Elsevier; 2012. p. 415–438. doi:10.1016/B978-0-444-53349-4.00216-8

[18] Inagaki M., Kang F., Toyoda M., Konno H. Advanced materials science and engineering of carbon. Beijing, China: Elsevier & Tsinghua Univ. Press; 2013. 434 p. doi:10.1016/B978-0-12-407789-8.00003-X

[19] Whitener K.E., Sheehan P.E. Graphene synthesis. Diamond and Related Materials. 2014;46:25–34. doi:10.1016/j.diamond.2014.04.006

[20] Stankovich S., Dikin D.A., Piner R.D., Kohlhaas K.A., Kleinhammes A., Jia Y., et al. Synthesis of graphene-based nanosheets via chemical reduction of exfoliated graphite oxide. Carbon. 2007;45:1558–1565. doi:10.1016/j.carbon.2007.02.034

[21] Bustos-Ramírez K., Martínez-Hernández A.L., Martínez-Barrera G., Icaza M.D., Castaño V.M., Velasco-Santos C. Covalently bonded chitosan on graphene oxide via redox reaction. Materials. 2013;6(3):911-926. doi:10.3390/ma6030911

[22] Kim J., Cote L.J., Kim F., Yuan W., Shull K.R., Huang, J. Graphene oxide sheets at interfaces. Journal of the American Chemical Society. 2010;132(23):8180–8186. doi: 10.1021/ja102777p

[23] Whitby R.L.D. Chemical control of graphene architecture: tailoring shape and properties. ACS Nano. 2014;8(10):9733–9754. doi:10.1021/nn504544h

[24] Hu Y., Song S., Lopez-Valdivieso A. Effects of oxidation on the defect of reduced graphene oxides in graphene preparation. Journal of Colloid and Interface Science. 2015;450:68–73. doi:10.1016/j.jcis.2015.02.059

[25] Zhao J., Pei S., Ren W., Gao L., Cheng H.-M. Efficient preparation of large-area graphene oxide sheets for transparent conductive films. ACS Nano. 2010;4(9):5245–5252. doi: 10.1021/nn1015506

[26] Chua C.K., Pumera M. Chemical reduction of graphene oxide: a synthetic chemistry viewpoint. Chemical Society Reviews. 2014;43(1):291–312. doi:10.1039/C3CS60303B

[27] Liang Y., Frisch J., Zhi L., Norouzi-Arasi H., Feng X., Rabe J.P., et al. Transparent, highly conductive graphene electrodes from acetylene-assisted thermolysis of graphite oxide sheets and nanographene molecules. Nanotechnology. 2009;20(43):434007. doi: 10.1088/0957-4484/20/43/434007

[28] Su Q., Pang S., Alijani V., Li C., Feng X., Müllen K. Composites of graphene with large aromatic molecules. Advanced Materials. 2009;21(31):3191–3195. doi:10.1002/adma.200803808

[29] Schniepp H.C., Li J.-L., McAllister M.J., Sai H., Herrera-Alonso M., Adamson D.H., et al. Functionalized single graphene sheets derived from splitting graphite oxide. The Journal of Physical Chemistry B. 2006;110(17):8535–8539. doi:10.1021/jp060936f

[30] Wu Z.-S., Ren W., Gao L., Zhao J., Chen Z., Liu B., et al. Synthesis of graphene sheets with high electrical conductivity and good thermal stability by hydrogen arc discharge exfoliation. ACS Nano. 2009;3(2):411–417. doi:10.1021/nn900020u

[31] McAllister M.J., Li J.-L., Adamson D.H., Schniepp H.C., Abdala A.A., Liu J., et al. Single sheet functionalized graphene by oxidation and thermal expansion of graphite. Chemistry of Materials. 2007;19(18):4396–4404. doi:10.1021/cm0630800

[32] Toh S.Y., Loh K.S., Kamarudin S.K., Daud W.R.W. Graphene production via electrochemical reduction of graphene oxide: synthesis and characterisation. Chemical Engineering Journal. 2014;251:422–434. doi:10.1016/j.cej.2014.04.004

[33] Guo H.-L., Wang X.-F., Qian Q.-Y., Wang F.-B., Xia, X.-H. A green approach to the synthesis of graphene nanosheets. ACS Nano. 2009;3(9):2653–2659. doi:10.1021/nn900227d

[34] Gao J., Liu F., Liu Y., Ma N., Wang Z., Zhang X. Environment-friendly method to produce graphene that employs vitamin C and amino acid. Chemistry of Materials. 2010;22(7):2213–2218. doi:10.1021/cm902635j

[35] Zhang J., Yang H., Shen G., Cheng P., Zhang J., Guo S. Reduction of graphene oxide via L-ascorbic acid. Chemical Communications. 2010;46(7):1112–1114. doi:10.1039/B917705A

[36] Tabrizi M.A., Varkani J.N. Green synthesis of reduced graphene oxide decorated with gold nanoparticles and its glucose sensing application. Sensors and Actuators B: Chemical. 2014;202:475–482. doi:10.1016/j.snb.2014.05.099

[37] Suresh D., Nagabhushana H., Sharma S.C. Clove extract mediated facile green reduction of graphene oxide, its dye elimination and antioxidant properties. Materials Letters. 2015;142:4–6. doi:10.1016/j.matlet.2014.11.073

[38] Suresh D., Nethravathi P.C., Nagabhushana H., Sharma S.C. Spinach assisted green reduction of graphene oxide and its antioxidant and dye absorption properties. Ceramics International. 2015;41(3):4810–4813. doi:10.1016/j.ceramint.2014.12.036

[39] Suresh D., Kumar M.P., Nagabhushana H., Sharma S.C. Cinnamon supported facile green reduction of graphene oxide, its dye elimination and antioxidant activities. Materials Letters. 2015;151:93–95. doi:10.1016/j.matlet.2015.03.035

[40] Akhavan O., Ghaderi E., Abouei E., Hatamie S., Ghasemi E.. Accelerated differentiation of neural stem cells into neurons on ginseng-reduced graphene oxide sheets. Carbon. 2014;66:395–406. doi:10.1016/j.carbon.2013.09.015

[41] He D., Peng Z., Gong W., Luo Y., Zhao P., Kong L. Mechanism of a green graphene oxide reduction with reusable potassium carbonate. RSC Advances. 2015;5(16):11966–11972. doi:10.1039/C4RA14511A

[42] Akhavan O., Azimirad R., Gholizadeh H.T., Ghorbani F.. Hydrogen-rich water for green reduction of graphene oxide suspensions. International Journal of Hydrogen Energy. 2015;40(16):5553–5560. doi:10.1016/j.ijhydene.2015.02.106

[43] Georgakilas V., Otyepka M., Bourlinos A.B., Chandra V., Kim N., Kemp K.C., et al. Functionalization of graphene: covalent and non-covalent approaches, derivatives and applications. Chemical Reviews. 2012;112(11):6156–6214. doi:10.1021/cr3000412

[44] Hirsch, A. The era of carbon allotropes. Nature Materials. 2010;9(11):868–871. doi: 10.1038/nmat2885

[45] Georgakilas V. Covalent attachment of organic functional groups on pristine graphene. In: Georgakilas V., editor. Functionalization of graphene. Weinheim, Germany: Wiley-VCH; 2014. p. 21–58.

[46] Spinato C., Ménard-Moyon C., Bianco A. Chemical functionalization of graphene for biomedical applications. In: Georgakilas V., editor. Functionalization of graphene. Weinheim, Germany: Wiley-VCH; 2014. p. 95–138.

[47] Ju H., Zhang X., Wang J. NanoBiosensing: principles, development and application. New York, NY: Springer Science & Business Media; 2011. 586 p. doi:10.1007/978-1-4419-9622-0

[48] Georgakilas V. Addition of organic groups through reactions with oxygen species of graphene oxide. In: Georgakilas V., editor. Functionalization of graphene. Weinheim, Germany: Wiley-VCH; 2014. p. 59–94.

[49] Luong N.D., Sinh L.H., Johansson L.-S., Campell J., Seppälä J. Functional graphene by thiol-ene click chemistry. Chemistry–A European Journal. 2015;21(8):3183–3186. doi: 10.1002/chem.201405734

[50] Ferrari A.C., Bonaccorso F., Fal'Ko V., Novoselov K.S., Roche S., Bøggild P., et al. Science and technology roadmap for graphene, related two-dimensional crystals, and hybrid systems. Nanoscale. 2015;7(11):4598–4810. doi:10.1039/C4NR01600A

[51] Kemp K.C., Cho Y., Chandra V., Kim K.S. Noncovalent functionalization of graphene. In: Georgakilas V., editor. Functionalization of graphene. Weinheim, Germany: Wiley-VCH; 2014. p. 199–218.

[52] Pavlidis I.V., Patila M., Polydera A.C., Gournis D., Stamatis H. Immobilization of enzymes and other biomolecules on graphene. In: Georgakilas V., editor. Functionalization of graphene. Weinheim, Germany: Wiley-VCH; 2014. p. 139–172.

[53] Moench I., Meye A., Leonhardt A. Ferromagnetic filled carbon nanotubes as novel and potential containers for anticancer treatment strategies. In: Kumar C.S.S.R., editor. Series: nanotechnologies for the life sciences vol. 6: nanomaterials for cancer therapy. Weinheim, Germany: Wiley-VCH; 2007. p. 259–337. doi: 10.1002/9783527610419.ntls0068

[54] Georgakilas V. Functionalization of graphene by other carbon nanostructures. In: Georgakilas V., editor. Functionalization of graphene. Weinheim, Germany: Wiley-VCH; 2014. p. 255–282.

[55] Zhang M., Yin B.-C., Wang X.-F., Ye B.-C. Interaction of peptides with graphene oxide and its application for real-time monitoring of protease activity. Chemical Communications. 2011;47(8):2399–2401. doi:10.1039/C0CC04887A

[56] Lu F., Zhang S., Gao H., Jia H., Zheng L. Protein-decorated reduced oxide graphene composite and its application to SERS. ACS Applied Materials & Interfaces. 2012;4(6): 3278–3284. doi:10.1021/am300634n

[57] Zeng Q., Cheng J., Tang L., Liu X., Liu Y., Li J., et al. Self-assembled graphene–enzyme hierarchical nanostructures for electrochemical biosensing. Advanced Functional Materials. 2010;20(19):3366–3372. doi:10.1002/adfm.201000540

[58] Liu J., Li Y., Li Y., Li J., Deng Z. Noncovalent DNA decorations of graphene oxide and reduced graphene oxide toward water-soluble metal–carbon hybrid nanostructures via self-assembly. Journal of Materials Chemistry. 2010;20(5):900–906. doi:10.1039/B917752C

[59] Liu S.-J., Wen Q., Tang L.-J., Jiang J.-H. Phospholipid–graphene nanoassembly as a fluorescence biosensor for sensitive detection of phospholipase D activity. Analytical Chemistry. 2012;84(14):5944–5950. doi:10.1021/ac300539s

[60] Fang M., Long J., Zhao W., Wang L., Chen G. pH-responsive chitosan-mediated graphene dispersions. Langmuir. 2010;26(22):16771–16774. doi:10.1021/la102703b

[61] Kouloumpis A., Zygouri P., Dimos K., Gournis D. Layer-by-layer assembly of graphene-based hybrid materials. In: Georgakilas V., editor. Functionalization of graphene. Weinheim, Germany: Wiley-VCH; 2014. p. 359–399.

[62] Yang K., Feng L., Shi X., Liu Z. Nano-graphene in biomedicine: theranostic applications. Chemical Society Reviews. 2013;42(2):530–547. doi:10.1039/C2CS35342C

[63] Wang Y., Li Z., Wang J., Li J., Lin Y. Graphene and graphene oxide: biofunctionalization and applications in biotechnology. Trends in biotechnology. 2011;29(5):205–212. doi: 10.1016/j.tibtech.2011.01.008

[64] Nyanhongo G.S., Steiner W., Gübitz G.M., editors. Biofunctionalization of polymers and their applications. Berlin Heidelberg: Springer-Verlag; 2011. 288 p. doi:10.1007/978-3-642-21949-8

[65] Wikimedia Foundation. Biofunctionalisation [Internet]. February 8 2015. Available from: http://en.wikipedia.org/wiki/Biofunctionalisation [Accessed: May 7 2015]

[66] Rodríguez-González C., Martínez-Hernández A.L., Castan~o V.M., Kharissova O.V., Ruoff R.S., Velasco-Santos C. Polysaccharide nanocomposites reinforced with graphene oxide and keratin-grafted graphene oxide. Industrial & Engineering Chemistry Research. 2012;51(9):3619–3629. doi:10.1021/ie200742x

[67] Amieva E.J.C., Fuentes-Ramirez R., Martinez-Hernandez A.L., Millan-Chiu B., Lopez-Marin L.M., Castaño V.M., et al. Graphene oxide and reduced graphene oxide modification with polypeptide chains from chicken feather keratin. Journal of Alloys and Compounds. 2015;643:S137–S143. doi:10.1016/j.jallcom.2014.12.062

[68] Schrooyen P.M., Dijkstra P.J., Oberthür R.C., Bantjes A., Feijen J. Partially carboxymethylated feather keratins. 1. Properties in aqueous systems. Journal of Agricultural and Food Chemistry. 2000;48(9):4326–4334. doi:10.1021/jf9913155

[69] Das A., Saikia C.N. Graft copolymerization of methylmethacrylate onto non-mulberry silk-Antheraea assama using potassium permanganate–oxalic acid redox system. Bioresource Technology. 2000;74(3):213–216. doi:10.1016/S0960-8524(00)00020-1

[70] Khalil M.I., Abdel-Fattah S.H., Kantouch A. Manganese (IV)-initiated graft polymerization of vinyl monomers on polyamide fibers. Journal of Applied Polymer Science. 1975;19(10):2699–2708. doi:10.1002/app.1975.070191005

[71] Mostafa K.M. Graft polymerization of acrylic acid onto starch using potassium permanganate acid (redox system). Journal of Applied Polymer Science. 1995;56(2):263–269. doi:10.1002/app.1995.070560217

[72] Moharana S., Mishra S.B., Tripathy S.S. Chemical modification of jute fibers. I. Permanganate-initiated graft copolymerization methyl methacrylate onto jute fibers. Journal of Applied Polymer Science. 1990;40(3–4):345–357. doi:10.1002/app.1990.070400304

[73] Zhang L.M., Chen D.Q. Grafting of 2-(Dimethylamino) ethyl methacrylate onto potato starch using potassium permanganate/sulfuric acid initiation system. Starch-Stärke. 2001;53(7):311–316. doi:10.1002/1521-379X(200107)53:7<311::AID-STAR311>3.0.CO;2-A

[74] Sarac A.S. Redox polymerization. Progress in Polymer Science. 1999;24(8):1149–1204. doi:10.1016/S0079-6700(99)00026-X

[75] Martinez-Hernandez A.L., Velasco-Santos C., Icaza M.D., Castaño V.M. Grafting of methyl methacrylate onto natural keratin. e-Polymers. 2003;3(1):209–219. doi:10.1515/epoly.2003.3.1.209

[76] Jyothi A.N., Carvalho A.J.F. Starch-g-copolymers: synthesis, properties and applications. In: Kalia S., Sabaa M.W., editors. Polysaccharide based graft copolymers. Berlin Heidelberg: Springer; 2013. p. 59–109. doi:10.1007/978-3-642-36566-9_3

[77] Martínez-Hernández A.L., Santiago-Valtierra A.L., Alvarez-Ponce M.J. Chemical modification of keratin biofibres by graft polymerisation of methyl methacrylate using redox initiation. Materials Research Innovations. 2008;12(4):184–191. doi:http://dx.doi.org/10.1179/143307508X362828

[78] Shon C.H., Gao J., Thomas D.A., Kim T.-Y., Goddard III W.A., Beauchamp J.L. Mechanisms and energetics of free radical initiated disulfide bond cleavage in model peptides and insulin by mass spectrometry. Chemical Science. 2015;6(8):4550–4560. doi:10.1039/C5SC01305D

[79] Fuhrmann J., Clancy K.W., Thompson P.R. Chemical biology of protein arginine modifications in epigenetic regulation. Chemical Reviews. 2015;115(11):5413–5461. doi: 10.1021/acs.chemrev.5b00003

[80] Aljawish A., Chevalot I., Jasniewski J., Scher J., Muniglia L. Enzymatic synthesis of chitosan derivatives and their potential applications. Journal of Molecular Catalysis B: Enzymatic. 2015;112:25–39. doi:10.1016/j.molcatb.2014.10.014

[81] Kim S.-K., Medis E. Bioactive compounds from marine processing byproducts – a review. Food Research International. 2006;39(4):383–393. doi:10.1016/j.foodres. 2005.10.010

[82] Hejazi R., Amiji M. Chitosan-based gastrointestinal delivery systems. Journal of Controlled Release. 2003;89:151–165. doi:10.1016/S0168-3659(03)00126-3

[83] Dash M., Chiellini F., Ottenbrite R.M., Chiellini E. Chitosan—A versatile semi-synthetic polymer in biomedical applications. Progress in Polymer Science. 2011;36(8):981–1014. doi:10.1016/j.progpolymsci.2011.02.001

[84] Xu G., Shi T., Li M., Yu F., Chen Y. Difference between the effects of modification graphene oxide with two biomass molecules: chitosan and cardanol. Research on Chemical Intermediates. 2015;41(11):8499–8513. doi:10.1007/s11164-014-1906-0

[85] Yang Q., Pan X., Clarket K., Li K. Covalent Functionalization of Graphene with Polysaccharides. Industrial & Engineering Chemistry Research. 2012;51(1):310–317. doi:10.1021/ie201391e

[86] Shao L., Chang X., Zhang Y., Huang Y., Yao Y., Guo Z. Graphene oxide cross-linked chitosan nanocomposite membrane. Applied Surface Science. 2013;280:989–992. doi: 10.1016/j.apsusc.2013.04.112

[87] Wang J., Gao X., Wang J., Wei Y., Li Z., Gao C. O-(carboxymethyl)-chitosan nanofiltration membrane surface functionalized with graphene oxide nanosheets for enhanced desalting properties. ACS Applied Materials & Research. 2015;7(7):4381–4389. doi: 10.1021/am508903g

[88] Singh A., Sinsinbar G., Choudhary M., Kumar V., Pasricha R., Verma H.N., et al. Graphene oxide-chitosan nanocomposite based electrochemical DNA biosensor for detection of typhoid. Sensors and Actuators B: Chemical. 2013;185:675–684. doi: 10.1016/j.snb.2013.05.014

[89] Rana V.K., Choi M.-C., Kong J.-Y., Kim M.J., Kim S.-H., Mishra S., et al. Synthesis and drug-delivery behavior of chitosan-functionalized graphene oxide hybrid nanosheets. Macromolecular Materials and Engineering. 2011;296(2):131–140. doi:10.1002/mame. 201000307

[90] Bao H., Pan Y., Ping Y., Sahoo N.G., Wu T., Li L., et al. Chitosan-functionalized graphene oxide as a nanocarrier for drug and gene delivery. Small. 2011;7(11):1569–1578. doi:10.1002/smll.201100191

[91] Depan D., Pesacreta T.C., Misra R.D.K. The synergistic effect of a hybrid graphene oxide–chitosan system and biomimetic mineralization on osteoblast functions. Biomaterials Science. 2014;2(2):264–274. doi:10.1039/C3BM60192G

[92] Wang J., Zhao G., Jing L., Peng X., Li Y. Facile self-assembly of magnetite nanoparticles on three-dimensional graphene oxide–chitosan composite for lipase immobilization. Biochemical Engineering Journal. 2015;98:75–83. doi:10.1016/j.bej.2014.11.013

[93] Zuo P.-P., Feng H.-F., Xu Z.-Z., Zhang L.-F., Zhang Y.-L., Xia W., Zhang W.-Q. Fabrication of biocompatible and mechanically reinforced graphene oxide-chitosan nanocomposite films. Chemistry Central Journal. 2013;7(39). doi:10.1186/1752-153X-7-39

[94] Justin R., Chen B. Strong and conductive chitosan–reduced graphene oxide nanocomposites for transdermal drug delivery. Journal of Materials Chemistry B. 2014;2(24): 3759–3770. doi:10.1039/C4TB00390J

[95] Depan D., Girase B., Shah J.S., Misra R.D.K. Structure–process–property relationship of the polar graphene oxide-mediated cellular response and stimulated growth of osteoblasts on hybrid chitosan network structure nanocomposite scaffolds. Acta Biomaterialia. 2011;7(9):3432–3445. doi:10.1016/j.actbio.2011.05.019

[96] Mohandes F., Salavati-Niasara M. Freeze-drying synthesis, characterization and in vitro bioactivity of chitosan/graphene oxide/hydroxyapatite nanocomposite. RSC Advances. 2014;4(49):25993–26001. doi:10.1039/C4RA03534H

[97] Li L., Luo C., Li X., Duan H., Wang X. Preparation of magnetic ionic liquid/chitosan/graphene oxide composite and application for water treatment. International Journal of Biological Macromolecules. 2014;66:172–178. doi:10.1016/j.ijbiomac.2014.02.031

[98] Fan L., Luo C., Li X., Lu F., Qiu H., Sun M. Fabrication of novel magnetic chitosan grafted with graphene oxide to enhance adsorption properties for methyl blue. Journal of Hazardous Materials. 2012;215–216:272–279. doi:10.1016/j.jhazmat.2012.02.068

[99] Li L., Fan L., Luo C., Duan H., Wang X. Study of fuchsine adsorption on magnetic chitosan/graphene oxide. RSC Advances. 2014;4(47):24. doi:24679–24685

[100] Hegab H.M., Wimalasiri Y., Ginic-Markovic, Zou L. Improving the fouling resistance of brackish water membranes via surface modification with graphene oxide functionalized chitosan. Desalination. 2015;365:99–107. doi:10.1016/j.desal.2015.02.029

[101] Cheng W., Wang M., Yang Z., Sun Y., Ding C. The efficient enrichment of U(VI) by graphene oxide-supported chitosan. RSC Advances. 2014;4(106):61919–61926. doi: 10.1039/C4RA09541C

[102] Maleki A., Paydar R. Graphene oxide–chitosan bionanocomposite: a highly efficient nanocatalyst for the one-pot three-component synthesis of trisubstituted imidazoles under solvent-free conditions. RSC Advances. 2015;5(42):33177–33184. doi:10.1039/C5RA03355A

[103] Rajesh R., Sujanthi E., Kumar S.S., Venkatesan R. Designing versatile heterogeneous catalysts based on Ag and Au nanoparticles decorated on chitosan functionalized graphene oxide. Physical Chemistry Chemical Physics. 2015;17(17):11329–11340. doi: 10.1039/C5CP00682A

[104] Liao K.-H., Lin Y.-S., Macosko C.W., Haynes C.L. Cytotoxicity of graphene oxide and graphene in human erythrocytes and skin fibroblasts. ACS Applied Materials & Interfaces. 2011;3(7):2607–2615. doi:10.1021/am200428v

[105] Ling Y., Li X., Zhou S., Wang X., Sun R. Multifunctional cellulosic paper based on quaternized chitosan and gold nanoparticle–reduced graphene oxide via electrostatic self-assembly. Journal of Materials Chemistry A. 2015;3(14):7422–7428. doi:10.1039/C4TA07160C

[106] Song Y., Liu H., Tan H., Xu F., Jia J., Zhang L., et al. pH-Switchable electrochemical sensing platform based on chitosan-reduced graphene oxide/concanavalin a layer for assay of glucose and urea. Analytical Chemistry. 2014;86(4):1980–1987. doi:10.1021/ac402742m

[107] Li X., Zhou H., Wu W., Wei S., Xu Y., Kuang Y. Studies of heavy metal ion adsorption on Chitosan/Sulfydryl-functionalized graphene oxide composites. Journal of Colloid and Interface Science. 2015;448:389–397. doi:10.1016/j.jcis.2015.02.039

[108] Gruskiene R., Deveikyte R. Ricardas M. Quaternization of chitosan and partial destruction of the quaternized derivatives making them suitable for electrospinning. CHEMIJA. 2013;24(4):325–334.

[109] Han D., Xiao P., Gu J., Chen J., Cai Z., Zhang J., et al. Polymer brush functionalized Janus graphene oxide/chitosan hybrid membranes. RSC Advances. 2014;4(43):22759–22762. doi:10.1039/C4RA02826K

[110] Pan Y., Wu T., Bao H., Li L. Green fabrication of chitosan films reinforced with parallel aligned graphene oxide. Carbohydrate Polymers. 2011;83(4):1908–1915. doi:10.1016/j.carbpol.2010.10.054

[111] Justin R., Chen B. Characterisation and drug release performance of biodegradable chitosan–graphene oxide nanocomposites. Carbohydrate Polymers. 2014;103:70–80. doi:10.1016/j.carbpol.2013.12.012

[112] Sayyar S., Murray E., Thompson B.C., Chung J., Officer D.L., Gambhir S., et al. Processable conducting graphene/chitosan hydrogels for tissue engineering. Journal of Materials Chemistry B. 2015;3(3):481–490. doi:10.1039/C4TB01636J

[113] Han D., Yan L. Supramolecular hydrogel of chitosan in the presence of graphene oxide nanosheets as 2D cross-linkers. ACS Sustainable Chemistry & Engineering. 2014;2(2):296–300. doi:10.1021/sc400352a

[114] Wei H., Han L., Tang Y., Ren J., Zhao Z., Jia L. Highly flexible heparin-modified chitosan/graphene oxide hybrid hydrogel as a super bilirubin adsorbent with excellent

hemocompatibility. Journal of Materials Chemistry B. 2015;3(8):1646–1654. doi:10.1039/C4TB01673D

[115] He L., Wang H., Xia G., Sun J., Song R. Chitosan/graphene oxide nanocomposite films with enhanced interfacial interaction and their electrochemical applications. Applied Surface Science. 2014;314:510–515. doi:10.1016/j.apsusc.2014.07.033

[116] Abdelhamid H.N., Wu H.-F. Multifunctional graphene magnetic nanosheet decorated with chitosan for highly sensitive detection of pathogenic bacteria. Journal of Materials Chemistry B. 2013;1(32):3950–3961. doi:10.1039/C3TB20413H

[117] Luo Z., Yang D., Qi G., Yuwen L., Zhang Y., Weng L., et al. Preparation of highly dispersed reduced graphene oxide decorated with chitosan oligosaccharide as electrode material for enhancing the direct electron transfer of *Escherichia coli*. ACS Applied Materials & Interfaces. 2015;7(16):8539–8544. doi:10.1021/acsami.5b00297

[118] Li M., Wang Y., Liu Q., Li Q., Cheng Y., Zheng Y., et al. In situ synthesis and biocompatibility of nano hydroxyapatite on pristine and chitosan functionalized graphene oxide. Journal of Materials Chemistry B. 2013;1(4):475–484. doi:10.1039/C2TB00053A

[119] Feng X., Wang X., Xing W., Yu B., Song L., Hu Y. Simultaneous reduction and surface functionalization of graphene oxide by chitosan and their synergistic reinforcing effects in PVA films. Industrial & Engineering Chemistry Research. 2013;52(36):12906–12914. doi:10.1021/ie402073x

[120] Ko T.Y., Kim S.Y., Kim H.G., Moon G.-S., In I. Antibacterial activity of chemically reduced graphene oxide assembly with chitosan through noncovalent interactions. Chemistry Letters. 2013;42:66–67. doi:http://doi.org/10.1246/cl.2013.66

[121] Pan Y., Bao H., Li L. Noncovalently functionalized multiwalled carbon nanotubes by chitosan-grafted reduced graphene oxide and their synergistic reinforcing effects in chitosan films. ACS Applied Materials & Interfaces. 2011;3(12):4819–4830. doi:10.1021/am2013135

[122] Yang X., Tu Y., Li L., Shang S., Tao X.-m. Well-dispersed chitosan/graphene oxide nanocomposites. ACS Applied Materials & Interfaces. 2010;2(6):1707–1713. doi:10.1021/am100222m